GOD GAMES

What do you do forever?

Neil Freer

ISBN 1-885395-39-6

Published by:

The Book Tree
c/o Post Office Box 724
Escondido, California 92033

Call for Our Free Catalog
1 (800) 700-TREE

Cover Art By:

Ursula Freer

Edited By:

Paul Tice

Layout and Design by:

Tédd St. Rain

COVER IMAGE: A depiction on an Assyrian clay tablet of Ninhursag the Mother
Goddess, and Enki genetically engineering the first human is juxtaposed with the
image of ourselves we attached to the Pioneer 10 space craft, our beginning and
our future among the stars.

CONTENTS

ABOUT THE AUTHOR

NEIL FREER is a researcher, writer, lecturer, philosopher and poet living in Santa Fe, NM. Neil and his wife, Ursula, have lived in the Eldorado area since 1994. Neil holds a BA in English and did graduate work in Philosophy and Psychology at the New School for Social Research. He has taught college courses in Philosophy and History of Religion, gives private and public seminars and lectures and has done over one hundred forty radio and TV interviews. Neil is the author of *Breaking the Godspell*, a book which explores the ramifications of the archaeological, astronomical and genetic proof for our being a genetically engineered species and presents the ramifications of this new paradigm of human nature that resolves the Creationist-Evolutionary conflict.

Neil's work appears as part of the symposium, *Of Heaven and Earth* (Book Tree, 1996) which includes the paper he presented as invited guest speaker at the Zecharia Sitchin Day special event at the International Association For New Science conference in 1996 in Denver. An essay "In The Middle of Whose Ship Are You Standing?" is included in David Pursglove's *Zen And The Art Of Close Encounters*. Neil published his first full collection of poetry, Neuroglyphs, in 1994 Please visit Neil's website for a more detailed overview of his work and publications at *http://www.concentric.net/~freer1*

INTRODUCTION

by

Zecharia Sitchin

The search for Truth takes many routes and has varied beginnings. A certain Greek philosopher looked for it with a lantern in the dark streets of Athens. A certain Hebrew was summoned, a millennium earlier, to a burning bush in the Sinai wilderness. Mine began when I was reprimanded by my Bible teacher for questioning why a certain word in Genesis chapter six – *Nephilim* – was deemed to mean "giants" and not, literally, Those Who Had Come Down, (in context, from the heavens down to Earth).

From that reprimand grew an obsession to find out who the Nephilim were and why the Bible referred to them as the sons of the gods (in the plural) – an odd reference in a sacred canon uniquely devoted to monotheism. From that there grew the puzzlement regarding who had been the *Elohim* in the Genesis creation segments who, speaking in the plural, said, Let *us* make The Adam in *our* image and after *our* likeness. After a lifetime of research that encompassed myth and mythology, history and prehistory, geology and geography and biology, archaeology and astronomy, the answer was found in ancient texts and in pictorial depictions bequeathed to us by the Sumerians, a people who had flourished in what is nowadays southern Iraq (the biblical Mesopotamia) and who, some 6,000 years ago, attained suddenly and as if out of nowhere a high civilization whose components and knowledge are still part and parcel of modern civilizations.

Piecing together the information provided by the Sumerians, and treating their texts as records of actual events and not as myth, the series of my books known as *The Earth Chronicles* began, in 1976, with the publication of *The 12th Planet.* Its central conclusion, that there is one more planet *in our own solar system* (called Nibiru by the Sumerians, Marduk by the Babylonians, and – as I have shown in *Divine Encounters* – Olam in the Old Testament) from which intelligent beings

5

had come to Earth in the distant past; used genetic engineering to bring about Homo sapiens by mixing their genes with that of harsh-evolved hominids; gave Mankind knowledge and civilization at times that correspond to the arrival of Nibiru in our vicinity; and became, in the end, the gods of ancient religions. The Sumerians called them *Anunnaki,* meaning "Those Who From Heaven to Earth Came;" the Bible called them Nephilim.

It was, by any measure, a daunting task to convert fragmented data and varied 'myths' into a coherent tale that chronicled the prehistory of Mankind and its planet. "Mind-boggling" was the word that was most used in reviews and comments on *The 12th Planet* and its sequels.

But then came Neil Freer and undertook a different kind of mind-boggling task. If all that I had concluded was true, he said, what does it all mean not to the human race and the planet in general – what does it mean to the individuals, to each one of us? He tackled the task adroitly in his first book, *Breaking the Godspell.* Now, when so many advances in all fields of science and thought have corroborated the ancient knowledge, he transfers his gaze from the Past to the Future; and in so doing he challenges not just the scientific, religious and political establishments. He challenges, no less the Anunnaki.

He titles his new book *God Games.* But, if all the above is the Truth, it is *not a game.*

Zecharia Sitchin

New York, April 1998

PREFACE

My first promulgation of the new paradigm, in *Breaking the Godspell* in 1987, already seems like a very long time ago. In it I explored the ramifications of the startling realization that, in light of the overwhelming archaeological evidence as presented by the Sumerian scholar, Zecharia Sitchin, we are a genetically engineered species. I realized then that the thesis that I was advancing was far ahead of the curve, taking too many large chunks of information as factual to be easily accepted by the public in general, and academicians and scientists in particular. I made the decision to simply go ahead and say it all without reservation or shrewd academic political considerations. The results have been quite gratifying. I have done over one hundred radio talk-show interviews, most with live call-in segments and roughly eighty-five percent of the response has been positive to very enthusiastic. The *Godspell* is in its third printing. What has been most interesting is the fact that at least some scientists have actually been willing to discuss the topic due to the fact that they have been intellectually honest enough to recognize that the evidence has developed to a point where, even if all of them do not interpret it as I do, it is at least recognized as significant evidence. It is the amassing of evidence at an accelerating rate that has caused my attitude and approach to gradually change since the publication of the *Godspell*. I was very "close" to the topic at that time, it's import literally so new and its force requiring so much comprehensive, mind-boggling thought, that it shows in the convoluted sentence structure. The new paradigm, if true, meant that we had to reevaluate every facet of human existence, and the process of literally rethinking the planet was a bit overwhelming. At that time the seminal evidence came from the brilliant archaeological interpretations of Zecharia Sitchin and clues and reinforcement from a variety of sources. Since 1987, however, many vectors of reinforcing data have come from a number of sciences. The implications for religion, philosophy, science and the New Age have become clear and even integrated by some. The work of thinkers like Joseph Campbell and Julian Jaynes has been put into perspective.

I have had the opportunity to think about the topic a great deal and the good fortune to discuss, develop and debate it with many keen intellects. I have been the host of a private electronic conference on the Well (an excellent bulletin board computer system based in Sausalito, California) for over three years which is devoted to a critical pro and con discussion of both my work and Sitchin's.

The ramifications of the new paradigm have come to the attention of many people and their responses have reinforced my conviction, as has audience reaction at the many lectures I have given. So my approach has evolved: I am oriented to putting the facts as I see them out as accurately as possible, as soon as possible. I don't think we should wait to explore it in depth until everyone is completely ready to look at it. We could wait a very long time. *God Games*, to those who have read *Breaking the Godspell*, may seem a bit more articulate, the metaphors a bit more succinct — and my attitude a bit brasher. But that is a result of the sharpening effect of discussion and debate and conviction based on the plethora of factual evidence.

OVERVIEW

In the elational planetlight
Of genetic enlightenment
We shall overcome
The ancient godspell,
The slave blindness,
Parent god-fright,
Totemtaboo,
Babel-factoring our genetic genius
Into negative quotients.

Neil Freer
Neuroglyphs

This book is about what to do after you get genetically enlightened. By "genetic enlightenment" I mean what it's like once you have broken the godspell, the effect of the ancient, subservient master-slave attitude that is the deepest dye in the fabrics of both Eastern and Western culture, the result of having been created as slave-animals by the Anunnaki/Nefilim, the transcultural "gods" (alien humanoids from the tenth planet, Nibiru, in our solar system) known to the ancient civilizations. This book is about how we may live, how we will conceive of reality after we fully real-ize the fact that we are a genetically engineered species. How we will operate as our own "gods," according to our own genetic credentials, play our own god-games according to our own transcendental choices, creating our own confident realities. It is about what it is like to be a member of a synthesized species which is going through a special case of evolution, a rapid metamorphosis, when the ramifications of that unthinkable thought are now about to bring about a turning point in our history. It is about the new civilization we are about to create out of the new planetary vision.

But, let it be said, at the beginning, the new meta-paradigm is not a call for a return to a paradise lost, some golden age of our racial youth, finding, creating or in any way suggesting a new religion. If that is your intention, please do not read this book. This is all about living beyond religion in any way we have conceived of that strange phenomenon before.

Do not read this book if you are looking for a justification for shrink-wrapping the universe on some sophomoric atheistic basic either.

What is the "godspell?" I use this word, previously part of the title of a Broadway play, as a precise way of characterizing the racial mind-set we humans have been infected with for some three thousand years. The godspell is the obsequious, subservient attitude toward the "gods" we were programmed and conditioned to when the Nefilim were here and which has evolved into the same slave mentality toward "God," Allah, Jehovah, the deity of your choice, which are all sublimations of the ancient Anunnaki/Nefilim.

9

The godspell mentality is, unquestionably in my mind, the most fundamental flaw in our planet's culture. It is the most pervasive, most fundamentally deep element at the root of war, conflict, division, and misunderstanding among humans. It is such, precisely, because it is essentially a complete falsehood and the use of "spell" in the word godspell is nearly literal. It is an ideology based on millenniums of fear, deep conditioning, further and further separation from the physical reality of our inventors and the truth about our invention, and sublimations on top of sublimations of the Nefilim into a Supreme Being. It is an elaborate, false, hollow, intellectual construction and an emotional trauma that still forms the skeletal framework of our planetary cultures, manifesting as religion and theology, and reinforced by strong emotional evocation. It cannot be proven reasonably and, therefore, there is "faith," the belief in that which cannot be seen, usually justified on the basis that that which is believed in is transcendental and/ or ineffable. The most charitable judgment on the phenomenon of the godspell mentality is that it was probably inevitable, given our original status, the Nefilim departing the planet, depriving us of the direction and teaching we had known under their domination, thereby forcing us to do for ourselves.

We have been explaining, adjusting, arguing among ourselves, those who claim to know, killing each other over what the god(s) really want from us and what we are supposed to be doing as humans for some three thousand years now and the process has always been stultifying, very counterproductive and horribly tragic in the vast number of humans who have lost their lives in the name of one god or another. We have come through the stages from slave to serf to Savior and are now regaining some sense of self. We are long overdue for a liberated, truly human existence. To this point, the last three thousand years of our history may be characterized as a prolonged proprietary polity fight over who holds authority over the absolute version of reality, who represents and interprets the will of some "god." We kill each other over claims as to what human nature really is and by what rules it should operate. We have treated the sociobiological event of our beginning as a species as if we could never be sure if it ever really occurred. . . In a time when we are required to deal with the politics of non-overlapping alien realities, we are not able to resolve the separations caused by our overlapping intra-species realities. We are theologically inane, philosophically naive and scientifically cramped. But we will not be for long.

The planet is on hold; in a perverse ecology we recycle outmoded primitive paradigms. We shuffle our feathers-and-molasses confusion between hands, creating governments peopled largely by the cynically devolved, sincere, perhaps, but myopic, in a time when sincerity is tragically inadequate. Domestically we are a dead poet's society paralyzed and waiting perennially for the futant we have probably already terminated at the stake, in the courtroom, the boardroom, the lab, the dean's office, the classroom for violating a taboo.

We still matriculate our young, these amazing parallel processing, relativistic, quantum jumping, multi-dimensional consciousnesses, semi-illiterate and naive, for fear of them questioning our shambling senilities. In a time when we need to stretch our historical sense to allow for the visitation of our planet by alien species from before our origins, we teach them drum and trumpet mammalian history fleshed out with desiccated parochial political platitudes. And expect that they will somehow be ready to step into stellar society. Yet everybody knows that there is something grand about to happen. Everybody knows that we are becoming a new race; not just a somewhat improved one, but we are moving to a new plateau of racial existence.

There is a new human emerging. Our children show all the signs; they are underwhelmed and overqualified. We feel it. And the signs are everywhere. There is the ongoing philosophical effort to provide the ideational context, the adequate maps and metaphors, the comprehensive unifying epistemology and clear picture of the way it might be if we had already reached a status beyond war, want, primitive competition, and the inevitability of death. But the solutions offered, however sincere and correct, are partial pieces of the puzzle.

There is the growing trend in psychology toward healing, integrated personal growth, the recognition of other than ordinary states of consciousness. But there is not yet a full context in which to bring together both the good in the old approaches and the valuable elements in an overarching concept of our racial psychological development that is a complement of individual development.

There is the scientific thrust to make life easier, to eliminate disease, to provide longevity and, indeed, immortality. We are about to read the entire human genome, the complete set of genetic instructions that make us the way we are. Science is on the verge of providing the synthesis for a unified field theory, to explore the parapsychological with precision, to provide for the wants and needs of the entire planet. Until now the context in which to integrate all those elements has been lacking.

There is the New Age phenomenon in all its multifaceted, multicolored charm, speaking of a world without war, with a harmonious ecology, concern and caring for all, love among all humans and life as just plain fun. But how to take that grand conspiracy to celebration is, after almost a quarter century, still awkwardly uncertain.

There is the broad sweep of a gradual reintegration of the profound philosophies of the East and the West driven by the intuition that there is a more profound level of truth below the usual consensual understanding of them. But there is still lacking a clear conception of a common source for these seemingly very different views of reality.

There is the groundswell rising to a new politic. The worldwide direction is clearly in the positive direction of individual human dignity and rights, toward individual opportunity. Peristroika and glasnost, representative government, civil and equal rights, freedom. But, however powerful the thrust of the demand for the recognition of the individual's dignity and rights, there is still not a clear and global understanding of the fundamental nature of the human which is required to ground it.

There are the strident calls for a completely new economic structure that will meet the needs and honor the dignity of all, foreshadowed by Buckminster Fuller's *World Game*. But, although it is not by bread alone, it is by bread by which we live. We do, indeed, need an economic context and system befitting the dignity of all humans and for all humans, and there are some scenarios that are particularly enlightened in that respect but they must, by their nature, be a part of the much grander vision.

And there is a dark side. There is a slinking cynicism, an often unspoken, viral attitude in human society that holds that there is no chance for such a new plateau of human existence. In that view it is impossible to get out of the criteria vacuum in which religion, philosophy and science rattle around with no way to initialize a common ground; impossible to get past the communicatory barriers of belief, taboo, turf and custom. Death is the pervasive conditioner, the source of the enervation of all effort toward transcendent expansion and an excuse for not even trying. We get here without a manual and the struggle for survival prevents even beginning to write one for oneself. That desperate cynicism holds it impossible to arrive at an overview of the human situation comprehensive enough to subsume all familiar worldviews and unite humans in a planetary paradigm.

I categorically, unequivocally and emphatically reject that point of view. Our most advanced thought from the ancient maps of consciousness to quantum mechanics points to the fact that we are capable of and do, indeed, create our own reality. It is clear that we have already attained a degree of knowledge and racial experience that allows us to begin to take charge of and responsibility for the direction of our own evolution. The vectors of our progress point inexorably in that direction. It is the godspell, ultimately, that gives rise to passive, debilitating cynicism.

With the totems and taboos of our racial adolescence dispelled, Prometheus can get off his rock and reach genetic satori; Job can get off his dung heap and complete his EST training; Buddha can open his eyes and reach genetic enlightenment. The new paradigm says that we are Homo-erectus Nefilimus, a genetically engineered species going through a rapid racial metamorphosis, a special case of evolution under the impetus of our more advanced Nefilim gene component. Transcendental experience is a natural neuropsychological phenomenon, a product of the rapid drive toward greater self-knowledge, and expansion into greater and greater physical/psychological dimensions of the universe on the part of the individual and the race. The essential paradigm is far more than even the sum of the multiple vectors it subsumes.

The archaeological key to the new paradigm is the work of Zecharia Sitchin: the gods, the Anunnaki/Nefilim were flesh and blood humanoids from the last planet in our solar system who invented us as slave animals and eventually accepted us as limited partners. That is the true "natural history of the gods" that Joseph Campbell attempted to write and only succeeded in resolving to a significant psychological component of our puzzling nature. The word from station DNA is that these "gods" wear designer genes. If you want to get half an idea of what an alien looks like, look in the mirror. If you want to know what the new human looks like, look again. The huge volume of archaeological and historical evidence and proof for the new paradigm is corroborated and reinforced by the evidence for Planet X/Nibiru indeed being in our solar system as the ancient records describe. NASA and the Jet Propulsion Laboratory are working with an observatory in Black Birch, New Zealand, searching the southern skies where its clear gravitational pull on Uranus and Neptune would indicate where it now is. It may have already been discovered by the IRAS satellite in 1984. Corroboration and reinforcement comes also from the mitochondrial DNA research undertaken to determine the time and place of our species' genesis; the place, Central Africa, and the time, around 250,000 years ago, closely matches the place and time, 300,000 years ago, given in the ancient records. The latest findings of Anthropology concerning our existence quite possibly preceding that of Neanderthals also comes closer and closer to the details we already have from the recovered information on clay tablets and stone. The collective information is overwhelming: eventual, obviously belated, acceptance into the academic arena is only a matter of time.

The elements, the characteristics of the new civilization, are a collective, consensual awareness of our generic humanity as a planetary unity, a recognition of a common, known history from the beginning of the race as a genetically engineered species, a globally common conception of human nature and its evolutionary direction, a multi-dimensional sociobiology capable of reflexive self-analysis, a psychology devoted to fostering the positive evolutionary development of the individual over the entire spectrum of consciousness, a unified scientific field expressed through a profound natural language based on self-referential consciousness, an ecological, non-competitive, ubiquitously helpful economics, an integral systems approach to the management of the material realm, with the primary focus on the transcendental as the essential human process.

The characteristics that mark the new human are an unassailable personal integrity, relativistic epistemology, profound compassion, robust depth of informational data, understanding of the universe in terms of a full unified field, broad-spectrum competence, transcendental competition, facility in dimensional shifting, preference for dyadic operation, a profound ability to enjoy, to play the games most enjoyable and satisfying to generic "gods," an expanded capacity to literally have great fun creating new realities, with the primary focus on the multi-dimensional. And to play those god games in the context of relative immortality and, eventually, habitual four-dimensional consciousness.

The turning point will be reached when we real-ize it; we will real-ize it when we reach it. That is not a devious Zen saying but a precise statement of fact, for the nature of the change is of the nature of a reality creation by us and we will make it real as we do it. The quintessential character of the change is determined by the fact that we have, for the first time, reached a point in our evolution where we can actually determine our own evolutionary direction — and we can and shall do that. That's the nature of human reality creation. It may be objected that such a planetary unity of understanding of our nature will limit human freedom, stultify creativity and diversity and take the mystery out of life. Not if it is correct; if it, indeed, is in keeping with the deepest elements of our genetic makeup it will do just the opposite. It will relieve survival pressures, honor and foster individuality, create the greatest latitude for multiple realities and support and accelerate our expansion into greater and greater dimensions.

My point is extremely simple: unless we know who we are and how and where we came from we will remain at odds with each other over the definitions we give as to what we are — and therefore, what we are about. Until we can overcome those barriers and know how we are generically human, we will not attain the positive common planetary ground that is the essential context needed to support the greatest degree of individual freedom and complex variety of human activity, exploration and expansion. All the political, economic, social, scientific, New Age, evolutionary variables are only symptomatic; the genetic level is where the profound realization, enlightenment, must blossom. In addition, because it restores our true history to us, the new paradigm brings a previously unattainable, unassailable integrity, individually and as a race, which will be essential to entering into direct contact with alien species — either in the future or with one that is already here on the planet. For that appears quite clearly to be the event that will cause the turn to a new civilization if we do not do it for ourselves.

As curious, strange, fantastic, stunning — or stupid — as any of the concepts advanced in this book seem to any reader, the fact is that there are already a small number of humans who have already assimilated and surpassed them as theory, and actually live as what I have called the "new human." I assume the minor premise that evolution and, particularly, our special case of it, progresses through individuals extending and living according to their unique potential's fullest expression. Demanding that the individual wait for any obstacle or condition facing the race to be overcome or perfected before exploring the next phase is counterproductive and not to be imposed on anyone.

We have at hand the common understanding of our racial beginnings, of our bicameral nature, of our common history and expansion over the face of the planet, of our relationship to the Nefilim and to each other. The awareness and comprehension of this fundamental, generic, racial self-knowledge I have called genetic enlightenment. It is the basis for a profound unification of all humanity. Not some superficial, social homogenization of peoples, cultures and philosophies but a unification that frees us to be one race and explains and enhances our diversity of adaptations and cultures and contributions.

With the godspell broken, the Babel factor canceled, we can finally enter stellar society as a mature race which knows who and what it is, what is good for it and what is not, with whom or what it could interact, with whom and what it would be dangerous to make contact, with the minimum of preconceptions as to how things should be. I invite you to participate in an exploration and mapping out of the post genetic enlightenment context of collective and individual existence in terms of the unique genius, sometimes disconcerting accelerations, and precocious proclivities of our bicameral genetic heritage.

Santa Fe, New Mexico

May 1, 1998

PART I

A BRIEF HISTORY

OF OUR HISTORY

CHAPTER 1

TAKING ARCHAEOLOGY OFF "HOLD"

IT'S HARD NOT TO BE A "CHILDE" OF YOUR TIME

> . . .intellectually, Archaeology is in deep trouble, trouble that has been brewing for more than a century and has now reached crisis proportions.

Robert C. Dunnell
Hope for An Endangered Science
Archaeology Magazine, J/F 89

> This is the end-game of an age, be certain.
> The masque of the hapless hero
> With a thousand hang-ups
> Is over.
> Mark it well.
> Done, done,
> In one revolution
> Of that unknown, so familiar planet.

Neil Freer
Neuroglyphs

The rediscovery of the evidence for our being created as a race by the Nefilim through genetic engineering comes primarily to us through Archaeology. It is well, therefore, that we reexamine Archaeology as a science and listen to the opinions of archaeologists themselves about their discipline and its potential contribution to the new paradigm. It will be revealing in spite of itself. Archaeology is a unique science in a strange sort of way. Although it does not deal with human nature as Psychology does, it, nevertheless, has a bit of the humanity of sociology and the romance of history. Seldom do you meet a scientist in any other field whose eyes twinkle as an archaeologist's sometimes will when speaking of a find of ancient human artifacts telling of the daily drama of existence in a lost culture in the remote past.

Archaeology in Crisis

The wealth of recovered history grows, seemingly, at an ever increasing rate as advanced techniques from satellite imaging of hidden sites to sophisticated dating and even DNA analysis are implemented. By the admission of no less than many respected experts in the field itself, however, Archaeology has a serious problem. Although possessing masses of data and elegant techniques by which to analyze that mountain of evidence, Archaeology has no adequate context, no overarching, unifying schema in which to integrate and explain the bewildering complexity – and sometimes seemingly contradictory nature – of its data bank of facts. How bad is it? As if it was not awkward enough that there is a whole spectrum of theories, there is a tendency to believe that theories are all that there should be. And this in a scientific arena that is founded on the assumption that there is an objective order of reality that is there for the discovering.

It is the discoveries, rather than the philosophizing, of the ordinary archaeologist which changes our minds and culture. The ordinary archaeologist's mind has been set in graduate school in the traditional context and most exhibit little inclination to become deeply involved in the fundamental philosophical issues or even the "big questions" such as the rise of civilization, or even the origins of agriculture. Von Daniken has pointed out classic examples of this lack of depth. A single example: the archaeologist, when asked for whom our ancestors constructed colossal temples, always answers "for the gods." When asked to state for what "gods," the answer is always something like "the gods in nature" (lightning, thunder, the sun, etc.). But when asked how the "gods in nature" could therefore speak and give commands, detailed instructions, blueprints for temples, the answer is close to totally incoherent. The honest scholar, Julian Jaynes, when faced with this question, could only fall back on the explanation that all humans in those ancient times must have been schizophrenes, hearing and seeing hallucinations, not real gods.

Innovation Almost Always Comes from an Outsider

Von Daniken, long the center of a storm of controversy for his outspoken opinions that the artifactual archaeological evidence all over the planet speaks of alien visitation, has played a role very similar to that of one of the major founders – perhaps instigators is a better word – of Archaeology, Schliemann. Because of his sensational discovery of the ancient city of Troy, that had been considered, previous to his physically digging it up, mythical, Schliemann is, deservedly, included in the histories of the science of Archaeology. But Schliemann was not, at least by modern standards, truly an archaeologist. He couldn't have been one because, before his time, there was no science of Archaeology in the West. The Judaeo-Christian ethos precluded the consideration of man and civilization in ancient times due to the prevalent interpretation of the Old Testament. Although the Chinese were doing scientific archaeological excavating in 200 B.C. it took this free-thinking German man of commerce to go against the consensual grain and decide to actually go and search where the old maps and documents and legends said the fabulous city had stood. The pictures of Schliemann's beautiful wife wearing the stunning gold jewelry from the excavations of Troy splashed across the newspapers of Europe could not be ignored for long. Schliemann dug but did not excavate in the careful, measured and documented systematic form of the modern archaeologist and he unwittingly damaged or missed some evidence. But he, and other pioneers, opened the Pandora's box of the past – and eventually the cautious academicians, no longer able to simply ignore the facts, were forced to move in to claim the territory and make it a "science."

It is clear that Von Daniken has done the same in modern times. He may be faulted by some for his methods but he has probed and prodded and goaded, asking the uncomfortable questions that cannot be answered in the context of current archaeological theory and presenting sensational evidence for facts too awkward to fit consensual theory. In the eyes of the consensual scholar he is a trouble-maker. But just as with Schliemann, the core concepts will inevitably have to be assimilated by the scholars – and, perhaps, a new branch of science with a name suitably academic will be formed. I must emphasize that I have been speaking accurately I think, in general about the "ordinary" archaeologist and that I also recognize the many exceptional minds who are dedicated to such a necessary discipline. Among them are many who see the condition of their science (some even feel that it is not yet even a full science) as I have portrayed it here. Nevertheless, because of a private recognition of and frustration with this tendency to inconsistency and inadequate thinking, the archaeologist who would rather just dig and investigate, tends to leave the arguments to the philosophers of history or science. And how do they deal with the fundamental problem? By dealing with its symptoms, not its cause. It is generally assumed, in professional Archaeological circles, that the crisis of Archaeology is caused by a lack of quality of interpretation. The problem is almost always formulated in a context which would tend to pit the traditional, interpretive methodology in Archaeology against the modern hi-tech precision techniques of measurement and investigation. Within the field, this condition has brought some to blame the classical attitude (cast, often, as an inadequate, outmoded "humanist" modality which must give way) for the perceived deficiencies, a lack of precision and resultant inaccuracy in archaeological explication.

Their proposed solution: a scientific method, based on an evolutionary context recognizing both genetic and cultural trait transmission, implemented by refined existing and emerging techniques of dating, material analysis, magnetometry, resistivity, photogrammetry, deep-probing radar, computer-assisted information analysis, etc. The adoption of this scientific approach is seen as overcoming the acute problem threatening the essential vigor of the discipline by furnishing a rigorous basis for interpreting the archaeological record.

Those who would not relinquish the value of the broader classical investigative approach caution us not to lose the overview and opportunity for appropriate synthesis while also advocating the use of the most progressive and advanced techniques to implement the work. Of course, the skeptical wags are heard claiming the impossibility of any real scientific method of duplicable experiments under controlled conditions. By the nature of the subject, they say, the only rigor is in the digging and one might as well leave essential theory to the science fiction people. And besides, the refrain goes, archaeologists are naive adolescents when it comes to clear philosophical methodology. There are those who, just as vigorously, espouse an intermediate position, articulated well by Stephen Jay Gould, as the best solution to the methodology problem. They hold that there is no one "scientific method," only methods appropriate to individual disciplines, that there is a crucial distinction between fact and theory. "Einstein's theory of gravitation replaced Newton's, but apples did not suspend themselves in mid-air pending the outcome." (S.J. Gould, Discover magazine, May 81).

Ultimately, the common goal is to discover law-like generalizations that are conducive to description, covering explanation and prediction. There is a serious concern, voiced by some in the field, that there are also deeper epistemological problems manifest in archaeological theory. But the authors of the generally respected text *Archaeological Explanation: The Scientific Method in Archaeology* (Watson, LeBlanc, Redman) make it quite clear (while acknowledging that the fundamental epistemological problems focused on by the philosopher as such are How we know and How do we know we know, etc.) that the archaeological discipline begins with the frank assumption that there is a discoverable reality which we can know. In this view, the most acute potential problem hinges on the concern as to how one attains accurate knowledge of the past and what the justification for wanting to do so might be.

It has even been suggested by some that "this 'critical theory' foment could lead to 'terminal skepticism' due to the objection by some who feel that every attempt to reconstruct the past can only be viewed as a projection of the present, designed to serve the interests of the present." But, in spite of the tendency to what Clifford Geertz has called "epistemological hypochondria," the apples continue to fall from the tree, the trowels continue to carefully remove the soil. The concern is more for the practical law-like generalizations drawn from the social sciences which can be taken as givens so one can get on with the digging. I submit, however, although all these perspectives have undeniable merit, they are only symptoms of a far more fundamental problem. The root cause of Archaeology being on "hold" does not lie at the relatively superficial level of methodological dispute, the need for technological implementation, a lack of interdisciplinary communication and cooperation, or even epistemological concerns or lack of philosophical sophistication. Ultimately, the hope, although often professionally and ambiguously disavowed, is that there will be discovered a synthesis, an overarching coherent paradigm which will constitute a sort of unified archaeological field theory. Most fundamentally the cause of the intellectual crisis in Archaeology is a socio-cultural bias, a mind-set that permeates Western culture and, therefore, Western Archaeology and which precludes a comprehensive paradigm, already extant, which is precisely the contextual overview being sought.

The history of that mind-set is fascinating. Although treated generally only as a nuisance, the perennial Scopesian struggle of the evolutionists with the Creationist ideologues in the academic and legal arenas is a far more significant formative agent of the public and intellectual character of Archaeology as a discipline than we would like to admit. We are still battling the Creationists' political power plays from such a militant posture that intramural questioning tends to get squelched as an internal security threat. This habitual defensive posture is most destructive in that it tends to foster preclusive thinking. Over the last century, in the heat of the battle "to overthrow the dogma of separate creations" (C. Darwin, *Descent of Man*), which forces the acknowledgment of a chronology of eons rather than a 6000 year existence of the earth and establishes the fact of humans existing in very ancient times, we espoused Darwin's concept of evolution. It is, I suggest, an alliance which has proven to be flawed not because the general concept is erroneous but because of a particular interpretation of it.

Even though we have been arguing about the theory of it's driving mechanism since Darwin's time, we have held to the concept of "the fact of evolution" as a continuous, lineal progression from more primitive forms to us so tenaciously that our thinking has become a preclusive block to any evidence for an alternate interpretation. By this I do not intend some sort of reconsideration of the specious ideology of Creationism, I mean an interpretation of our specifically human form of evolution different from the currently consensual in that it is understood as partially non-continuous with preceding forms, has a radically unique rate of development from its inception approximately 300,000 years ago and has dimensional characteristics that are unique, even novel. I suggest this is the refined model of human evolution that should be adopted as the working context of the new Archaeology on the basis of the following interdisciplinary evidence.

The Biochemical Contribution

Douglas Wallace of Emory University began studying the mitochondrial organelle DNA in the cytoplasm of the cell in the 1970's. The mitochondria is a tiny, bubble-like energy producing element in the human cell. It is thought to have been symbiotically incorporated millennia ago.

It had been established previously that, although the mother and father contributed randomly to the nucleus DNA, the mitochondrial organelle's DNA came exclusively from the mother. This furnished a unique, elegant tool for tracing the most ancient gene sources. Wallace's conclusions were that, depending on the technique used, the source could be considered either Asia or Africa. Most importantly it could be determined that the time for the appearance of human genes was approximately 200,000 years ago, plus or minus 50,000 years.

Rebecca Cann, working with placental tissues from all over the world, studying the mutations identifiable in the mitochondrial DNA, has found the same evidence as Wallace but would hold for an African origin as would Wilson of Berkeley. The inescapable conclusion is that we are a genetically very young race with a beginning around 250,000 years ago. This area of investigation does not address any interruption in the evolutionary progression of species as such; it is focused on the data for indication of time and source of inception. But the conclusions are significant to the argument developed here as shall be seen.

The Anthropological Contribution

In the intense discussions concerning the theory as to how the process of evolution actually works, there has been a gradual shift of interpretation from the strict Darwinian emphasis on the struggle for reproductive success within any given population toward adaptation at the level of the individual, i.e. toward behavioral ecology.

The Observable Fact: Human Uniqueness

A greater and greater gulf has developed in our thinking about the differences between ourselves and previous hominid forms. The blurring of the differences between the previous conceptions of "primitive" and "civilized," coupled with our having to retreat from the concept of progress within evolution, tends to redirect our thinking to a consideration of our species being basically the same from its inception: human progress is unique in its speed of development and precocity relative to anything observable in any other species of this planet. These three vectors taken together mutually reinforce. I suggest that the interpretations are changing because we are experiencing more and more difficulty integrating specifically human evolution into the slow patterns of evolution as they have manifested themselves over time measured in millions of years. It took Homo erectus 140,000 years to progress from rough flaked stone tools to smooth ones and some use of fire. If we have moved from square one to traveling vicariously to Mars in 1992 in a brief 250,000 years as a new species, there is an obvious radical difference in speed of evolutionary progress. And the way in which we do it clearly is adaptive, precociously so. If the current reasonable and intense discussion in the professional arena were limited to these variables only, things would be just fine. But the crux of the matter hinges around a cluster of factual vectors, the import of which is almost unthinkable, which we are precluding from the intellectual arena a priori, on the basis of a socio-cultural emotional mind-set rather than any reasonable examination in a fact forum. If you do not believe this, simply read the next paragraph.

It is not that the particular interpretation of the fact of evolution as an unbroken linear process from more primitive forms to humans causes us some slight handicap. This concept of our inception as a species is particularly detrimental because it precludes us from dispassionately and seriously reconsidering and re-evaluating the two major premises of all the ancient civilizations of the Middle East: the "gods" were real and created the human race. (I was correct about the visceral reaction was I not?) We all enthusiastically agree with Margaret Conkey "The appeal of Archaeology lies in its ability to think the unthinkable and to imagine all sorts of human pasts. Every decade yields not only new finds and techniques but new ways of thinking about the past as well." But to do it, ah, that's the visceral rub. Why should we bother to consider or reconsider these two central principles which those civilizations went to such pains to reiterate, record and preserve for the future concerning themselves? Because the reasons we have given ourselves previously that they couldn't (does the word ring any bells?) be so have been so badly eroded over time that they no longer are substantive or valid. Because a cluster of interdisciplinary data vectors has developed over the last one hundred years, the resolution of which not only demands the reconsideration of these unthinkable premises, but attest to their literal truth.

Consider first the reasons why we always "knew" the "gods" were unreal, fictional, myth. I do not mean here the local nature spirit projections and unscientific guesstimations of simple peoples; not the political creations of local shamans; not the theological institutionalized fictions thrice removed centuries or millennia after; not the symbolic codifications of specific psychological states; not the cosmic fu-fu of some modern lazy minds. Campbell and others lump all those diverse elements together in a confusing amorphous blob creating serious confusion. I speak here of the "gods" of "the world's first great trans-regional civilization, the one that cannot have been influenced by already developed civilizations", that "vast nexus of urban societies that centered on the Nile and the Mesopotamian rivers" (Henry T. Wright), the Anunnaki/Nefilim. (It is informative, in itself, that it is only the second dictionary definition of "myth" meaning fictional, unreal, as opposed to the primary meaning of "ostensibly historical content whose origin has been lost" that is always applied.) The belief that the "gods" (the Anunnaki of the Sumerians, the Nefilim of Genesis) must have been fiction because the technology and weapons ascribed to them, i.e. atomic bomb-like or laser-like forces of destruction, their ability to fly through the atmosphere and space, to communicate over long distances, to create humans, etc. were fantastic (to the minds of very recent generations) have been ruled out by the simple fact that our current level of technology is at least equal to any of that described in the ancient documents. Nagasaki and the Star Wars project speak clearly enough in themselves. The belief that the "gods" must be fictional because it is impossible to travel in the "heavens" has already been destroyed by our astronauts hitting a golf ball around on the moon.

Mythology as the Greatest Myth

The belief that the "gods" must be myth because the "divine powers" that enabled them to perform the "superhuman" feats such as living for thousands of years and restoring the slain to life were unbelievable is negated by the relatively near-term potential for us to do precisely that through nanotechnology (see Drexler, Eric: Engines of Creation) and genetic engineering – just as the possibility of genetically engineering an entirely new species of humans is in our grasp. The ability to create humans is not just something ascribed to them as a mythic power but is in our newspapers as a 5 billion dollar program to read our entire genetic code and in our laboratories in the ability to clone, to create. Which is precisely the crux of the matter. But isn't it enough to negate the entire consideration that those ancients claimed that the "gods" came from a planet in our solar system that we "moderns" have not even discovered yet? Not if you can give credence to NASA, JPL and reputable astronomers. Ten years ago, a ten page article in Astronomy magazine (Oct. 88) detailed the considerable evidence deduced from telltale effects on the path of the orbits of known outer planets, Neptune and Uranus concerning this Planet X, Nibiru, and the opinions expressed by various astronomers within and without NASA that it exists. It is difficult to locate because of an extremely elliptical, comet-type orbit that takes it close to the inner solar system only in a periodicity that they estimate at 1000 years.

There is an unfortunate closed loop *pas de deux* being executed between astronomers and archaeologists at this point. If the astronomical material were to prove out, the archaeologists might revise their criteria and their position. If the archaeologists were to revise their position then the astronomers would feel free to take a look at the archaeological material. Meanwhile everybody is shouting back and forth over their professional back fences about how much we all need interdisciplinary communication but afraid to move their peer-petrified duffs. If the consequences were not so vital the situation would be as humorous as the Abbot and Costello "Who's on first?" skit. It's about the same intellectual level . . . No longer can we deal with this most unquestioned, most unproven of assumptions in a way epitomized by Gordon Childe: "Perhaps these gods were projections of ancestral society . . . But the gods, being fictions, must have had real representatives, nominally their specialized servants, who must have done much to give concrete form to the imaginary beings, and by interpreting must have invented their desires." (*What Happened In History*, p.100). (Note the word "must").

I suggest that Hemple, favorite philosophical authority of the authors of *Archaeological Explanation,* in their development of a model scientific methodology and thinking, would cringe at the circularity of such sophomoric postulates and logic upon which the interpretation of those ancient societies has been based. How many times have we heard specious "logic" like the following: The gods were fictions. Men must have invented them. Men that invent fictions are unreliable in what they say. Therefore when those men said that the gods were real they could not be relied on and were propagating fictions which they were inventing. Yes, the Sumerians spoke of them, sculpted, drew them, treated them, as physical, imperfect, real anthropoid beings with whom they could converse and from whom they could receive even written instructions – but after all, "gods" are supernatural like God, right? And the supernatural went out with the Bible and Moses in Darwin's time didn't it?

And since we know from Darwin that we are just another animal in a long line of animals then our beginnings must have been in savagery, right? And savages in the "state of nature" always war on each other and are simple-minded and project fantasies. Yes, the Sumerians had every institution we have now in their civilizations but they must (that word is a classic clue) have been half-savages just turned nouveau urbanites and naive fantasizers who imagined "gods." Therefore how can you even consider that the "gods" were real when we know that they were the creations of simple-minded fantasizers whom we know were fantasizers because they imagined "gods"? And besides, we all know that physical evidence is more reliable than written (a piece of chamber pot from Jefferson's privy is more accurately informational concerning early U.S. society than the Constitution?) because the writers were priests and priests were the primary propaganda inventors of "gods," working from self-interest, right? At least we know they were priests because "temples presuppose priesthoods." If the "gods" were real they would have been palaces but we know the "gods" weren't real so they must have been temples. And, besides, even though it all stacks up we don't have any interplanetary Nefilim trash . . . And on, and on, and on. . .

The convoluted tangle of thinking and projections we have inherited related to the primary principles of those civilizations is clearly the most god-awful mess of self-referential "logic" in all of western culture. (Pun intended.) It's hard not to be a "Childe" of your time. I submit that there is not a single valid postulate which can be advanced as even the beginning of a proof that the Nefilim were fictional. All that is left is a residual non-scientific emotional bias. And, indeed, over time the objections have mutated into positive indicators, vectors pointing at a new, if unthinkable, synthesis.

But to recognize the possibility for an extraordinary interruption, a break in the natural evolutionary process with an act of genetic engineering, even to recognize that all the "reasons" we have given ourselves for that central belief that the "gods" were unreal no longer hold, is only to clear the arena. Among this cluster of interdisciplinary vectors pointing to a shift toward a constructive reexamination of those unthinkable premises upon which an entire civilization was based there are, I submit, archaeological components which have preeminent positive significance.

The newest discoveries continue to swell the volume of information enormously. Only several years ago an Iraqi team of archaeologists from the University of Baghdad under the direction of Dr. Walid Al-Jadir discovered a trove of clay tablets at the ancient site of Sippar. In a sealed room measuring 6 by 9 feet they found 2500 intact tablets, neatly arranged on shelves. (It is unfortunate that Saddam Hussein will obviously not allow this material out to the world.) Over 400 tablets have been extracted and processed at this writing, giving the complete texts of important documents which we have only known from damaged or incomplete texts as well as new information about our past and beginnings. Already the history of our creation has been found repeated in concise detail. It spells out, among many other events, the work strike of the lower echelons of the "gods" that forced the creation of humans to take their place. These documents also speak in detail of a flood as a pivotal event in the "gods'" history, even more so for humans in that it was used as a deliberate attempt to wipe out the human experiment. It should be noted that the theory that the entire Antarctic ice cap could slide off on a layer of built up slush with sufficient perturbation of the earth, due to the configuration of a number of geological factors, was published in Scientific American at least a decade ago. Such a monstrous event would be relatively short-lived but cataclysmic in its effect. (Current scientific estimates indicate that even a warming of the planet by a greenhouse effect could trigger such an event in slow motion.) Putting the interdisciplinary scientific data together is not difficult; the implication being that such perturbation occurs with the periodic passage of the "god's" home planet through the asteroid belt region and was anticipated and used by the "gods" as a means to wipe out the human experiment.

That the ancients knew unequivocally of that planet – and indeed of all the other planets in our solar system – is clearly demonstrated by the artifactual archaeological evidence. A single example: a 4500 year old Akkadian cylinder seal in the State Museum in East Berlin. Catalogued VA/243, it depicts the size (5 magnitudes of the earth) and position of that planet in our solar system, as well as the size and position of all the known planets around the sun. I respectfully submit that imaginary auditory hallucinations could not have given them that kind of quality information. Zecharia Sitchin's thesis integrates all of these vectors into a coherent interpretation that gives a positive answer to the simple question "What if, having been forced to acknowledge their possibility, those unthinkable premises so fundamental to those vast and sudden civilizations were literally true?" Having opened the door, he is able to give plausible explanation for enigmas previously resisting all efforts. Although Henry T. Wright has expressed some doubt that "such a synthesis will probably be realized not by a single brilliant scholar" I suggest that the contribution of the cross-disciplinary scholar Sitchin fills the bill (although Sitchin is the first to acknowledge the contributions of generations of scholars whose work paved the way for his thesis). I suggest that the Sitchin phase, resulting in a master synthesis based on nothing but the archeological evidence, is the contextual "critical theory" that Archaeology needs to take it out of the holding pattern which is the source of the threat of "terminal skepticism." I am confident it will constitute the operational platform for and form the real character of Archaeology in the 21st century, showing the origin of the suspected extraterrestrial thesis to be from our solar system, explaining the creation of man as a hi-tech genetic engineering process, revealing the nature of the "temples," ziggurats and pyramids and generally correcting and making precise what had been previously only inaccurately guessed. (See: Sitchin, *The Twelfth Planet*; *The Stairway to Heaven*; *The Wars of Gods and Men*; *The Lost Realms*; *When Time Began*; *Genesis Revisited*, Avon paperbacks.)

The scholarship and linguistic abilities that Sitchin exercises on this material form a solid foundation for the conclusions he has reached and the facts enable him to untangle the knots of previous scholarly dilemmas and mistakes. He is one of 200 people in the world who can read the Sumerian on the ancient clay tablets or carvings. If he is as correct as I believe him to be he should receive the Nobel prize.

The museums, libraries and collections of the world have accumulated well over 500,000 pieces – probably much closer to a million at this writing – of artifact, art and documents. The tremendous amount of information now available to us may be grasped through a few examples. From the 4000 year old city of Nineveh we have recovered the better part of the famed library of Ashurbanipal. This tyrannical but intellectually sophisticated ruler who claimed that he had learned from the professional scribes (copiers, document writers) how to read the documents written in the older languages, had collected a vast library (the section on astronomy and astrology alone was the equivalent of 25,000 of our modern volumes) of those writings and history from before the Flood (a historical dating marker used throughout the ancient world). His huge library burned down and fired the clay tablets into a sturdy ceramic that preserved them, fortunately, for us.

Sitchin's conclusions may be summarized using a paraphrase of the statement of objectives made in his prologue. Using the Old Testament as an anchor, and submitting as evidence nothing but the texts, drawings, and artifacts left us by the ancient peoples of the Near East, he proves that Earth was indeed visited in its past by astronauts from another planet just as the ancient peoples themselves said that superior beings "from the heavens," the ancient "gods," came down to Earth. The Nefilim, mentioned in Genesis (chap. 6), whose name in the ancient Sumerian translates correctly to "those who were cast down upon earth," were an advanced race who landed here approximately 450,000 years ago to set up a colony with the purpose of mining gold. The pictograph sign, term, for the lords in the Sumerian ("god" or "gods" was not in their vocabulary) was a two-syllable word: DIN.GIR. It can only be translated as "the righteous ones of the bright, pointed objects" or, more explicitly, "the pure ones of the blazing rockets." There can be no doubt that they were "living beings of flesh and blood, people who literally came down to Earth from the heavens." The Anunnaki, as they were called in the Sumerian, were the Nefilim of Genesis. They were aliens but from within our solar system, not from another star system or from another galaxy.

He identifies the planet from which these astronauts came as a tenth member of our solar system. Known in the popular press as Planet X but to the ancients as Nibiru, it is now being searched for by the Naval Observatory and NASA. The existence of this last planet in our solar system is suspected by a number of astronomers because of the purely scientific, observational clues that have been discovered over time.

The similarities between the data being accumulated by NASA and the information translated and interpreted by Sitchin from the ancient records concerning the home planet of the Nefilim are clear and impressive. The ancient texts plainly show that this planet is 5 magnitudes of the earth, orbits in an extended elliptical path that is tipped up from the plane of the other planets by nearly 45 degrees. Whereas our NASA astronomers theorize perhaps a 1000 year long orbit, the ancient records clearly speak of a 3600 year orbit. This difference may well be due to the fact that the ancients knew that Nibiru orbits in a direction opposite (clockwise) that of the other planets, passing through the region of the asteroid belt in its closest approach to the sun and then moving outward in an extended elliptical orbit far beyond Pluto. The implication being that the physics of Newton's and Bodes laws would require a quite different periodicity if this planet does indeed travel in such a unique path.

He lays bare ancient reports of a celestial collision, the result of an intruding planet being captured into the Sun's orbit in the very early stages of development of our solar system, and shows that all the ancient religions were based on the knowledge and veneration of this tenth member of our planetary system since it was recognized as the home planet of the Nefilim, of the gods. (The ancient way was to count the ten planets plus our moon and the sun as the principle members of our solar system). He submits recovered texts and celestial maps dealing with the artifactual space flights to Earth. He describes the Nefilim and shows how they looked and dressed and ate, glimpses their craft and weapons, follows their activities upon Earth, their loves and jealousies, achievements and struggles, and unravels the secret of their "immortality." He traces the dramatic events, an actual revolution of the lower echelons of the Nefilim who were fed up with their onerous mining tasks, which prompted their geneticists to engineer a creature to take their place that led to the "Creation" of Man, and shows the advanced methods of genetic engineering by which this was accomplished. The first attempts combined the genes of a number of animals with those of Homo erectus and the results were generally bizarre and frustrating; the results also indirectly gave rise to part of what has traditionally been considered mythology, tales of creatures that were sometimes strange combinations of very different animal and hominid characteristics. Finally the decision was made by the Nefilim to combine their own genes with Homo erectus and we were the successful product. They apparently had advanced skill because they were able to control and predict the sex of the individuals produced.

The first successful creatures were called Adaba in the Sumerian –the Adam of Genesis. He then follows the tangled relationship of Man and his deities, and throws light on the true meaning of the events passed to us in the tales of the Garden of Eden, the Tower of Babel, the Deluge, the rise of civilization, the three branches of Mankind. The Garden of Eden tale in Genesis is a condensed version of detailed accounts in Sumerian texts of how Man, when first "invented," could not procreate as is the usual case with mutants; how that ability –knowing" in the sense of "Adam knew his wife" – was given to us; how the genetic experiment, as a result, got out of the Nefilim's control leading to the expulsion of the new creatures "to the east," the outback, out of the area where they were kept as slave-servants, the area reserved to the Nefilim. Over an extended period of time, our ancestors acquired knowledge, skills, experience and eventually began to copy their masters the Nefilim.

The Tower of Babel tale is, again, a condensed version of the historical episode concerning the discovery, by the Nefilim, of the fact that men had gone so far as to try to duplicate one of their rockets – we do not know, in detail, how accurately – to the extent that the Nefilim agreed that, if men could progress that far, they could do, eventually, whatever they really set their mind to do. As a result the Nefilim deliberately brought about a divisions of languages, on the basis of divide and conquer or, at least, control effectively. The story of the Deluge marks a point when the Nefilim had become so dissatisfied with the experiment, because of the proliferation of the new creatures, that they were ready to destroy all Men. The evidence shows that the Nefilim were apparently aware, as we have only recently learned in theory, that the Antarctic ice sheet could build up an underlying slush layer that would allow it to slide into the ocean under the influence of a sufficient disturbance. They resolved that, when their planet returned to the vicinity of Earth again causing a strong perturbation of this planet, they would let the ensuing flooding destroy all humans. It is possible that, if it were not for one of them, Enki, who was sympathetic to the humans, who secretly instructed Noah, one of his favorites, in the construction of a craft this book would not be written, and you, dear reader, would not be around to read it. We came very close, at that point, to a deliberate extinction.

After the Flood and our near annihilation it was decided to perpetuate our kind for our practical value and we eventually became partners with the Nefilim. When humans became too plentiful the surplus was apparently literally expelled into the regions, the "out-back," surrounding the Nefilim centers. An understanding of the differences and the similarities between the culture of the humans in the Nefilim dominated centers and the culture developed by humans on their own is the key to the common elements and traditions found throughout the cultures of the world that has so intrigued and puzzled scholars for decades.

The character of the major cultures of the world can be seen to be the reflection of the personality of the Nefilim rulers of the geographic area in which the culture developed. We will examine in a later chapter the development of those differences in character as exemplified by the male-dominant Semitic culture of Yahweh's area as contrasted with the female-dominant cultural character of Ishtar's India. Sitchin then shows how Man – endowed by his makers biologically and materially with "instant" civilization, knowledge, and science as a result of the deliberate choice to preserve Man as a useful tool after the Flood – ended up crowding his gods off Earth.

Apparently the Nefilim phased off the colony between 1250 to 200 B.C. The decision to leave us on our own on the planet is analogous in many ways to the decisions struggled with by NASA in their projections of how to contact or not contact an intelligent species we may eventually discover on some planet around another sun when we begin to travel among the stars. He shows that Man is not alone and that future generations will have yet another encounter with the bearers of the Kingship of Heaven. The politics of that encounter, in light of the strong independence developed in our precocious metamorphic evolution, may be critically significant to our future and should be given much thought before the fact. The degree of planetary unity required for the most advanced and intelligent determination of our common posture in that event does not yet exist. The genetic enlightenment afforded by the ramifications of Sitchin's resolution of the last critical pieces of the puzzle of our beginnings and our history is the key to that unity.

Sitchin's thesis, at this point in time, is controversial. It is not accepted, at least publicly, by the archaeological community. But it apparently is enough of a "threat" to the consensual view of the archaeological establishment to provoke a typical reaction. The editors of Archaeology magazine saw fit to publish an article by Stephen Williams of Harvard and the Peabody Museum ("Fantastic Messages from the Past," Sept./Oct. 88) which is a poorly disguised attack on Sitchin himself. The potential of his thesis is apparently enough to warrant bringing in some great archaeo-communicator to warn innocent professionals and simple layfolk about a "rogue scholar." Preferring *The Clan of the Cave Bear*, Williams dismisses 40 years of research, which "all seems to fit; it flows and the evidence seems to be there" as a "bit of von Daniken warmed over." Substituting ad hominem innuendo for scholarship, using Bigfoot, channeling, buzzwords and buzznames as props, he brands Sitchin's thesis as "impossible" – while acknowledging it plausible enough for even him to believe contingent only on seeing a piece of Nefilim trash.

If Williams savaged a consensus-establishment scholar, right or wrong, in such patronizing, shoddy and cavalier fashion he might well be censured by his professional society or employer. But no need for decency if you judge the victim does not have enough consensual clout; get him quick and dirty on the way to the fact forum. The archaeo-McCarthyism of "I hate these messengers who cannot, or will not, tell truth from fiction" is near theological. He might more correctly have called the course on Fantastic Archaeology he was teaching at that time in Harvard Heretical Archaeology and his "debunking shop" (!) an office of the Archaeological Inquisition. It is interesting and significant that Williams, who has set himself up as a "debunker," a sort of self-appointed Amazing Randy of anthropology, declined my solicitation for a concise and professional contra Sitchin article to be included in a pro and con symposium on the grounds that he was not expert in Middle Eastern Archaeology as his specialty is the North American area! This type of non-professional attack is typical of situations of impending radical change in more than just the academic community. I have accepted Sitchin's thesis after analyzing it for a number of years. I am not claiming that his thesis is the last definitive word in every detail, only that the work should be included in any serious fact forum and that his thesis and conclusions are substantially true. The weight of the evidence he presents and the supporting data from the sciences is far too strong to be ignored. There are some individual members of the archaeological, astronomical, and academic communities who have expressed interest in it but have difficulty when faced with the decision to recognize what has been traditionally considered unreal myth as simple historical fact. Peer pressure and the radical nature of the change required make it difficult.

Sitchin's synthesis of the information from Archaeology, history and the Old Testament unites the mainstream interpretations with that of the dissenting school of thought simply by correcting the deficiencies of each. The breadth and detail of Sitchin's work can only be summarized very briefly here; there is much to be gained by a thorough study of his writings. Although the implications of his work are perhaps not fully realized as yet, I predict that his work will have the relative impact in time as Darwin's *Origin of Species*. The significance of the evidence presented by Sitchin should be immediately, overwhelming convincing. Yet the fact that its ramifications do not seem apparent to many persons exposed to them is a classic example of how widespread conditioning and programming can seriously cripple human thought. The Sumerians and subsequent ancient Middle Eastern peoples based their entire civilizations and cultures on two fundamental principles; the "gods" were real, flesh and blood humanoids with whom they could associate and communicate and those humanoids created the human race. It is the reasons we, following long-standing traditions, have considered these humanoid colonizers of our planet to be mythical, fictional, that we will consider next. Few seem to have noticed that those "reasons" have become totally invalidated by our current scientific and technological knowledge and capabilities.

Any evidence that the Bible is anything other than what the various religions based on it claim it to be is rejected a priori. Any major contradiction of the prevailing doctrine concerning our direct evolution from prior anthropoid forms tends to be rejected by the scientists most concerned with that theory as well as the proponents of the various theories concerning the even earlier evolution of life on earth and where it came from. Astronomers and geologists and astrophysicists who have laboriously worked out theories of the beginning and evolution of our solar system tend to react negatively to a gratuitously given, detailed account of the actual historical facts of that development coming from a time when they assume humans did not have the capabilities of discovering it. But, as with any new information, we can only hope that everyone will approach it with an open mind and that their private context will not be too threatened. But think about it: have we not been telling ourselves for some time that the real story could be that way in our science fiction? Have we, indeed, not been projecting a hope of such discovery – Tillich has called it our "ultimate concern" – as if from some deep source in our collective unconscious? Is it not beyond coincidence that, when the facts are made known, we recognize something startling yet strangely very familiar as if almost programmed in our genes. . .?

We should not be embarrassed about the naiveté that we have exhibited collectively over centuries; if it were not for the fact that we have reached the level of technology we now possess we would probably still be in the same position, under the godspell. It is critical to appreciate the fact that unless we had attained a grasp of the fundamentals of genetic engineering, of rocketry and space flight, of geology, of the nature of our solar system etc. as we only recently have acquired them, and reached the level of scientific achievement we now enjoy, we could not even have understood the real nature of the events about which Sitchin is writing. He has succeeded in slipping into place the last pieces of the puzzle of human origins, setting in proper perspective, once and for all, an important historical event on a medium size, ordinary planet orbiting an ordinary star located in the outer regions of an ordinary galaxy: our creation as a species and a race. Since the event involves all of us intimately we call it important; in the overall perspective of the universe it is, most probably, relatively trivial. It also seems reasonable to allow the possibility that we, being the successful and enduring product of a planetary experiment, may be something of a curiosity to whatever beings are capable of studying or learning about us. The interest, on the part of whatever imaginable or unimaginable entities could simply be curiosity that makes us a local tourist attraction or it could be a more serious sociological study. It is even reasonable to assume that there is an ongoing observation and monitoring of this planet by the Nefilim, providing that their race has not met with some catastrophic disaster.

We may have been able to remain blasé to the fact that pioneering explorers and archeologists have repeatedly uncovered the physical reality of cities and civilizations that formerly were considered to be only the content of legend or myth; previously, may have been able to hold in skeptical suspension our judgments concerning the reality of gods, demi-gods, kings and dynasties and high cultures the sophistication and frequent splendor of whose material remains and records now lie in our museums for all to witness; may have been able to previously marvel, without drawing conclusions, at the stupendous feats of monumental engineering and organization written large across now remote or barren landscapes; been awed, in unresolved puzzlement, over the contradiction of mature law, trade, education, travel, economics, and advanced medicine, science and technology clearly evident in remote times the scholars have insisted were primitive; may have been able to labor as students and scholars, docilely submissive to the righteousness of "authorities," naively viewing the bulk of the history of the world previous to the Greeks as somehow the irresponsible and unreliable figments of innocent, ignorant, even primitive peoples probably somehow less human than ourselves; may have been able to sidestep the multitude of discovered artifacts, ooparts whose tangible logic speak eloquently of our predecessors being technologically and socially like ourselves rather than the grunting savages the cloistered savants would have us believe; formerly taken refuge from the responsibility and risks of personal education and evaluation behind the robes of ecclesiastical or institutional dogmas acknowledging as unquestioned inspired authority the very texts parts of which were claimed to be the product of less developed minds and to be taken metaphorically; formerly, been able to leave to the "experts" the explanation of the gross inconsistencies of our species' developmental patterns when viewed against the known sequences of previous species; capitulated to the hive pressure not to judge for ourselves what we could glimpse over the academic and religious barriers.

Our racial childhood, however, is now rapidly coming to a close and we are coming, typically reluctant and turbulently, out of our racial adolescence faced with the responsibility of self-determination and mature action. The point of critical-mass of information about our unique genetic creation and who we really are is inescapably upon us. The last pieces of the puzzle have fallen into place. The primary immediate effect of Sitchin's thesis allows us to recognize that there is an emerging world paradigm, an overview of the human condition coming into focus, which subsumes and resolves both the perceived problems in Archaeology and the "big questions" as symptoms. If the intensely introspective characteristic of our rapid evolutionary/metamorphic progress has tended to foster a sense of who we are individually, the new paradigm tells us fundamentally what we are. It revives us from a sort of racial amnesia and restores our true history to us. In providing that resolution it both defines the natural role of Archaeology and fulfills it and goes beyond the immediate benefit to Archaeology when we consider its ramifications. The new perspective reveals Archaeology as the discipline preeminently responsible for its conception and central in its future development and implementation. The restoration of our true history has far reaching implications for the planet since it reveals the common racial root that subsumes the superficial Babel factors of tribe, language, nationality.

Tenure Tetanus

But, ah, the visceral twitch. The remembered causticity of the loaded buzzwords. von Danikenism (his questions were correct if his methods were questionable); pyramididiot (if we share Brian Fagan's excitement, "the greatest revelations will result from months of patient detective work with the tiniest of clues. . ." then we should be in awe at Sitchin's elegant detective work (*The Stairway to Heaven*, chapter 13) proving Vyse's claim that Cheops built the pyramid we know by that name a fraud); "bullshit Archaeology" (boringly obscene); "alternate Archaeology" (now there's an archaeologically inquisitional one); "astronaut or astro-Archaeology"; "lunatic Archaeology" – all contributed by Glyn Daniel on a single frothing page of "passion and polemic" indiscriminately lumping everything, from the risible to the profound potential, acerbically together. Williams can only add "fantastic Archaeology" to the list as the name of his "debunking" college course in which he lumps Sitchin's scholarship with Bigfoot and Mu (Archaeology, S/O 88). Ah, the collegiate colitis, the tenure tetanus, the collective academic catatonia. . . There has already been objection raised that Sitchin has offered "nonstandard" translations and interpretations. How can a new view not be "non-standard"? Being so or not is no criterion of its accuracy; it is either correct or it isn't.

But there is a clear, developmental, evolving pattern of translation/interpretation from the time of the publishing of Eberhard Schrader's *Die Keilschriften und das alte Testament* (1872), George Smith's the *Chaldean Account of Genesis* and L.S. King's *The Seven Tablets of Creation* down to the present with the publishing of Sitchin's first volume, *The Twelfth Planet* (1976). And all these works are within the established scholarly context. We now have the capability to recognize a clear ascending spiral pattern in those questions considered fundamentally problematical in Archaeology. This insight affords a potential key to the resolution of those problems and exposes the even more troublesome socio-cultural mind-set that prevents us from reconsidering the new paradigm with anything approaching objectivity. We can easily agree with Brian Fagan that the "Archaeology of the 1980's will seem remote and arcane by the year 2050." Our advantage is that of hindsight. How many expert authorities do we have to see parade through the spotlight of our history's stage making retrospectively preposterous, definitive pronouncements that this or that could not be, before we recognize the necessity for more intelligent due process? I submit that we are at a point in our racial development where we can understand the pattern of our collective psychology sufficiently well to discern the mechanisms of and reasons for our preclusive thinking when confronted with such a thesis.

The Godspell Effect in Archaeology

This chronic indisposition of Western culture is neither subtle nor unfamiliar. The difficulty with bringing it into focus is that we experience it as a familiar condition rather than understanding it as a handicap. In its dreary generic symptoms it manifests as turfish preclusivity. It's root cause, for the sake of mnemonic dramatization, I call the godspell. For it is theological, philosophical, psychological and only incidentally scientific and a solipsistic effect of the very phenomenon it tends to prevent us from acknowledging. It is the ancient master-slave relationship, the godspell of near craven subservience, subliminally influencing our cultural attitudes. Modern scientific archaeologists influenced by old theocultural thinking!!?? If the shoe pinches, don't ignore it. . . We tried breaking the godspell with Darwin and only got it half right. (And, if a voice out there is heard to say "Well, thank God for Darwin regardless of how imperfect his work was," many of us will still not catch the irony in that statement.)

The Galileo-Bruno Syndrome

The problem of preclusivity: we deny the possibility of some information being evidence, rule it out in advance, before it is even considered as such in any fact forum. As we have seen, the first phase of the problem began with the advent of the Western Judaeo-Christian ethos. Its dominant character was theological; the doctrines of Christianity precluded investigation, archaeological and geological evidence, and imposed a Gallilean intellectual house-arrest on dissidents. No one wanted to be first to denigrate the Bible and Genesis. It could even be dangerous to your health. Witness: Jordano Bruno burnt at the stake – only 36 years before the founding of Harvard. In retrospect, we marvel at the power of a mindset which, by precluding the subject itself, prevented the very inception of scientifically oriented Archaeology in the West until a mere 150 years ago. Even in the 19th century it took 25 years for the new geological evidence to even begin to take effect.

But is it not a textbook truism, as Glyn Daniel states flatly in his *Short History of Archaeology*, that it was with the publishing by Lyell of *The Principles of Geology* in 1830-33 that science was freed from Moses and the Bible and marked a new era? Was not myth dispelled and true history finally on the ascendant? Hardly. Hindsight shows clearly that all we did was move from an Inquisitional style frying pan into a semi-secular philosophical fire. If the first stage was the result of theological crudity the next was even more clumsy, philosophically, causing the continuation of problems in Archaeology that are still with us. If the preclusive mentality of the first stage is clinically clear to us, that same attitude with regard to the linear, continuous interpretation of evolution applied to the human species is not. But there is no difference between the theological and the academic preclusive syndrome when it renders authorities indurate, those subject to their influence intellectually paralyzed, their discipline stultified, and licenses otherwise outrageous "passion and polemic" in the name of science.

We should not be too hard on ourselves since it seems clear that, in general, individually and collectively we have been sincere if incorrect. It is also clear that sincerity is not enough. If the psycho-sociological component of unquestioned authority enforcing unquestionable dogmas, preclusive in its attitude and pro-nouncements, first manifests in the Gallilean syndrome (we will not look through the telescope of your evidence because we know that we will not see what we know cannot be there) it leads rapidly to the outright persecution of those holding non-consensual opinions. The psychological component manifest by the dogmatizees is epitomized by the Cuvier mind-set. Brilliant enough to be the founder of vertebrate paleontology, with the evidence on his desk, Cuvier never-theless denied the possibility of fossil man in antiquity as something that could not be because the prevalent theology said it could not. Other scholars simply kept silent, protected themselves, or fell into a kind of intellectual catatonia.

I suggest that this condition is not to be taken for granted as "normal" in the process of evolution but that it is a direct symptom of our unique situation as a metamorphosing mutant species. It is a left-over, a secular continuation of that highly resilient mind-set which made part of the work of the early pioneers of Archaeology not just discovery but a proof that what had been considered myth or at least legend was physically real and true. It created the oversimplified con-flict between Catastrophism and Uniformitarianism, and even now prevents us from moving on to the next decisive step past recycled Scopesian melodrama. Practically, it is a gag order on the archaeologist and the thoughtful scholar. It becomes clear that the current tendency to "terminal skepticism" is a direct result of the old preclusive mind-set.

Robert Foley has emphasized that, in the last analysis, it is the way we con-ceive of human evolution that is critical to how we interpret the fossil evidence and the circumstances in which they are found. The early oversimplified Darwin-ian survivalism and insistence on linear continuity with primitive forms, forced the view of humans of even those founding civilizations as driven by blind sur-vival mechanisms, having somehow just stepped out of animalistic savagery, char-acterizable even as delusional schizophrenes. If we could not believe what they said about the most important, central facets of their lives and history, we had cut ourselves off from any criterion and source of accurate information. And now, the weight of the evidence from biochemistry, genetics, geology, astronomy, anthro-pology, space science and rocketry, those very interdisciplinary scientific sources which we have begun to look to for quality interpretation, has revealed the possi-bility of all the unthinkable, seemingly impossible things that we used as reasons to believe that view.

To compound the dilemma, we have developed, over time, an acute reflex-ive awareness of ourselves getting caught in this kind of professional bind, unwit-tingly straight-jacketing our thinking with preclusive postulates so many embar-rassing times in the past, we have begun to even doubt our ability to see anything of the past as it was and even, perhaps of the present.

The Generalist's Contribution

The generalist's interdisciplinary contribution is to point out that, eventu-ally, every discipline seems to evolve to a stage where it becomes appropriate to turn its own methods and data back on itself in a sort of reflexive self analysis. We have already seen the concept of the history of history formulated and written about profitably. We already have histories of Archaeology. Archaeology has matured to the point where a psychology of Archaeology as it has evolved is not an unreasonable study. And it should not surprise anyone if Archaeology has evolved generically similar to other disciplines.

Any discipline is founded, develops a character, a purpose, a methodology, a data base, a mode of interpretation, a language form of communication, interactively determined by the perceived need/value to know and the nature of the subject under investigation. It dynamically evolves under the driving force of that reciprocal engine in a process of gathering information and the forming of successively more comprehensive and adequate models of the reality assumed existing and discoverable. It matures to where there is developed a set of inherent parameters which are a function of the assumptions concerning the nature of the subject matter and the natural limitations of the methodology employed. The slowing of interpretive potential due to this circumscription is eventually perceived as a severe frustration and the first attempt to correct it is by an improvement of methodology. But the improvement of methodology being constrained by the more primary assumptions, usually unwittingly preclusive, about the subject itself, causes an inevitable deep intellectual crisis. The holding pattern causes a variety of individual reactions depending on genetic predisposition, imprinting, conditioning, intelligence, quality of training, etc. Typical reactions range from blaming predecessors, through peer savaging, bombastic dogmatic pronouncements, retreat into the displacement activity of preoccupations with details, to seeking interdisciplinary resource clues, deepening of philosophical perceptions, and, finally, courageous questioning and revamping of consensual assumptions and a redefinition of the subject, even of the discipline itself. Often the key to the turning point has to come from outside because no one in the professional group will "go first." We have seen it happen in other disciplines. I suggest the pattern is clear in Archaeology.

We are at a point where, having become enlightened sufficiently through the data of Archaeology and history itself, we can not only identify and evaluate the patterns of our processes of identification and evaluation but use that knowledge to avoid the errors, the unscientific emotional sets, the cultural traps, the preclusivity. We should even be able to use those known processes to predict and accelerate our scientific progress. We no longer need be restricted by the painfully immature psychological mechanisms identified by Kuhn in his analysis of scientific revolutions. In summary: we (Western culture) have passed from the Christian domination which was able to preclude and, therefore, suppress archaeological and geological investigation itself, through a time of partial transition in which we adopted the principles of Darwinian evolution while retaining a preclusive attitude with regard to the physical reality of the Anunnaki/Nefilim, the "gods." This preclusive position is partially due to our lack of technological development and understanding which would allow us to entertain the possibility of those capabilities in the past, and an over-simple overreaction to anything even mistakenly associated with the supernatural or the "divine." To get Archaeology out of the deep intellectual crisis it is in currently we need to make a second, interdisciplinary, transition from a preclusive posture to allowing Sitchin's thesis into the fact forum for rigorous examination, consideration, debate. And in that transition we can and should improve the nature and quality of humanistic-scientific due process. I support Eric Drexler's opinion that Karl Popper's standard for distinguishing facts from values is useful: a statement is considered factual (whether true or false) if, in principle, an experiment or observation could disprove it, if it is refutable.

Even if the 21st century may be the "century of the small object" in Archaeology it should not be the century of the small mind. In the light of the transformative nature of the new paradigm, however, the function of Archaeology will not be preoccupied with only microartifacts. The intellectual crisis of Archaeology is the symptomatic warning displayed by the bellwether discipline of our culture. The release that comes with the new paradigm takes Archaeology off hold and designates it as the preeminent discipline to provide and analyze the voluminous information and artifacts concerning this central fundamental event of our history as a race. The potential is already available to realize the expectation of Brian Fagan for the time when "the past will serve the present and the future."

The new paradigm is the context in which the broad classical approach and the new approach with its emphasis on hi-tech techniques can be united and subsumed. Even more critical to realize is that the context of the new paradigm is essential if many older and newer discoveries are to be interpreted at all. It is the context in which the entire field of Archaeology can be unified; it is the key to the synthesis hoped for by sincere professionals in the field for so long. The new Archaeology can make profound sociological contributions. If we have instinctively turned to anthropology more and more for the answers as to "how we should live" the new paradigm elucidates the nature of its contribution. We are returning to a time when scholars in all disciplines, from fieldworkers to theoreticians, even in the most esoteric areas are becoming aware that their work has a profound influence on the focusing and molding of human vision. Even the layman is becoming aware of how our knowledge and interpretation of the past influences our daily lives.

Archaeology as Sociobiology 1A

The new paradigm reveals that Archaeology contributes, in a far more profound way than currently envisioned by its most ardent proponents, to world unity. We are planetarially, generically human with a common origin. Taking Archaeology off hold in this way empowers Archaeology to contribute to taking the planet off "hold." Once the door is open to the possibility of there being an advanced alien culture on the planet in relatively ancient times we will begin to examine past history and archaeological discoveries with a new expectation and in a new perspective. An integrated archaeological world view is now possible. The new paradigm provides the context that fulfills the classical broad investigative, interpretive approach and eliminates the "epistemological hypochondria" and "terminal skepticism" and recycling of outmoded paradigms. It adds a profound humanity to the classical humanism. This will lead to a strong focus of technological search using the most advanced technology because we will know what we are looking for and where to look for it. The recognition of our generic humanity and common origin should promote a degree of cooperation among, in particular, archaeologists, that has not been seen before. The new context clarifies the objectives, focuses the worldwide search, gives cohesive vital meaning to our history, affords us a critical racial self-knowledge that is liberating, maturing and unifying. A critical realization will be the understanding of the relationship between the part of the new human race living in the Nefilim controlled, basically urban centers and those humans who lived in the "outback" (having been originally thrust there as surplus and forced to fend for themselves and develop an adaptive culture of their own).

The common tradition of origin, the preservation of fragments of the traditions and knowledge of the cultural centers by those moving farther and farther from those centers, the attenuation and corruption of history at greater and greater distance from its source, the tensions, interchange, and sometimes the conflict between these two groups as the result of their respective orientations I suggest is at basis of the enfolding of our real political and sociological history on the planet. It provides the context in which we can easily see the answer to the puzzling fact that the same "myths" and pantheon of the gods can be found all over the planet, how the American Indians can have forty two versions of the Flood story, why there is a common thread of general culture throughout the world settled by the migratory segments of the human populations.

Without too much exaggeration, Archaeology will now become not just a source of "raw materials" for written history but the fundamental ground for Sociology 1A. A few simple examples of the practical effects the change in interpretation should bring about will provide the reader with an indication of the way Archaeology should move.

Sitchin has already pointed out the location of the original Nefilim sites in the Tigres-Uphrates area as being geographically located precisely on radial points focused as a landing grid on the ancient spaceport as well as the same precision of radial location of Jerusalem with regard to the pyramids and other key points for the same purpose. There is indication of a worldwide grid system having been laid out by the Nefilim and a search for the marker points should prove fruitful probably by using these known sites as a beginning. The new paradigm provides sufficient context to integrate information from the finds of ancient human sites in the old and the new world. Having recognized our beginnings 250,000 years ago, the archaeologist should be relieved of the prevalent assumptions of a much more recent beginning that causes difficulties in reconciling the obviously well developed cultures whose traces lead us back into a much more remote past. If, indeed, Shamash was the main spaceport of the launching of the Nefilim spacecraft we should look most intensely there for possible high-tech remnants of their technology. It is true that metallic objects do not endure too long in that type of climate but perhaps some things could have lasted through favorable conditions.

If, indeed, the ancient site of Baalbek with its colossal foundation stones weighing a thousand tons was a landing place with counterparts in other places the archaeological investigations should focus on these features and a logical extension of that investigation would center on a determination of the significance of the sites for their favorableness for space launch and landing by correlation with the knowledge acquired by NASA and JPL in our time.

If, as reported, the excavations in Northern India where the Nefilim established their third settlement on the 30th parallel show that some of the buildings' thick stone walls are turned to glass to a depth of ten or twelve inches, we should indeed examine them for the effects of atomic radiation. The Vedic and other texts speak of just such a type of conflict being waged by the gods. The area of the Sinai peninsula, where Sitchin has pinpointed the destruction of a Nefilim spaceport by two of their members with devices resembling atomic weapons causing devastating fallout that killed many humans and almost killed two of their own kind, should be examined with this scenario in mind for traces that apparently still remain.

The island of Leucadia has always been held to be the vacation spot of the gods. If it was literally so, the interpretation of the archaeological finds there should take on a whole new light. The reasons for towns existing there with elaborate palaces surrounded by varying types of dwellings and specialized shops constituting a hierarchical support structure yet without any fortifications or defenses will become clear. In view of the "unthinkable" scope and nature of this new paradigm, Henry T. Wright was perhaps far more profoundly correct than he realized when he said "The archaeologists of the future, however, will have a responsibility to go beyond innovation in techniques and forge new ways of thinking about the earliest civilizations."

The unthinkable is what we need to think about. Without taking the model for the reality, let us consider Sitchin's thesis nevertheless with intelligence and good humor, that critical ingredient of a true fact forum. But the way the philosophers, scholars and scientists have handled it so far is the best way to illustrate how awkwardly things stand at present. No more painful example can be found than the Jaynes-Campbell syndrome which we examine next.

CHAPTER 2

EVOLUTION REVISITED

The intellectual life of man, his culture and history and religion and science, is different from anything else we know of in the universe. That is fact. It is as if all life evolved to a certain point, and then in ourselves turned at a right angle and simply exploded in a different direction.

Julian Jaynes
The Origins of Consciousness in the
Breakdown of the Bicameral Mind

The clear, dominant trend of the study of evolution in general and the study of Man particularly is in the direction of more and more precise explanation of social, cultural, and evolutionary phenomena in terms of more and more minute determinism on the molecular level.

Edward O. Wilson

When we look at process we are looking at our selves looking at process: that is a tautology that is difficult to escape.

Phil Johnson
Darwin on Trial

The real meaning of the gradual rejection of evolution is not just as an unsatisfactory or false theory unsupported by the research and hard evidence: we are rejecting it because we are Darwinian defectors, our consciousness is telling us that, whatever the developmental process is, we are capable of determining the nature of it as we go.

Frank Clinton
Diary Of A Bewildered Politician

To call the current situation we find ourselves in, as a race, with regard to the fundamental questions:

Where did we come from?

What are we?

a puzzle is an understatement; for those who have considered it in any depth, it is a dilemma of the first order. The disagreements are not about details and fine distinctions; the disagreements are about the fundamental issues and they are radical.

For centuries the answers to the questions Who are we? and Where did we come from? came mostly from religious sources. Even after Charles Darwin published his *The Origin of Species* and his ideas began to take hold, the only evidence we had was fossil – and a single one from Kenya for some time at that. What was important was the fact that evolutionary theory – of any kind or accuracy – was accepted scientifically. In order to fully grasp the depth of the conflicts over evolution and to integrate the concept itself into the new paradigm we should be familiar with the evidence as background.

The Remote Background of the Species Homo

Until about twelve years ago it was generally held from the fossil evidence that the direct ancestors of man split off the evolutionary tree about twenty million years ago. Then the advent of bio-chemical techniques suddenly began to change that view. In 1967 Allan Wilson and Vincent Sarich compared the degree of reaction of human antibodies (the compounds produced by the body to neutralize or destroy anything it recognizes as a threat to its health) to the blood albumin of various primates as a benchmark standard. Their fundamental assumption was straightforward: the greater the human antibody reaction the farther away would the primate species be from the human species. The standard indicated that the split occurred only about 5 million years ago and that the chimpanzee was much closer although not that close to the human and the baboon was quite distant. The "split" they were referring to assumed an evolutionary process and therefore a developing line of creatures moving in the direction of greater intelligence and complexity. In the picture they envisioned, therefore, first the baboon, then the chimp, then humanoids, had split off, diverged, from an ancestor common to all three types.

The determination of species has always been accomplished from the fragmentary skeletal remains on the basis of anatomy and brain case size. The anatomical characteristics of skeletal remains indicate posture, the type of food capable of being eaten, etc. Brain case size and, therefore, brain size is taken as an indicator of intelligence and sensory requirements and some subtle skull formation indicators can be at least used as indicator of brain development in certain specific areas. Associated tools, ornaments, traces of dwellings, fire, industry, burials, etc. have furnished clues to social structures.

But the controversial evidence was now beginning to come from sophisticated biochemical genetic investigative techniques. Because of these findings being in conflict with then current opinion, the reaction on the part of the scientific community was typical. Either the data was ignored or, as David Pilbeam did, rejected on the basis that the genetic clock was not that regular. So Sarich proposed a test: if the clock was accurate then the evidence should be reinforced by the fossil evidence. And, in fact, it was validated by the 1980's.

Then John Alquist and Charles Sibley at Yale made the clock more precise by extracting and combining human DNA with chimp and baboon DNA to see how relatively compatible they were. The less compatible they were the earlier in time the split between the particular species had taken place. The results showed that the split of human from chimp was around 7 million years ago by their method of determination and about 10 million from the baboon, generally reinforcing the findings of Sarich and Wilson. Pilbeam, a respected English scientist, still maintained his position that the "clock" standards being used could not be trusted and Steven Jones voiced the same opinion.

In 1974 Donald Johannsen had discovered 40% of a female Australopithecus Afarensis skeleton in Ethiopia; the now famous "Lucy," as she had been nick-named, walked upright as could be ascertained by the anatomy of her pelvic and leg structures. Mary Leakey also had found 3.7 million year old footprints in Leotil, Tanzania clearly indicating upright walking by adult and child.

Randall Sussman and Jack Stern did electrical muscle tests to compare the chimpanzee and human walking. When they compared the anatomy of Lucy with their findings concerning chimp and human they found the critical pelvic characteristics resembled the parallel legs in chimp rather than the knock-kneed human. Lucy's anatomy indicated that her species was halfway between man and chimp. So if Lucy was halfway from chimp to man and was dated to 3.7 million years ago this significant finding reinforced the 5-7 million estimates since 3.7 was very close to just half of 7 million years for the complete process of splitting anatomically and otherwise from the chimpanzee. Sarich's proposed test had been met: the genetic, molecular evidence was supported by the fossil.

The gradually accumulating paleontological evidence had begun to show that, after the split off from the evolutionary tree by pre-human types 5 to 7 million years ago, 1.9 million years ago Homo habilis, the first tool maker, ap-peared. The precursor of habilis is still debatable. Some experts hold that Homo habilis evolved from Australopithecus afarensis or africanus and some experts hold that all early Homo types were australopithecine. Because of the general physical characteristics of Homo habilis many experts would hold that habilis is the first of our genus Homo and that habilis eventually gave rise to Homo erectus 1.5 million years ago. Homo erectus was a user of fire and a traveler. One mil-lion years ago there are clear signs that Homo erectus began migrations from Africa to many parts of the world, including Java, China, etc.

Although the open questions regarding the exact ancestors of Homo habilis are certainly important, their resolution will not change the way we think of ourselves substantially. But there are questions which continue to surround the immediate precursors of us, Homo sapiens sapiens, that are profoundly critical to the discussion of our true beginnings. The simple fact is that there is no real agreement among the experts, at this stage of the investigation, concerning the genesis of Homo sapiens. One theory says early Homo sapiens (of which many experts consider us, sapiens sapiens, as a subspecies) arose directly from Homo habilis; Homo erectus, also arising from Homo habilis, was a dead end. Another school of experts holds that Homo habilis gave rise first to Homo erectus and Homo erectus, in turn, gave rise to Homo sapiens.

If that is not enough, there is the even more perplexing questions surround-ing Neanderthal man. It had been thought, until recently, that Homo sapiens neanderthaliensis, a stocky, muscularly powerful species evolved as a subspe-cies from an early form of Homo sapiens probably as long as 200,000 years ago. But there are three different interpretations of the Neanderthals relationship to us. Some experts say they were a species of Homo that went extinct some 30,000 years ago. Some judge they were a subspecies that went extinct. Some say they were a direct precursor of Homo sapiens sapiens, us.

Even within current evolutionary theory, the crux of the matter with regard to precisely how we arose as a species, therefore, focuses on whether we are directly evolved from Homo sapiens neanderthaliensis or Homo erectus. This is not a trivial consideration. Homo erectus, in one interpretation, was our ancestor; in another not at all. In one interpretation Homo erectus gave rise to Homo sapiens Neanderthaliensis; in another erectus had nothing to do with the rise of Neanderthaliensis.

The matter is incidentally complicated by the discoveries of Neanderthal grave sites showing a consideration for the dead, Trinkaus' pathology studies at Shanidar indicating that a crippled person had been supported for a long time, etc. But these indications of a level of intelligence and sensitivity well above the brutish have only tended to distract from the critical issue; it may well indicate whether there was sufficient developed potential for us to have developed from a particular species but it does not furnish any proof we did.

The State of Current Theory

We are actually seeing a reversal of the consensual opinion concerning the relationship of Homo sapiens sapiens and Homo sapiens neanderthaliensis in recent times. Early discoveries of paleontology seemed to show that Neanderthals came first and were later displaced by Cro-Magnons (an early form of Homo sapiens sapiens). The remarkable fact is, however, that the most recent work and refined techniques have shown that Cro-Magnons could have preceded Neanderthals. The two groups could have coexisted for a time and even possibly mated; some evidence is claimed for hybrids. The two major disciplines contributing to the decipherment of these ancient puzzles are paleontology and genetics. Breakthroughs in genetic testing to determine hereditary lines, soon to begin using samples of recovered DNA from fossils, has given a greater magnitude of precision to the research.

In December of 1990, Christopher B. Stringer (The Emergence of Modern Humans, Scientific American, December 1990, p. 98) propounded the "out of Africa" explanation. In essence this model proposes modern humans evolving in central Africa as a racially undifferentiated stock, eventually spreading throughout the world. Racial characteristics, the variations of adaptive racial features of color, hair type, etc. developed afterwards and locally. The mitochondrial DNA research is taken by Stringer as reinforcing the paleontological evidence: central Africa as the locus of our genesis, racial differentiation becoming obvious by about 100,000 years ago. The "out of Africa" scenario postulates a generic beginning in the region of Africa of the Nefilim gold mines and their laboratory where we were first invented. In this context, racial characteristics are understood as adaptations to local conditions; skin color, stature, hair and eye type are all relatively superficial characteristics that can develop in the relatively short time of some 10,000 years under the pressure of survival. Modern humans seem, in this model, to be a line distinct from Neanderthal, reinforcing the Sitchin scenario.

Current Theory and Sitchin's Thesis

Problems still to be solved are very instructive with regard to the new paradigm of our genetic genesis. Sitchin's thesis explains those items which may not be able to be verified or explained by the out of Africa model.

Since the current paleontological and genetic position is still working against a concept of our evolution as a continuum from less evolved species, a progenitor species is assumed. A major item is that even our best science has not been able to find definitive indication as to what the immediate progenitor of modern humans was. But the problem, however, of finding one may exist because there was none in the sense that scientists expect and for which they are looking. Our instant genesis from the merging of the gene code of two hominid species, one indigenous and one alien would leave, literally, nothing to look for, nothing to find. We, conceivably, might find – and may well be finding – the fossils of the indigenous species. It appears doubtful that we will find the remains of Nefilim since they apparently were in a mode evolved enough to control their own lives, had overcome death, and were capable of advanced medical repair of injury and even sophisticated resuscitation of those who had been killed.

A further condition of verification of the out of Africa explanation propounded by Stringer, the discovery of natural conditions that would have accounted for the advent of modern humans in central Africa, may be simply impossible to determine. If we were literally invented for pragmatic purposes, our genesis was obviously not prompted by a confluence of local natural conditions but by the arbitrary decision of the Nefilim.

Natural conditions of some sort are anticipated, by those holding the out of Africa hypothesis, for the proliferation of humans out of Africa and over the world. Since the Nefilim controlled humans and their distribution, no "natural" conditions may ever be discovered. Detailed accounts tell of how the newly created human slaves, invented and being used in the gold mining operations in Central Africa, became coveted by the Nefilim in the area now known as Iraq. They actually went to Africa and physically took them away from their Nefilim counterparts and brought them to the Middle East. This was not a migration of the sort conceived of by the paleontologists, it was an artificial transportation. Once we had become too numerous after being given the ability to procreate we were pushed out of the Nefilim centers in the Middle East into the outback towards the East. A gradual spread over the world from there seems to be the picture being drawn by even current scientific investigation, but it will not be able to explain in purely natural terms other events that were determined by the Nefilim such as the transportation of humans, quite possibly by air transportation, directly to South America (the tin and gold mining operation they set up in Peru) for example.

An explanation is sought for the replacement of less evolved hominids by modern humans in terms of mental or behavioral or cultural superiority. Some scientists even hold that the lesser hominids were replaced even before any clear cultural or behavioral advantages are seen in the records. If our entire existence was artificially controlled by the Nefilim, explanations in these terms may not exist. In the most radical of out of Africa interpretations modern humans would have replaced Neanderthals without any interbreeding between the two types. Although there is some evidence of contact due to the fact that Neanderthals probably changed their tool-making techniques to parallel those of modern humans in some areas of Western Europe, there is only a slight indication of hybridization. This would seem remarkable due to the widespread and sustained potential contact between the two groups if we consider them simply as two hunter-gatherer cultures as the paleontologists do. In the context of the new paradigm, however, where at least some humans were always directly in contact with at least portions of the advanced Nefilim culture from the beginning, 300,000 years ago, the contrast between the two groups may have been far more than is generally conceived. Even the humans who were not in touch with the Nefilim on such a direct basis because they had been, or were the descendants of those who had been expelled from the Nefilim centers, may have retained such a tradition of their beginnings and status that they saw themselves as very different from the Neanderthals.

The sequence of development may then be understood as progressing from the successful production of the original humans who were unable to procreate around 300,000 years ago to Neanderthals (able to procreate and migrate) around 250,000 years ago, and finally to our type of human.

All of these pending elements, significant or major difficulties in the view of current paleontological science, are easily explained in the context of Sitchin's model and may be taken as positive reinforcement of its correctness. Stringer feels that it is possible that we may be on the brink of a comprehensive explanation that would unite all of the archaeological, genetic, fossil and even linguistic elements of evidence. It would be unfortunate if such a fundamentally sound hypothesis as the out of Africa model was discarded in frustration because it could not answer these questions satisfactorily because its proponents would not consider Sitchin's thesis and information. It will become clear to even the most skeptical of scientists and investigators over time that many of these questions will not find an answer outside of the context of the new paradigm.

The Creationist-Evolutionist Conflict in Light of the New Paradigm

I emphasized, in *Breaking the Godspell*, that, in the light of the new paradigm, both sides of the conflict between the Creationists and the Darwinian evolutionists, as witnessed in the popular press, are partially correct. I pointed out that, if the new paradigm is correct, there was a creation of Homo sapiens sapiens but it was a genetic synthesis of two species in a laboratory; that there seems to be some process of evolution on this planet but it was not completely continuous from single-celled creatures to us but had been interrupted by the Nefilim. To make the point that just about all previous paradigms concerning our beginnings and nature were partially correct and now could be subsumed by a larger worldview in which a planetary consensus and unity could be achieved, I put it simply and, in substance, it was and is an accurate statement. I also emphasized how not only Creationism but Evolutionism was and is a product of the ancient godspell mentality in the expression of its conceptual basis and in its social projections – and especially in the way its proponents have promulgated and defended it over time. Everyone is aware that there are conflicts over evolution between the creationists and the evolutionists. Fewer are aware that even some scientists contest the very notion and existence of an evolutionary process itself. The state of the topic is in such flux at this time that it is worth clearly setting out the details of both the Creationist and Evolutionary points of view because the popular images of both in the press – and even in many scholarly treatments – are incomplete and, sometimes, inaccurate. Too simplistically, Creationists are all portrayed as fundamentalists holding for a literal interpretation of the Biblical account of the creation of the Earth, man, animals and plants. Too often evolutionists are all portrayed as atheistic materialists.

Even more fundamentally, there has been persistent and serious conflict between those known as Creationists over the interpretation of the Bible in this regard, and there has been persistent and serious questioning of whether there is any such thing as the Darwinian type of evolution by natural selection or any other type of evolution on the part of scientists who have been brave enough to even raise a question.

The Creationist Camps

There are two basic differences among Creationists. The difference best known to the general public is between those who hold that the creation of the earth, humans, plants and animals actually took place in six, twenty-four hour days as we know a day as measured by one rotation of the earth around its axis, and those who hold that the "days" spoken of in Genesis, the first book of the Biblical Old Testament really meant ages of time, longer periods of gradual developmental progress.

There is also a difference of opinion between some Creationists regarding whether evolution could actually be considered the means by which the Creator brought about the creation of plants and animals and man. There were a number of important preachers and evangelical ministers and theologians who, early in the confrontation with Darwinian evolutionary theory, did indeed interpret the Creation story in Genesis as meaning a creation over ages or eons, and some even allowed that it was possible, if evolution was a fact, that it was the very means through which the Creator achieved His purpose.

Darwin's Theory: A One Hundred Forty Year Old Problem

Darwin's theory of evolution, about how species have come into being on this planet through what he saw as a process of survival of those individuals who were lucky enough to be able to adapt to ambient conditions, is about one hundred and forty years old. It is still contested, not only by Creationists, but by some scientists. If it were a matter of a straightforward choice between natural evolution of species according to some clear mechanism and Creationism of whatever variation, life would be simple in that regard. But the matter is far from simple. In the broadest overview, very complex continuing arguments range from whether there is any process of evolution at all in the way that Wallace, Darwin or Lamark or anyone conceived of it or, if there is, what the mechanism is by which the real evolution operates. If there is not a process of evolution as described by any of the various theories, what process are we actually witnessing and experiencing? Most of those arguments are at least recognized by scientists and Creationists regardless of which variant of position they take. Again, if it were only a matter of resolving the controversies on the basis of some criteria agreed on by all parties on both sides, things would be still simpler.

Much of the difficulty in even communicating among the various factions stems from the fact that it is not the interpretation of the details of the evidence that is the cause of the fundamental differences of opinion but the dichotomies between the primary criteria used by the various parties to the dispute. There is no common criterion against which to make judgments. The scientist would make the scientific method the criterion of truth. The Creationist or theologian would make the Bible taken as divine revelation the criterion. The philosopher tends to apply syllogistically logical reasoning to judge the validity or either position and sometimes adds to the list of theories. Furthermore, there is obviously no meta-criterion acceptable to the adherents of all these schools of thought by which to judge whether their individual criteria are correct. Because religion, science and philosophy work from fundamentally different definitions of what a human being is, emotionally charged factors distort and cloud the issues and fuel the arguments.

The Creationists demand that Creation Science be taught in the classroom.

Darwinian evolutionists claim there can be no creation science, that it is religion, not science. The Creationists counter that they want the evidence that would throw doubt on Darwin's theory at least brought out in the school. The majority of evolutionists would have it that evolution is a "fact" and not open to contest. Scientists in other areas of study tend to accept anthropological evolutionary theory because it is the respectable scientific thing to do.

Doubts about Darwinism

Yet some scientists and reasonable thinkers have grave doubts about the evidence for a Darwinian type of evolution. They point to the lack of fossil evidence of intermediate, transitional stages between species and the predecessor species from which they were thought to arise.

Darwin himself indicated that serious lack of evidence of a multitude of transitional forms would be fatal for his theory. They question how complicated, specialized structures such as an eye could have arisen by natural selection. According to that theory the separate parts of an eye, having no constructive purpose in themselves, would have to build up in nature in organisms over time until they finally came together in the form of the eye in an appropriate organism. But, according to that same theory, natural selection would have tended to eliminate the beginning stages because of their not being immediately promotive of survival. Because of the difficulties brought to light by these arguments, convinced Darwinists have moved from the scenario of continuous development through many transitional forms to "punctuated equilibrium," long periods of stable form punctuated by bursts of rapid change and development in the face of lack of evidence for transitional forms in the fossil record. Even some scientists who favor an evolutionary view but who have doubts about the evidence say that evolution has become a sort of religion not to be questioned among scientists.

This seeming unscientific phenomenon can be puzzling to a layman not completely familiar with the history of the theory. The problem lies in the fact that those espousing Darwin's theory and other forms of evolutionary explanation did and do so for mixed motives. Parallel to and, sometimes, overriding the acceptance of evolutionary theory because of its perceived accuracy in explaining the fossil record was and is the opportunity to defeat the theological, religious, Bible based Creationist authority and power. So the options open to the scientist who would be courageous enough to reject the theories of evolution available are severely limited and there is, literally, nowhere to move. To attribute some sort of purposeful determining factor, some teleological guiding element, that was working in nature to develop and preserve these remote precursor forms (of an eye in the example above) would be to go exactly contrary to the random working of the mechanism of survival of the fittest that canceled out any notion of a God. So what does even an honest scientist do? Certainly not walk back into the Creationist camp and surrender without a great deal of reluctance. In fact what happens is that most scientists who are directly occupied with evolutionary theory such as paleontologists, anthropologists and biologists, simply keep looking in quiet desperation for more fossil evidence to reinforce the theory that is assumed true, rather than test the theory itself as would be the usual scientific procedure. Scientists and academics who are not directly involved with evolutionary theory tend to simply follow the lead of those who are because peer pressure forces them to follow the party line. And yet, even when the strictest of scientific methodology is applied, the scientific materialist is sometimes dismayed and feels trapped by the conclusions.

The Scientific Darwinian Dilemma

At this point in time the scientific community and our general intellectual thought is caught in a dilemma with regard to the concept of specifically human evolution. Beyond the relatively shallow conflicts between those of evolutionary and creationist persuasions there is a fundamental problem caused by the acceptance of the theory itself, by its ramifications. This dilemma is articulated well by E.O. Wilson (*On Human Nature*, Harvard University Press, 1978). He is one of the few who faces the ramifications of the current position of science concerning Darwinian-type evolution clearly and fully.

The problem is this: science, in accepting evolutionary theory and jumping from that to a too simple and absolutistic conclusion of complete determinism of the human being, has boxed itself into a very unsatisfactory position. In the minds of those who influence contemporary consensual scientific and intellectual thought, the rejection of the creationist position leaves genetic variations and chance as the only other candidate for the principle that created human nature. Most scientists leave the topic at that. But Wilson's thought is an excellent touchstone for an understanding of the situation in the scientific and academic communities because he will not leave the subject at that point.

Wilson points out that the elimination of God as the creator of the human species leaves, in his thinking and that of others like him, only "genetic chance and environmental necessity" as the maker of man. He emphasizes that the acceptance of genetic variation and chance as the only cause for the existence of human kind, infers a complete biological determinism and negates any possibility of "soul" or spiritual component in human nature. He admits that this causes a sense of "hesitancy and even dread" when approaching the subject of man's essential nature for it reduces man to a completely biologically determined being and "our souls cannot fly free."

This is not simply the opinion of one respected scientist; it is the consensual opinion of the general scientific community except for those professional scientists who espouse a religious Creationism view. Wilson represents those who hold this "unappealing proposition" in a very honest and articulate way, the "legacy of the last century of scientific research." He is careful to point out that this position does not rule out a Deity who made the fundamental particles of the Universe and set it in motion, but not at the level of the creation of the human species. It is his deep conviction and the focus of his work that everything human can, and will, eventually be explained by genetic, bio-chemical programming. Free will is an illusion in this context.

Those disagreeing object that Wilson's type of reasoning and conclusion on this point is oversimplistic. They argue that, just because the Creationist scenario of a cosmic God performing a special case creation of the planet, plants, animals and man in an apparently short period of time has been negated by scientific research, it is not even logical to jump from that to an assumption of complete biological determinism. They would argue that, if, even according to Wilson's logic, a cosmic God may still be allowed as the Creator of the universe, could not that universe, from the quantum mechanical level up to the level of man, be structured in such a way that there was inherent freedom and the possibility of dimensional expansion at any level? In other words, could not the universe be complex enough so that, even though completely determined, we were programmed to determine our own determination? They claim that there is a strong hint of a preclusive selectivity of conclusion in the usual logic, a too simple dialectic operating and a clear prejudice toward following Darwin in his intention to "eliminate Moses and the Bible" once and for all, too eagerly. Those questioning the logic, it is important to note, are not arguing for or against the existence of a God, but simply objecting to what they consider to be poor logic as such on Wilson's own terms.

From a psychological perspective, one may ask Why should the conviction that we are completely determined biologically, genetically, make us feel so bad if, indeed, that is the real truth? If we have been relatively comfortable with our apparent, even if illusory, degree of freedom and ability to evolve so far, why should that knowledge suddenly destroy our existence? What is the real cause of the uneasiness and fear we experience when considering that imputed absolute determination? Wilson himself acknowledges that we have reached a point where we can consciously influence the course of our evolution. If we are, indeed, genetically determined to determine our own evolutionary development, is there not a degree of freedom inherently involved, and therefore a significant paradox?

It is well to remind ourselves, as Wilson does, that this position of science concerning evolution is the "legacy of the last century of scientific research." To be more precise, it is the legacy of late 19th century thinking in science and religion. Darwin's theory of evolution was published in 1859, about one hundred and forty years ago, and it has been disputed in general or in detail since then. The Lamarckian versus the Darwinian versus the Durkheim/Radcliffe-Brown controversies are seen to be over specific mechanisms rather than overarching explanations. There has been no major counter theory to replace it so far. It is the thrust of this chapter to show that not only does the new paradigm do so, but adds an additional dimension to our knowledge of our beginnings and, as a result, to our present and future.

The unease and dread engendered by the seemingly inescapable logical conclusion that human nature is reduced to a completely biologically determined mechanism coupled with the fact that Darwin's original work has been disputed since it was published may be taken as strong clues that something is still lacking in the consensual theory of evolution as it applies to human nature. When the recently determined, puzzling sequence of development, which sees Homo sapiens coexistent with Neanderthals and even preceding them, is taken into consideration the clues are reinforced: the basic premise of natural selection is contradicted if the more evolved species precedes the less evolved. Over time the evidence has gradually made the natural selection principle more and more difficult to support according to the criterion laid down by Darwin himself. But the fact that all current explanations are received as unsatisfactory, unsettling and unacceptable to one or another segment of the population (and even the scientific explanations cause "fear and dread"), simply reinforces the fact that we recognize that none of them are completely correct. The fundamental bias of both sides in the argument is not based primarily on strictly scientific fact but on religious or anti-religious preconceptions and motivations. No other conflict of interpretations more clearly and acutely manifests the ancient godspell effect operating in the common psyche.

Against the context of the new paradigm, "we keep returning to the subject (man's ultimate nature) with a sense of hesitancy and even dread" because "the philosophical legacy of the last century of scientific research" that forces a choice between God and genetic chance and environmental necessity as the maker of the species is in error, not because it rejects the Judaeo-Christian concept of God but because it does so for the wrong reason. And we have known it was in error through our dread and hesitancy below reason but have been, to this point, unable to refute it. The new paradigm will allow us to return to the subject of man's ultimate nature with confidence, freedom to radically experiment and explore – and humor. The same reductionistic science that would honestly insist that we are totally determined by an infinitely regressive loop of ultimately quantum mechanical events is already telling us that we are determined to transcend in an infinitely progressive loop of fractal-like expansions. This is accurate whether we see ourselves strictly as a Darwinian resultant of ice age vectors or the precocious leading edge of a synthesized species metamorphosing rapidly, driven by the engine of our Nefilim gene component. It is more comfortable in the latter case – the truth, regardless of its content, tends to set one free. And the new paradigm also is totally consistent with the seeming anomalies of acceleration, ooparts (out of place artifacts) and syncretisms so apparent in the human situation, and is a reasonable explanation why "biological evolution is always quickly outrun by cultural change."

Forbidding Archaeology

To this point in the discussion we have considered the differences between the various schools of thought within the scientific establishment with regard to whether there is such a thing as evolution in the first place and, if there is, by what mechanism it may be driven. And we have compared those various positions to the various Creationist viewpoints.

There is another body of information that calls the specifically Darwinian thesis of natural selection into question from an archaeological and anthropological point of view. The Darwinian thesis works from a fundamental postulate that the "higher," more complex organisms, species, develop from the lesser complex. That is an assumption that compels the Darwinian to hold to that apes would come before human types, more primitive and lesser developed human types would have to evolve before more developed human types. Homo habilis would have to arrive on the scene before Homo erectus, Homo erectus before us.

Michael Cremo and Richard Thompson, devotees of a Vedanta guru, Bhaktivedanta Swami Prabhupada, who suggested that they do the research since the Vedic tradition speaks of humans on this planet in the very distant past, have published an exhaustive study of archaeological, paleontological and anthropological evidence that contradicts the evidence currently accepted by establishment science regarding the age of humans on this planet. In *Forbidden Archaeology*, a nine hundred page tome meant for the layman, they deal with four main categories of evidence:

1> tools, from crudely chipped pebble types to superbly crafted flint blades dated to remote times and normally attributed to advanced human types not assumed to have existed in those times.

2> skeletal remains of human types dated to past ages when human types that advanced are not assumed to have existed.

3> simple ooparts, artifacts like mortars and pestles found in strata of an age in which the human capabilities to manufacture and use them are not supposed to have existed.

4> advanced artifacts dated to times when the technology was not supposed to have existed.

The thesis of Cremo and Thompson is very straightforward: The collective evidence is strong enough, much of it being collected and analyzed by highly skilled and respected professionals, to demonstrate that advanced human types existed in very remote times, dated in many millions of years in some cases, alongside of primitive human types or even previous to some of them. If the evidence is accurate then the Darwinian explanation that advanced types always arise from more primitive types is contradicted and to be rejected. As part of their exposition they document many instances where the evidence is ignored, denigrated and even suppressed or destroyed because it contradicts the current establishment view. They give examples of the shameful way very competent scientists who have been willing to risk publishing their honest findings of this type have been literally drummed out of the field and their reputations and careers destroyed for their efforts.

The work of Cremo and Thompson forces a very critical question: If their evidence for advanced types of humans in extremely remote times is true, does it in any way contradict Sitchin's thesis that we, specifically Homo sapiens sapiens as we know ourselves, are a genetically engineered cross between Nefilim and Homo erectus? I have carefully studied all the evidence they present and I do not find that it does.

Scenario 1: Anatomically Modern Type Man Co-existent with Homo erectus?

Let us take the most radical scenario suggested by the evidence they present: humanoid types, ranging from the most primitive, near-anthropoid species to types anatomically similar to modern humans have existed from the most remote times alongside of each other and some of the more primitive types may very well still exist in remote areas on this planet. Although clearly negating the Darwinian thesis, this does not negate the possibility of the Nefilim choosing one of these humanoid types for their specific purpose of adding their genes to it to make it a satisfactory worker type. If, indeed, they truly existed at that time. Ultimately, in this matter, I believe that we must give priority to the actual recorded and reported history known to all of the earliest civilizations, taking it with the hard anthropological evidence to determine precisely what humanoid type the Nefilim chose to upgrade for their practical purposes.

The Nefilim selected a humanoid type with which to fuse their genetic code or part of it, enough to raise the humanoid species to a level of intelligent competence to work as gold miners. Sitchin has indicated that he believes that the humanoid type chosen was Homo erectus. Even if there were more advanced, physiologically nearly identical human types contemporary with Homo erectus 250,000 years ago, it does not, by that fact, contradict Sitchin's thesis. It does raise a question as to whether the Nefilim might have actually used a more advanced human type much closer to us to merge their genes with rather than Homo erectus if, indeed, one was extant.

Scenario 2: Repeated Advanced Civilizations in the Remote Past?

Why would the Nefilim have even bothered to invent us if there was a species already up to standards that they could have enslaved – or hired? It is difficult to postulate that a human type equal to the ordinary modern human was indeed present and flourishing on the planet some 250-300,000 years ago with a culture as highly developed and sophisticated as any we have identified in that time period because the hard evidence for that is lacking. But it is not beyond possibility that advanced human civilizations could have risen and fallen in the very remote past. The existence of an advanced human civilization in the remote past that was wiped out or degenerated for whatever reason, even many times, would not contradict Sitchin's thesis. If the situation, at the point when the Nefilim arrived here and eventually needed a slave creature, was such that the most advanced human types were of sufficiently low caliber to need upgrading even to do mining as the records indicate, there is still no contradiction.

The precise question at issue is, if there were anatomically modern humans already existing on Earth, identical to us both physiologically and mentally, even to the point of being self-aware as we identify ourselves to be, when the Nefilim are said to have created us, could that mean that some contemporary humans are from purely indigenous stock and some are from the Nefilim/humanoid cross? It is at least conceivable but logically not probable: Why would the Nefilim even bother genetically engineering us into existence if there was a species already up to standards that they could enslave – or hire?

Sorting Out the Genetic Strands

Contemporary humans, possessing self-awareness or the potential for it, can and do exist in a feral state. Koko the gorilla, who speaks by signing, shows all the characteristics of being at least rudimentarily self-aware. Self-awareness or intelligence level or tool utilization or degree of cultural complexity do not provide us with sufficiently fundamental criteria to determine if, indeed, there are at least two strains in the human population, one indigenous and one half-Nefilim. Only very detailed genetic investigation could determine that and probably only with great difficulty because, if it were so, the two strains may have melded so much by this time that the threads would be hard to entangle.

Could the Nefilim have Chosen a Lesser Species?

Another theoretical possibility presents itself: even though there might have been anatomically modern human types available, the Nefilim could have chosen a lesser species, say Homo erectus, as a base on which to create a slave animal for ethical or pragmatic reasons. We were described as eating the grass of the fields and drinking from the ditch when we were first invented. But, again, this does not seem logical on the basis of why bother when a more advanced but not up to standard type was available. And, in the last analysis, the records say nothing whatsoever about any human type then that would have been equal to us now.

Cremo & Thompson: A Remarkable Contribution

The work of Cremo and Thompson does present evidence that contradicts the Darwinian type of evolution and robust evidence for the existence of anatomically modern humans side by side with primitive types of humans and apes from the very remote past to the present. In doing so, they have also opened up the possibility that we may have underestimated the tool-making and using capabilities of the more primitive human types as well as their intelligence and cultural complexity. It may be completely erroneous to reason that, since a flint spear point is of a certain level of complexity, symmetry and elegance, it "must," on the basis of assumed level of required dexterity and brain capabilities required for conceptualization and perception of symmetry, practical efficiency of design, etc. have been made by an anatomically modern human. It may also even be quite incorrect to reason that because the anatomy of a human equates to ours that the mental capabilities "must" be the same as ours. It all comes down to a matter of pure genetics at that point rather than criteria based on sophistication of tools, evidence of social interaction, etc.

Is Evolution Still a Viable Concept?

A cardinal question then: Is there any version of evolutionary theory, Darwinian, Lamarkian, or modern variation that is still viable? and, if so, into which can we fit our own unique bicameral genetic metamorphosis? The resolution lies in a clear discrimination between how we understand the processes of change and development in biological organisms in general; how we view humanoid development up to Homo erectus; how we understand and define our own quite unique, rapid, metamorphosis; how we conceive of how we will change going into the future.

Redefining the Human Developmental Process

I believe the generic concept of evolution is certainly still viable if we think in terms of a generic developmental process as we witness ourselves experiencing – or at least some of us do. If we reject the term "evolution" for whatever reason, we are almost forced to invent another term to describe the developmental process we witness our peculiar species undergoing. I continue to use the term evolution here precisely because of that fact. It is difficult to claim that we are static creatures when we see ourselves already determined to eliminate disease, social injustice, crime, and eventually to master immortality and habitual four-dimensional consciousness. I am convinced that we are experiencing a racial and personal, developmental process. When we are free enough of both the Creationist and Evolutionist mind-sets to define that process for ourselves we see that it is a process of expansion, both physical and conscious, and becoming more and more reflexively directable by us. It is robust, sometimes rapid and novel enough to be almost startling. Perhaps we will, eventually, invent a new word for the human developmental process to replace "evolution" because of the various definitions and nuances of meaning previously attached to it.

If we continue to use the term "evolution" then the Darwinian mechanism must be modified and reduced to proper perspective. It certainly seems to apply in the specialized adaptation of species to local conditions. The specialized modifications of finches to food sources is a well known example of adaptation of physiology even if we cannot agree on the precise details of just how it comes about. As we examine adaptation and genetic selection operating in more and more complex species, the effects may be somewhat subtler: we do not find apes with one hand grown into a sharp sickle for cutting heavy food foliage. The tendency seems to be toward more comprehensive and generalized capabilities.

Stuart Kauffman's Self-determined Criticality

There is also a concept being advanced that requires serious consideration and which says that the advent of new species, change in species, or the extinction of a species comes about at the most fundamental level in a radically different way. Stuart Kauffman, a creative thinker attached to the Santa Fe Institute, has been working with the concept of self-organized criticality for some time.

In its barest, simplest form, this theory states that when all the conditions are right, a species will emerge which fits the niche created pretty much full blown. When it is great ape time it will great ape. When it is advanced human time it will advance human. The conditions for both might be simultaneous or contemporary: the natural, adaptive, survival type changes postulated by the Darwinian theory are negated partially in that there is no need for a more evolved species to be evoked from a lesser evolved species, i.e., evolution does not have to, probably does not, move from the less complex to the more complex. Adaptive changes can take place on a Darwinian basis after a species has shown up but these would be on a more superficial basis than in Darwinian theory. The foundation for this theory is set in the context of chaos and complexity theory and has been reinforced by computer modeling of the projected scenarios. This description of the theory is oversimplified but sufficient to give accurate understanding of its principles.

The Objections to the New Paradigm

Direct objection to the new paradigm by the scientific community generally is on the basis that, if our fundamental conception and definition of human nature is to be altered, it must be brought about by the application of scientific evidence and methodology and not by some gratuitous dogma. Setting aside academic objections to the blatant circularity of that demand, I would only counter that, if archaeology, astronomy, anthropology, paleontology, bio-genetics and, possibly, linguistics are indeed sciences then the data derived from those disciplines over the last one hundred and fifty years is the evidence, and the only evidence, to which I appeal.

Establishment science and the new paradigm diverge at the point of explanation of how our species Homo sapiens came about. The new paradigm does not, of itself, contradict the Darwinian general assertion that there is a process of evolution operant on this planet, a generalization that has been accepted by scientists even if they have not always agreed, concerning the mechanisms by which it operates. It recognizes that there was and is some sort of indigenous evolutionary process operant on this planet that eventually produced Homo erectus. And it does not contradict or argue against the thesis that it could have been genetic chance and environmental necessity that brought about Homo erectus. Neither does it deny that there was or could have been a continuum of evolutionary development from less evolved humanoid species through Homo erectus. With regard to all these fundamental elements of humanoid evolution, up to the point of Homo erectus, their is no inherent contradiction or conflict. Establishment science and the new paradigm diverge at the point of explanation of how our species Homo sapiens came about.

The DNA Evidence and the New Paradigm

The mitochondrial evidence for our first woman ancestor coming from east central Africa – precisely where and in the same time frame as mentioned by the ancient documents – reinforces the new paradigm. Establishment science says that Homo sapiens evolved from Homo erectus, furnishing the genetic base from which we, Homo sapiens sapiens, were derived. The new paradigm says that only half of our genetic base was derived from Homo erectus (the other half being Nefilim) and that it happened not through an extension of the continuum of the natural developmental processes of genetic chance and environmental necessity but by a deliberate interruption of the continuum by genetic engineering. The new paradigm affords a totally new and empowering concept and understanding of the specifically human participation in that process and corrects and improves evolutionary theory, and then adds a completely new dimension to it.

The Potential of the New Paradigm for Modifying and Expanding the Definition of Evolution

Comprehension and acceptance of the new paradigm affords us the opportunity to correct and complete our understanding and definition of both evolution in general and our unique special case of it as a synthesized species. Having assimilated that redefinition and adjustment, we are freed and empowered to explore the dimensionality of consciously directing our own personal evolution and contributing to the determination of our species' evolutionary trajectory.

If, indeed, the continuum of the indigenous processes of evolution on this planet was literally interrupted at the point of Homo erectus by our synthetic genesis then we gain at least perspective and accuracy through knowledge of our real history and bicameral nature. The new paradigm allows us to adequately and comprehensively explain and integrate the elements of the phenomenon of religion as a strictly sociobiological phenomenon in terms of a sublimation of the original master-slave relationship and, with an appropriate and frank reductionism, cleanly differentiate that matter of solar system bio-politics from the "resplendent and multidimensional" exercise of the potential for self-conscious natural transcendence of the resultant product. Three very fundamental practical questions arise at this point.

Is there a way of detecting evidence in the human genome itself of such a merging of the genomes or, at least, the imposition of specific Nefilim genes, as the texts would indicate, on the Homo erectus genome increase of the intelligence of Homo erectus to usable status? I have discussed this with Douglas Wallace at Emory University and he says that he is doubtful that such an event occurred due to the fact that we are only, by current estimation, 5% away from the chimpanzees but that, until we read out our entire genome at least, he does not see any way to adequately technically determine it. He does not negate the possibility but is initially skeptical. I would judge that, if five percent or even one percent makes us that different from the chimps, there is room for a whole lot more. I suggest that the genome project be eventually expanded to do such an analysis. It may be subtle evidence if 300,000 years of additional adaptation has blended and "smoothed out" the rough spots.

It seems clear that the rapidity of our development over only some 300,000 years is due to the impetus of the more advanced Nefilim genetic component. That is quite a head start and different from the situation of a new species which begins just a tick or two more advanced than the species from which it descended. The recovered records indicate that it was a degree of intelligence enhancement which the Nefilim intended to add to Homo erectus. Perhaps, once we have deciphered the entire human genetic code and, possibly, begun to discover the indications of what was the Homo erectus contribution and what was the Nefilim contribution, the actual combination was largely Homo erectus and a select part of the Nefilim code controlling intelligence.

The second question is an objection which I raised early in my consideration of the validity of Sitchin's thesis: Given the existence of the Nefilim and their arrival here 450,000 years ago from a planet on which they adaptively evolved under very different conditions than on Earth, how could a merging of gene codes take place when the assumption easily could be that those codes would be quite disparate, even, conceivably, very different in their general nature, quite possibly even in their fundamental structures, or worse?

The solution to that problem offered by Sitchin is based on the facts of our solar system cosmology as taught us by the Nefilim (as recounted in the recovered document *Enuma elish*). The solar system formed according to the dust accretion theories we have determined but, at a point after nine planets had been formed (Earth was not yet formed), Nibiru was captured into the solar system from outside, eventually collided with a large proto-planet, Tiamat, located in orbit where the asteroid belt is now, part of the proto-planet shattering into asteroids, comets, meteorites and other debris seen splashed all over the surfaces of our system's planets and moons – the larger part recongealing into Earth in its present orbit. That collisional event, it is indicated, caused a cross-seeding of the fundamental building blocks, probably higher even than the amino acids since the residual debris that is still crashing into Earth in the form of meteorites often contain fairly complex organic residue, that could have given rise to similar evolutionary trajectories.

I can at least allow some form of that scenario as possible at this point. It might also be argued, it has been suggested to me, from the data available from genetics already developed, that even very different gene codes could eventually be merged together although the capability is not quite available as of now. But a common base from cross-seeding of the resultant Earth would eliminate that need.

The third question: Would not a planet that travels so far out into deep space be uninhabitable by the type of humanoid the Nefilim were described as being? The answer is not necessarily. The ancient records speak of Nibiru as being a very radiant planet, possibly meaning it had a high core temperature. Although it is a source of controversy even among astrophysicists, some would hold that a body the size of a planet in an elongated orbit such as Nibiru's tends toward a circular orbit and the result is a great deal of internal and tectonic stress creating heat. By the existential fact that the Nefilim did and probably do exist and came from there claiming it was their home planet and were here for huge amounts of gold to create a sort of molecular gold shielding in their planet's atmosphere we can infer that humanoid life was possible there on, or below, the surface. Possible, at the very least.

Intimate knowledge of our bicameral genome will be of tremendous help in determining the potential for our future evolutionary directions, but we should not make it into some new form of religion. To summarize this point: having reevaluated and discarded the theological and religious definitions of human nature and attained a common planetary consensus of what and who we are, we shall go about investigating the most obscure details of our genome to determine what are the Nefilim and Homo erectus components if they are actually distinguishable at this point in time. We will do this for both practical as well as theoretical reasons. A detailed knowledge of the structure of our genome in these terms should afford us great insight, obviously, into why we are the way we are in detail. It should also give us the keys to the solution of genetic diseases that may actually be the result of such a cross-breeding of two complex and different genomes as well as other practical advantages.

But we must be careful not to fall back on a strictly laboratory style, biological approach to the implementation of the new paradigm in terms of genetics. Science will be a tremendous tool and very helpful for certain information of a restricted type. But the overarching process of reinterpretation of the human condition and the application of that knowledge for the improvement of the present and the determination of our future trajectory should be our focus. Having established that exalted focus and become accustomed to taking responsibility for ourselves in determining our own racial and individual futures, we must be careful not to turn the process into an institution or some sort of religious quest (in terms of how we understand "religious" traditionally). It is interesting to again consider the thesis of E.O. Wilson as he grapples with the phenomenon of religion.

By recognizing the phenomenon of religion as an evolutionarily programmed, deep racial imprint, Wilson, unwittingly, has precisely identified the godspell syndrome.

E.O. Wilson, the father of Sociobiology, is one of the most ardent proponents of scientific naturalism. His fundamental conviction is that all of human action will, eventually, be explained and predicted in terms of genetic and biological programming. Yet he emphatically disavows that scientific naturalism should in any way be made a substitute for religion. Within that context, his evaluation of religion is almost startling. He sees religion as mythology and rejects it as such. But he also sees religious belief as a predisposition of the human neurological system, a result of millennia of genetic evolutionary programming (! the godspell). Without making much of a value judgment concerning it, he acknowledges it as an integral part of human nature, hard-wired in the brain after all this time, which should be honored and incorporated into, and modify scientific naturalism as a proven evolutionary gambit. But he also sees scientific naturalism as a mythology, but one which can inspire and motivate seeking high goals.

By the human species being a unique class and case of evolution requiring a dimensional expansion of evolutionary theory, Wilson's reduction of all human action and proclivities to genetic potentials and the de facto definition of religion as myth parallels the Darwinian attempt to eliminate "the Bible and Moses"; it is only half right. Relative to the new paradigm, his assigning of the phenomenon of religious belief to the category of neurological programming is correct but for the wrong reason. It was imprinted deeply into the human psyche not through the development of myths about unreal deities projected by the human mind in response to felt need, but by the unique and overarching real physical presence of and total dominance of the human species by the Nefilim.

So what is most fascinating and significant, with regard to the new paradigm, is Wilson's identification, as a scientist, of the phenomenon of religious belief as an evolutionary development, as a programmed disposition. With the recognition of religious belief as being a deep, evolutionary genetic imprint, "hardwired into the human neurological giving rise to the classic phenomena of consecration of personal and group identity, attention to charismatic leaders, mythopoeism, etc.," he is identifying and has defined the godspell with clinical precision. Reinterpreting his statements in the context of the new paradigm is not a large step. Since the Nefilim were the major dominant factor in the creation of the human species and were undisputed masters of this planet, whom all humans had to acknowledge while they were here, it is clear why what we know as "religious belief," in whatever sublimated form we witness it currently, is such an important and profound element in our psyche. Human existence, racially and individually, was dependent on the Nefilim for a very long time, long enough for it to be deeply imprinted in our genetic makeup. If that was not sufficient reason in itself, there is also the fact that we are half-gods, half Nefilim and half Homo erectus, and the Nefilim component was literally genetic from the very beginning.

With the elimination of the necessity for identifying it as myth there comes an opening for sorting out the elements that have been lumped together as "religion." We can, then, easily distinguish the actual historical fact of our creation by the Nefilim, resulting in our original subordination to and awe of them, from the gradual sublimation of that knowledge and relationship into what we currently experience as "religion" after they left the planet. We can further differentiate the innate tendency of the human mind to expand, to explore new dimensional comprehension and perception from both of these elements and understand its precocious development in the perspective of our rapid sociopsychological metamorphosis.

The Human Species as a Special Class and Case of Evolution

The new paradigm not only provides us with a precise and robust understanding of our real beginnings and true nature and how our unique type of development fits into the grand scheme of evolution, but it differentiates the mechanism by which our unique species evolves from that of other organisms. We choose, or have the ability to choose, to evolve, we choose our trajectory, our direction and our goals, and we choose the criteria by which we determine one direction of evolutionary development is better for us than another. This sets us at least two dimensions beyond any organism which is not self-reflexively aware and is at the mercy of genetic chance and environmental necessity.

But, over and above the levels of even creatures such as chimpanzees and apes and perhaps whales and dolphins, who can and do exhibit some fundamental degree of self awareness, we need to understand the fullness of our unique ability to control our own evolutionary development. We need to explore fully just how much of a special case of evolution we really are. It is critically important, during this time when we, as a race, are in the process of breaking the godspell, that we redefine and make precise the kind of evolution we humans are experiencing because it contributes directly to the unifying concept of generic humanity.

We are about to read the entire human genome. Feedback of that detailed knowledge into our consideration of our future evolutionary direction will be very important. It is critically important not only for our present adjustment but also for our understanding of how to choose our future direction. When we have finally recognized the conflicting theological definitions of man for what they are and discarded them, we will need to have an accurate, robust and mature understanding of the full range of complexity of unique human evolutionary development available to us. By human evolutionary development I do not mean simply a history of how early humans originated and developed, although that is an integral and important element. Beyond that we must understand in detail how we evolve and will choose to evolve as creatures aware that we are aware that we can direct our own evolution. We shall explore that potential in the next chapter. Before we enter that dimension of the topic there are important factors that must be considered that bear, or will come to bear shortly, on our understanding of ourselves and our evolutionary choices.

Will We Read Our Beginnings in the Genome Itself?

A major biogenetic development now coming to the fore will contribute to our detailed understanding of evolution in general and our own in particular. We are in the process of reading the entire human genome, the complete set of DNA genetic instructions down to the last molecule that makes us humans. We may soon be able to read it in sufficient detail to tell how and when the actual gene splicing that produced us was carried out in the first place, precisely, perhaps even with an identification of specific Nefilim and Homo erectus sequences. Or perhaps the two have fused so intimately by this time that no easy identification and discrimination is possible. But certainly a serious and intense effort should be made: the results may well provide resolutions to genetic defects, diseases and puzzles as well as give us a great deal of insight into how and why we are the way we are. We need to reexamine especially both the overall schema and fine details of our genetic code. Having learned from that intense study, we can apply that knowledge as a feedback component as we inevitably consciously determine our future evolutionary trajectory.

We must not turn the process of directing our own evolution into a rigid process with rules and authorities, sects and conflict. We could, unfortunately, and some inevitably shall, turn the process of determining our individual and racial evolutionary trajectories into a quasi-religion, a perversion of the essential freedom involved. That would be totally counterproductive. Neither should we rely on too rigid and sterile a form of "scientific method." It is clear that the essential nature of the process requires the greatest degree of freedom to attain the greatest latitude of experimentation to achieve the greatest degree of expansion. It is equally clear that, by predicating these conditions of freedom, I am also advocating and setting rules. It comes down to the individual being free and informed enough to be able to use the information to determine his or her individual evolutionary path and to contribute to the determination of the general species' evolutionary direction. But it will not be simply the application of lessons learned about what worked and didn't work, what caused disasters and what raised us to new plateaus, what seems good for us and what not, what our vision was and how we might implement or expand it in the future. Our self-awareness, our intense awareness that we are aware that we are aware of ourselves evolving and can influence that process creates a complexity that is both difficult to handle and yet is the key that unlocks an adequate future. Even as we think about and go about determining how we can and should take charge of our evolution, as if we could do that without actually beginning to do so, we are, by that very fact, modifying our evolutionary development and trajectory.

Humanity: A New Category in Evolutionary Theory

How does the new paradigm expand the concept of a species evolving? The new paradigm demands that we create two new categories in the theory of evolution: that of a synthetically engineered organism and that of an organism that directs its own evolution. Although, at first, new categories would seem only to complexify the controversies, the recognition of these crucial distinctions will, in truth, finally resolve and clarify the entire subject of evolution in general and our genesis in particular.

Traditional and current evolutionary theory is based on the concept that any given species became that species by evolving from a previous species. The only persons who would disagree with that fundamental premise are those who reject the concept of evolution entirely. Those who hold the theory of evolution valid most certainly do disagree on the details of how it works as a process sometimes, but they do not disagree on that fundamental premise. So the fundamental premise of evolutionary theory creates a single category, in the context of the theory, that of organisms that evolve from other organisms by some process of natural selection, cultural evolution, or a mechanism of your choice.

But what, then, of our synthetic genesis, our sudden – indeed, instantaneous – coming into existence through the arbitrary decision of the Nefilim to create a species that would be their slaves, designed primarily for their purposes? The idea of "natural selection" through adaptation to natural circumstances by those most fit to do so must be stretched very far here if it is to apply at all. If we take the Nefilim as the instrument of "natural selection" then the Nefilim must be understood as representing a "force" of nature acting on a species (two species in this cases, Homo erectus and their own) and their purpose as some sort of deliberate adaptation to ambient conditions for the sake of survival. But, since it was their adaptation to ambient conditions (invention of a slave race to free them from hard physical labor) rather than that of Homo erectus (who was just used as raw material) the conditions of the usual definitions do not apply. So we must separate the processes of evolution which are adaptive and selective, undergone by species that are "natural" (defined as not having been synthetically created as we were) in general, from our special case of it in particular. Rather than try to warp and force our unique genesis into the usual evolutionary process of adaptation and gradual change through the favoring of the genes of those individuals who are most fit, as we have been attempting to do so far, we should clearly assign ourselves to a new category: any species that is genetically created. A bit of reflection on that category should reveal that it is not so novel as it might seem. We are already at a level of technological capability at which we can and have produced completely new, simple living organisms in the laboratory for experimental or practical purposes. One only has to look to the organisms into which specific genes have been spliced to make them into efficient tools for consuming oil spills or producing insulin or preventing potatoes from being damaged by frost. The concept is not completely novel.

We have no difficulty recognizing organisms as synthetically designed and created by genetic engineering for a specific purpose. That they are engineered to self-destruct under certain conditions for the purpose of prevention of their unwanted proliferation when their specific task is completed is recognized and accepted. That we could design an organism purposely so that it would not evolve is conceivable and causes only casual interest. Nor is any of this, with regard to simple organisms which we can manipulate, so startling that it has provoked geneticists into considering it philosophically and assigning it to a new category.

Ought we to create another classification of evolutionary theory for organisms that are incapable of evolving? Or organisms that are deliberately created to devolve for the sake of experiment or self-destruction? It will not be too long before we can engineer, through the use of nanotechnology, specialized humans for living underwater at great depths or with wings to fly or with what would be relatively superhuman powers. Those kinds of potentials provoke immediate profound debate in themselves, as to whether we should do such things, whether it is ethical, whether such a being, if the alterations are carried to the extreme, is even to be considered as another species altogether. This last consideration comes closest to the present topic of discussion.

The question here is not whether we should do such things but, most fundamentally: Is this kind of creation of a new type of human to be considered evolution by any stretch of the traditional definition? As our control of genetic manipulation approaches total, the notion of natural selection or adaptivity begins to approach irrelevance in the usual sense. Our attainment of manipulative control of the basic mechanisms of life and evolution puts us on an entirely new plateau where survival, adapation, natural modification and adjustment, as we have previously conceived of those elements and their influence, do not apply or have to be reevaluated and re-thought. The beginning of the torturous discussions concerning simple cloning of animals and the possibility of cloning humans shows how awkward the subject is. We are being forced to redefine ourselves and life itself. We will gain consensus in due time concerning these intriguing and puzzling questions. I have explored their potential here only for the purpose of introducing the topic and using it as a background against which we may view our own redefinition. And, with regard to ourselves, it certainly is significant since it bears on the most fundamental questions of our existence and our future.

The question may fairly be asked Is a species such as we, which has reached a point of self-reflexive consciousness where it is aware of its awareness of its own evolution and the potential to choose its direction and control it, still evolving in the same way as it was previously by some mechanism of natural and/or cultural selection? In it's simpler form the question is this: Is conscious evolution still evolution by the criterion of some sort of selection process for survival as we have previously defined it? Quite clearly not.

We are aware that we are evolving and that radically alters the process. It is not Darwinian by any stretch of the definition, perhaps closer to self-determining criticality, but more than that. Human evolution has already been recognized by scientists as more complex than the evolution of simpler organisms because of our developed self-awareness. We are aware that we are evolving and that radically alters the process. We can, in a real and literal sense, interfere in that process.

We have begun, at least, to haltingly control the direction that we will take as we evolve. Inevitably we will even modify the determining genetic instructions that influence and determine our potential for evolving in different ways. Our self-awareness creates a modifying feedback into the process itself. It is certainly true that we are limited to a range of understanding of the process and a limited set of responses to our environment by the biological components of our nature. We could not simply determine, as example, that we were going to live in the middle regions of our planet's atmosphere as organized, self-aware, socially-oriented complex electrical fields without a radical effort of transmutation. But we can conceive of that possibility, perhaps even toy with some deep thoughts about nanotechnological transformations that just might do it, or even write a Star Trek script in which it was accomplished, plausible enough for even scientific consumption. Our awareness of our awareness of our evolution and the potential for actually controlling its direction magnifies the complexity of the concepts involved to such a degree that it seems quite legitimate to question whether consciously experienced evolution is evolution at all in terms of any previously advanced theory.

It may be argued that, even when we are fully aware of our evolution and determine its trajectory, that may be understood as selecting for survival adaptation and so matches the usual definition of natural evolution. But if one uses the concept of adaptation as a critical criteria then, perhaps, the answer must be negative in the case of the creation of a species ad hoc for a specific purpose, as the Nefilim created us. In our case it does not even fit the Lamarkian concept of evolutionary mechanism which works through the transmission of traits acquired by the parents in their lifetime to their offspring. Both the Darwinian biological and Lamarkian cultural models would define our ability to control our evolutionary direction as a biologically based capability developed through natural selection as a survival gambit. But genetic manipulation by us to supersede and improve on genetic chance and systematic modification of the environment to supersede and improve on environmental necessity takes us one more dimension past the purely reactive stage.

Our Unique Bicameral Genetic Imperatives

The bicameral nature of our genome, due to the fusion of two distinct genetic codes, Nefilim and Homo erectus, adds an additional degree of complexity to the redefinition of our evolution. We started out, as a species, as half Homo erectus, with hundreds of thousands of years of Homo erectus experience and wisdom of this planet.

It takes a bit more contemplation, however, to appreciate that we also started out as half Nefilim, already half of our genetic programming quite advanced and sophisticated with a vast time span of Nefilim experience and wisdom of the Planet Nibiru. The imperatives of our Homo erectus component and those of our Nefilim component may have been originally quite different, may have been, in the abstract, in conflict. But the integration, over time, of these powerful impulses in the human being has become quite complete. Their fused psychological residue may well be the source of the archetypes identified by Jung operating in the collective unconscious. So, over and above the expected adaptations, planetary conditions and the domination of the profound racial imprint as a slave species to the Nefilim working in our developmental process, we must take into consideration the interplay of our unique bicameral genetic components.

The Future as a Potential to be Determined

Our mythology rings hollow and childlike, our great literature sounds bombastic and lost, our traditions are not our history but a conflicting babel of parochial assertions, our religions hollow and rote claims to authorization to represent quaint and inadequate deities, our science a rigid two-step of investigation mandated by preconceptions of the universe which we are attempting to understand without preconceptions. The old arguments and posturing between the Creationists and the Evolutionists are stale and outmoded. Our politics and our public educational policies are heavily influenced by the godspell residue.

It is appropriate at this point to note the following: In *Breaking the Godspell*, I argued for a revolution in theology and religion that would bring about a gradual and graceful transition to the new paradigm in those institutions so that the radical changes would be as least traumatic as possible. I was aware, at that writing, that there was little hope or expectation that theologians and hierarchies of major and minor religions would be so enlightened as to easily accept the fact that religion as we know it is a cargo cult sublimation of the ancient master-slave relationship and gracefully relinquish either their claim to supernatural authorization – or power. Even less did I expect that they would begin to reexamine their entire history, context and traditions in a grand process of what could be termed, curiously perhaps, un-conversion. But, for the sake of least trauma, I made the case as strongly as I knew how. I am not surprised at all that, not only has nothing of the sort happened, but that the theologians and various clergies for the most part refuse to even consider the evidence. A few clerics and theologians who already were in the process of transition out of their respective religious contexts have become interested but they have been in no position of influence, obviously, to initiate any such movement within their contexts since they were being ostracized already in their situations.

I did expect that the ordinary layman, having little or no investment directly in power or influence within any of his or her major and minor religions, would be more open to the new paradigm. And this is, indeed, what has happened. In over one hundred radio and TV interviews, usually "live" call-in talk shows, all over the United States and Canada, eighty-five percent of the response has been positive. Typically, the caller, from whatever arena of life, acknowledges that they have been thinking about all the various components of the new paradigm for a long time. Although not able to put the pieces together, previously, they say that the new paradigm is the resolution for which they had been looking.

The new paradigm provides the base from which we can envision future evolutionary expansion and development. Those who think for themselves must be confident and brave enough to venture into that awesome plenum of potential. The new paradigm not only resolves, expands, corrects and, in some cases, eliminates, these difficult elements in current consensual evolutionary theory, it goes far beyond to provide a complete and detailed history of our species' metamorphic evolutionary development. It points the way to an expanded set of evolutionary classifications. In doing so it provides the base from which we can intelligently and consensually begin to envision and chart and explore the vast potential for future evolutionary expansion and development, individually and collectively. Those who would think for themselves must be confident and brave enough to venture into that awesome plenum of potential and freedom. We should not be startled by the message of DNA spoken there through our own voices and choices.

CHAPTER 3

REPOSSESSING OUR HiSTORY

GENETIC INLAWS AND OUTLAWS

History . . . six disciplines in search of a self definition.

Allan Bloom
The Closing of the American Mind

A people without history
Is not redeemed from time. . .

T.S. Eliot
Little Gidding

We have come, these last two incredible millennia,
In vulnerable suspension between identities,
Walking the eerie boundaries between ages,
Both equally ours, yet not quite ourselves;
Timidly murmuring precluded questions,
Hovering between the obvious and the unthinkable,
The delicate, evolving psyche
Palpated by the throbbing genetic dynamos,
Unripe defiance transmuting gradually
Into quiet detachment, yet avoiding
Premature disenfranchisement
In duplicitous cultures.
We have come, knowing that, somehow
We were supposed to know, ever less docilely,
Disowning our very history,
Stringing and unstringing the bow,
Denying the reality of the target,
Following the wrong gods home,
Down the uneasy valleys
Of our species' discontent,
Lately patting our pockets for the last few
Stereoarchetypes left to scratch dim light
Against the shadows and spectres
Of those petulant gods we have been
Trained to find peering, peevishly,

Through flaking scars in the silvering
Of the puzzled mirrors of our introspection.

Neil Freer
Neuroglyphs

What we need, in this time of racial transition from godspell to god games, is a completely new perspective from which to resurvey and conceptualize our history as a planetary-wide species. Our real history and heritage has not been fully in our possession for some two and one half millennia. That is a startling indicator of how radically we have become dissociated in our view of reality. But it was, most probably, as equally inevitable. We consider, in this chapter, not another recounting of our history in terms of dates and deeds and disasters but the broadest planetary view of human history from the time of our synthesized beginnings in the Nefilim laboratory. I am fully aware that many readers will see the word "history" in the title of this chapter and feel a cloud pass over their mind, a sense of dullness and disinterest due to some experience in school days that makes history a pathetic subject about wars, dates, and politics to be avoided.

I empathize and sympathize with that feeling, having known it myself for the better part of my academic life, for the same reasons. There was something very wrong with history it seemed, although I could not put my finger on it in my early school days. It was all drum and trumpet history, who conquered whom, very political the way it was taught, very primitive, close to, but lacking the sensationalism of, the nightly news. There was a profound sense of something askew, something very puzzling about the fact that there have been and are so many radically different interpretive realities. How could the same events be understood and translated as if the interpreters seemed to be seeing different planets? And how, in the face of such differences, could so many be so certain, so absolutely authoritarian, in their claims of being the only correct interpreters and, indeed, even the only representatives of some higher authority who they were ready to claim had put out their "official" version in the first place?

When we listen to the brooding verses of Sting's song complaining that we will learn nothing from our written history and that it only catalogues what amount to crimes, somehow those words speak to a deep feeling in all of us. The tremendously exciting fact is, however, that we now have the information to negate the despair of those lines. Something was missing, it's true, but now we have the keys to the past – and therefore to our future. And that turns out to be a very significant clue to what has been going on in the cultures of the world but especially in our Western culture, for all this time.

When I began to develop some ability to interpret for myself my first real evaluation was that the nerve had been deadened in the way history was taught for two very American reasons. The first was rather disgustingly shallow. It became quite clear that it was slanted to favor our country over others. Not quite propaganda, perhaps, as was obvious to even a schoolboy, but skewed. And then, eventually, there were the discovered lies, the "disinformation" in time of war, or to further the multinational corporations, cartels, even private interests. The sincere investigative scholar and journalist tend to become involved at the first level. Exposes, documentaries, Congressional investigations and cleanups result.

But, besides that "street" level, there is the tougher problem of the American classroom, Constitutionality. The Constitution of the United States is certainly, a most evolved political foundation, designed for a citizenry of plural religions, political views, cultural backgrounds. But the simple fact is that respecting all of those plural positions without contradicting anyone's philosophy teaching their children something contrary to their worldview is an agonizing, almost impossible dance for the sincere and conscientious teacher from nursery school through college. The fight over what shall or shall not be put in a textbook goes on all over the land. Sincere religionists and sincere scientists are often at the focal point of argument at the second level. One has only to recall the famous trial of the schoolteacher Scopes for violating Tennessee laws prohibiting the teaching of Darwinian evolutionary theory in the classroom in 1925. The most recent form of the conflict is between the scientists and the Creationists. It's almost like the Rocky movies; Scopes I, Scopes II, and probably III sometime in the not too distant future. But below the chessboards of local political "street fights" or the straightjacketing of the teacher lies a far more fundamental cause of the instinctual aversion to the "history" we teach in our schools we have all felt. They are only symptoms of the critical deeper fact that we simply do not know, agree on, what we are. We don't really know what human nature is and what it's origin was. So it is that, even until today, every generation experiences what Martin Marty, an historian of religion, has called a perennial polity fight over who is the official interpreter of reality according to the word of some deity. This may seem incredible in our so-called "modern" world but, as Julian Jaynes said in the 70's, perceiving the inevitability if not the nature of the impending planetary revolution,

> We, at the end of the second millennium A.D., are still in a sense deep in this transition to a new mentality. And all about us lie the remnants of our recent bicameral past. We have our houses of gods which record our births, define us, marry us, and bury us, receive our confessions and intercede with the gods to forgive us our trespasses. Our laws are based upon values which without their divine pendancy would be empty and unenforceable. Our national mottoes and hymns of state are usually divine invocations. Our kings, presidents, judges, and officers begin their tenures with oaths to the now silent deities taken upon the writings of those who have last heard them.

The ancient godspell will not yield easily. But the point of critical mass has been reached. The force of the evidence from archaeology, astronomy, bio-chemistry, anthropology and history will inevitably bring the new paradigm into the scientific arena and, eventually, preeminence. But the ancient master-slave relationship still influences our fundamental view of reality. We should be inspired, therefore, to relearn the salient features of our history from this new perspective. Since the evidence shows that every world-view has been partially correct we find ourselves in a position where "history" suddenly becomes vitalized and meaningful. Because its potential to resolve every facet of the most critical aspects of the recycled scenarios that have put the planet on hold has been restored, history becomes more a sociobiology than headlines from the Daily News. If the nature of our rapid metamorphosis as a species becomes the context of our racial drama it suddenly becomes us, not some puzzling group of warring robots, that it is all about. It becomes vitally exciting once our true history has been taken out of the hands of sectarian interests and restored to us. Within the context of the new paradigm our history undergoes a radical shift of perspective. The events held as most significant in the classroom, the shifts of power, rise and fall of nations, cataclysmic conflicts, crusades, jihads, constitutions and rulers and governments, the exploration of territories thought unknown, all are but symptoms of the far deeper process of healing and transformation of the collective human psyche as we come staggering out of the millennia-old night of racial amnesia and dissociation, waking to what we have never really forgotten.

We are ready to reclaim our heritage. Arriving at that point of resolution has not been easy – and may not even be acknowledged by the scientific community in general for some time. The general compartmentalization of the various scientific disciplines is clearly a handicap here. Because of the traditional, rather inane, game rules of professional behavior which tends to cover for the fact that scientists do not, as a rule, read the technical journals of disciplines other than their own and feel unqualified, one scientific discipline will not trespass on another's turf or fundamental assumptions or allow trespass on its own without extraordinary pressure.

The archaeologists say that the "gods" of the Sumerians were fictitious and, therefore, can give no consideration to the possibility that those "gods" could have interfered in the process of natural evolution and created us. Paleontologists, anthropologists and genetic researchers, the custodians of evolutionary theory, tend, therefore, to avoid and disdain any interpretation of that sort. Since they are not pressing such a possibility from their evidence the archaeologists, reciprocally, do not feel the necessity to reevaluate their position and assumptions. The loop is mutually and negatively reinforcing. Change usually has to come, and usually has come, from outside in such cases because of the self-induced paralysis. So, as we have seen, those sincere scientists who have devoted themselves to the study of our biological past generally restrict themselves to the concept of some form of evolutionary continuum from lower forms through us as the context within which to describe their findings. It is that concept which, paradoxically, as we have discussed previously, is the sticking point and which will have to be reevaluated with regard to our specific beginnings as Homo sapiens sapiens – or Homo erectus Nefilimus, as I think it is more appropriate to call us.

The Reluctant Contribution of Science

Yet it is from science, those sciences, that the indicators and data demanding the change in our understanding of our history are derived. The most recent scientific information we have obtained from the work in DNA "fingerprinting," the application of restrictive enzymes to DNA to pinpoint mutations, and the work being done in the area of both nuclear and mitochondrial DNA comparison show several findings of direct interest. The dating of our beginning as humans, regardless of how accurate the sophisticated bio-chemistry may prove to be, is not a direct proof of our Nefilim engineered beginnings; it can only indicate when we most likely came into existence as a species. But, even at its present stage of development, it does reinforce the statements in the ancient records concerning both time and place: Africa and 200-250,000 years ago. The DNA researchers are working within plus or minus parameters of 50,000 years at the time of this writing. The difference between the 300,000 years interpreted by Sitchin and the outside number of 250,000 years from the DNA work is probably only a matter of further refinement of both numbers to bring them together.

The Great Migrations

The current explanations offered by establishment anthropologists as to how the migration of humans took place over the earth assume some process of natural cause and effect. The reasoning is that there must have been pressures of climate or climate change, drought, famine, lack of raw materials, plague, change in food supplies, or some identifiable natural cause that stimulated humans to migrate even, eventually, to the tip of the southern hemisphere and the most inhospitable areas of the far North.

Again, there is clear evidence in the ancient records of deliberate manipulation by the Nefilim of whole human populations. The most obvious one is spoken of in the accounts of the early times and noted in the Old Testament as a significant event: there was an expulsion of humans out of the major center(s) when they became too numerous after being given the ability to procreate. The expulsion was toward the East, but it was meant clearly to force humans into the rural, outback areas. It marked the differentiation between the "in-law" (Nefilim center) and "outlaw" (outback) human groups. The immediate result, for the expelled, we may easily speculate, would have been of forced adjustment to a new environment, the necessity to create a lifestyle and culture adapted to that environment, the use of knowledge gained, originally, in the Nefilim centers and preserved in the outback to achieve those ends. In the long term the customs and knowledge preserved from the Nefilim centers would have merged with the new knowledge and customs of life in the outback.

How the important events of human history were effectively preserved even by the descendants of expelled humans at great distances from the Nefilim centers can be seen in the fact that forty-two versions of the history of the Flood were known by the American Indians. Since the potential existed for the superior specimens to survive and thrive, however, some of those pushed farther and farther into the outback and eventually migrated throughout the continents.

With adjustment to new climates, environments, the challenge of social adaptations, the precocious would have been capable, eventually, of forging and sustaining a higher culture. A critical fact that must be taken into account is that these humans were not starting from some primitive state or without knowledge of civilized living; they were expelled into the outback from the Nefilim city centers; they, or their ancestors, had been exposed to relatively advanced, organized, probably literate, civilization. We know from the ancient records that many had lived in close enough contact with the Nefilim themselves to even have begun to annoy their masters by their noisy sexual activities. We may safely assume that those humans originally expelled carried with them all that cultural knowledge, practical craft and technical skills such as metallurgy, advanced textile manufacture, leatherworking, animal husbandry, basic medicine, language and communication skills, knowledge of history and traditions, perhaps some even carried some basic science. Take it as a given that further anthropological and archaeological investigation, over time as the new paradigm becomes gradually accepted and the scientist can admit what he or she is looking for, will reveal more accurately what, precisely, they took with them as a legacy of civilization. I am confident that my assumptions are substantially accurate in this regard. It would seem that all contact with the Nefilim centers would not be completely cut off either, even after the surplus, or successive batches of surplus humans had been expelled. It would seem, speculatively, that even a strict mandate to cut off what are the deepest human ties, those of family, kinship, cultural traditions and bonds, would not succeed completely under the most severe enforcement by the Nefilim. If these reasonable assumptions that the humans expelled from the Nefilim centers as surplus took with them a substantial and diverse cultural tradition and were able to maintain some contact with the humans in the Nefilim centers is even half correct, we have a major breakthrough in comprehension. This is the key to the understanding of the puzzles presented by the advanced art and traces of sophisticated astronomical knowledge and technology found in very ancient sites. It is the key to why rudimentary to high culture is always found where humans are found no matter how far back we trace ourselves.

Half of the puzzle of ooparts, "out of place (only in the context of the old interpretation) artifacts" in history, is explained in this context of human cultural development. The artifacts were not out of place; our understanding has simply not been correct. The information gathered by von Daniken and Chatelain finally takes on clear and logical meaning and is set in proper context. Both have acknowledged the conclusiveness of Sitchin's synthesis. The other half of the puzzle of ooparts is explained by the fact that some actual traces of Nefilim technology may have survived. The entire area of archaeological and historical re-search and analysis becomes one of the most significant human endeavors and should be given the highest priorities. Archaeology becomes the textbook of human Sociobiology 1A. But first the archaeologist, the paleontologists, the geneticists, the historians and even the astronomers must reach the point where they will permit themselves to at least consider and debate the possibility that the Nefilim were real and created us as a synthetic species.

The "Outlaw" as Pagan

The concept of the "pagan" (original meaning: one who lives in a rural area; also one who practices polytheism), is a term of rejection in the Judaeo-Christian ethos referring to those who do not profess monotheism and/or who need "salvation." It is the "rural dweller" aspect that is intriguing when viewed against the background of the "in-law" – "outlaw" division of humanity from the most ancient times. It is not difficult to see how the concept could have evolved from the time when surplus humans were first banished into rural areas. Those "outlaw" humans would be the most prone to develop independence and drop their subservience to a local Nefilim ruler, to lose their monotheistic allegiance. Probably by the second generation after any individual or group of humans had been expelled as surplus or undesirable from a Nefilim center, a full adaptation to the natural conditions of the "outback" would be well underway. From basic survival from day to day, gathering and hunting food, building shelter and responding to social and psychological needs to the development of rules and customs and inter-group communication, regional and cultural traditions grew up. We had our indigenous Homo erectus genetic data bank to fall back on with regard to natural survival and adaptation and our Nefilim component intelligence to improve that base rapidly. It is quite clear that we developed adaptive cultures, in tune with the land and nature, sometimes quite peaceful, sometimes more belligerent.

But the other characteristic of our cultures the world over, is the influence of the residual tradition of the "gods," the Nefilim, regardless of how far we migrated from the Nefilim centers. The fact is that, simply because of the size of the planet and the fact that the Nefilim established centers in both the Eastern and the Western hemispheres, humans could not become very remote from the Nefilim influence. Before considering the orientation of those humans who actually moved into the most remote parts of the world it is helpful to eliminate some common misconceptions.

When I first began to consider the subject long ago I tended to view the primary focus of the Nefilim colonization efforts as being confined to the Middle Eastern locations which are commonly associated with them. But it has turned out that every known, major civilized center was founded by them. As an excellent example, we have thought for some time that humans migrated into the Western hemisphere by way of the Bering Strait in times long past and gradually populated the Americas all the way down to the tip of South America. It is implicit in this view that the great centers of civilization in Central and South America, therefore, were founded by humans who settled there. But it has become clear that these centers of great culture and organization were instituted by the Nefilim, from lake Titicaca to Central America and Mexico to the southwestern U.S. – if not to the Mound Builder societies in the central to southern part of the U.S. There is clear record of the transportation of humans as workers and technicians from the Middle East directly to Lake Titicaca in South America – quite possibly by air – in the year 3113 B.C. when the Nefilim set up their gold and tin mining operations there. It is clear that humans reached the various regions of the world by two means: migration and direct transport by the Nefilim. Understood in the new context, the legend of an Eskimo tribe that their first ancestors arrived at their territory in the far North "on the wings of an iron bird" takes on a very different character. The legends of the American Indian tribe concerning sky gods, visitors from across the seas, etc. can also be reevaluated. We tend to think of Australia as a relatively remote island continent but the aborigines of Australia speak of the Dreamtime (such a fine term for the remote past) when the "gods" mapped and marked the Earth.

From Stonehenge in present day England to the mounds of the Central United States, from Northern India to South America, indeed from Earth to Mars we find the same type of astronomically oriented constructions and monumental building. It seems quite clear that those astronomical constructions were focused on tracking the Planet Nibiru when it returns to the inner solar system every 3600 years. The important point here is that the knowledge of the presence of the Nefilim on the planet, in some cases even direct contact with them, has always been present to even those humans who have moved into the most remote and inhospitable areas of the planet. No doubt those traditions were carried there by the earliest arriving humans and preserved as important over time. Certainly that tradition has been distorted, misinterpreted and partly lost over long periods of time but it is usually recognizable. Since those sites are found throughout the reaches of the Earth on all continents, it is clear that, for humans, there were no "remote" and "isolated" places that were without knowledge of at least the existence of the Nefilim. In some cases the core tradition is clearer than the convoluted sublimation we have made of it in supposedly enlightened Western culture. With the realization of our true history the picture of humanity on the planet is altered forever. We can now see ourselves as a genetically created species, coming collectively and precociously of age as a planetary race, having spread over the face of the planet in the short span of some 300,000 years, adapting to climate and local conditions, preserving the ancient memories of our origins if not the full history.

Those of us who were directly involved with the Nefilim rulers in their centers of power were most traumatized by the departure of the Nefilim on whom we relied and whom we served. The profound adjustments we were forced to make in that vacuum of authority led, eventually, to the sublimation of the flesh and blood Nefilim ruler(s) into transcendental, absolute beings through merging their memory with metaphysical concepts. The traditions of servitude and ritual eventually were transformed into the institutional religions we know today. Those of us who adjusted to the variety of physical and social conditions in the remoter regions of the planet, removed from direct contact with the Nefilim rulers, did not feel the change of their leaving so sharply. We had a greater sense of self-reliance and independence and had begun to develop our own explanations of reality, of physical science and our relationship to the universe in general. But the memories and the stories and the traditions were preserved, mixed with local lore, partially lost, but still were like a faint echo in the background of the sacred traditions.

The Superficiality of Race and Place

Adaptations such as physical stature, skin coloration, eyelid configuration, hair type are, obviously, relatively superficial survival comfort adjustments to local climatic and sun conditions. Modern biological knowledge has shown that these adaptations take only 10,000 years to come about. Racial prejudices and similar abuses based on these superficial adaptations to local conditions have their source in the ancient godspell mentality and traditions and attitudes.

The Nefilim practices of using us as pawns in their power struggles and slave raiding and as guarded units of their political areas created separations that run deep. All these traumas will gradually fade with the elimination of the godspell syndrome. The panoramic view of our history afforded us by the new paradigm shows clearly the gradual evolution of the human population of the great and ancient cities, the former Nefilim centers of power and control, into human city states and eventually nations, their customs, governments, traditions and goals extensions and continuations of the Nefilim way we were taught or copied over time. That planetary view also shows the sweep of the other half of humanity, originally pushed into the rural wilderness, gradually spreading over the continents, adapting to local conditions both socially and physically, developing their own cultures and customs with a deepening sense of independence. In this perspective, over the centuries, we can better understand the relationship of the "in-law" to the "out-law" population. When the out-back tribes of independent "out-law" peoples came sweeping back into the city centers, our history books would have it that they were "savage" warriors, intimating an inferior culture of even some degree of depravity bent only on looting. But, if one sees these "out-law" cultures as independent tribes who were returning to the centers from which their ancestors had been expelled with a sense of reclaiming or appropriating their ancient heritage, the picture shifts radically. Modern, and probably less biased, anthropological and paleontological research has gradually revealed the cultures, so often portrayed as "primitive" or "savage," to be just the opposite. The tragic saga of the way the Amerindian peoples were regarded and treated is a classic example. In the context of the new paradigm, the descendants of those "in-law" humans from the Nefilim power-centers had developed religious cultures at root of which were the sublimated concepts of the Nefilim as a single transcendent God. Considering themselves the favored subjects of that stern God in whose name they were to "convert" the "heathen savages" who were often considered to be sub-human and without "souls," these "in-law" humans could only see their brothers and sisters waiting for them on the shores of other continents as inferior and exploitable. They even closed their eyes to the advanced development of the culture and knowledge of other city centers of Nefilim founding in the new world that, in some cases, easily surpassed their own in Europe. If the "out-law" tribes, called barbarians (bearded ones) by the "in-law" humans, acted harshly when they swept back into the citified areas, no less did the "in-law" types when they trespassed into the "out-law" human territories.

The end of the colonial empire phase of our history, through which this attitude continued, marks a turning point. The common consciousness has gradually awakened to some indications of the commonality and dignity of all humans. But the godspell mentality is the fundamental, traumatizing factor, at root of our religious, national and cultural attitudes and conflicts. Political, economic, philosophical and social differences are only symptoms at a relatively superficial level of the effect of this underlying and all pervasive influence. It has left deep and painful scars in humans and human society all over the globe.

The New Age movement has become, in part, a voice calling for the recognition of the cultures of the world, such as the native American Indian, for the value their traditions have for reestablishing harmony among peoples and with the planet. This is certainly a very valid and valuable contribution. But it could go further. In the context of the new paradigm, both the tribes themselves and their New Age and ecological supporters could work to reestablish their place in the restored history of humanity and reinterpret those legends. The most sacred traditions and histories focused on the "gods" can now be integrated into the world history of mankind and our involvement with the Nefilim. This is important and valuable for two reasons. It allows for a restored understanding of how the various cultures of the world, both "in-law" and "out-law," interacted with or were manipulated by the Nefilim. It also furnishes a precise base by which to differentiate that relationship to the Nefilim from whatever other contact may have been made by other alien visitors over time, whether casual or sustained. The interpretation of the "gods," the Nefilim, as mythological, has contributed to the confusing of all references and traditions of all cultures concerning the "gods" into an amorphous, nebulous muddle. But many of the "out-law" independent cultures, due to their remoteness from the Nefilim centers, had less tendency to dissociate and sublimate due to the trauma of the Nefilim's departure from the planet. Because of their more literal approach, they have, more often than not, been classified as naive, primitive, simple, even superstitious. But, in the light of our restored history, their traditions, in many cases, have retained, paradoxically, a more factual understanding of who the "gods" really were. The fact that their ancestors had been expelled thousands of years ago into the out-back and forced by circumstances to develop their own adaptive cultures caused these "out-law" human populations to be more in touch with nature than their "in-law" counterparts.

Reevaluating Our History

So the intuitive sense of revulsion to the way our history has traditionally been presented that has turned so many of us for so many generations away from the subject in school is vindicated. We have always known that we are more than the bickering, tribal, insular characters of our history books. We have always known that we were possessed of an intrinsic dignity and not the subjects of some peevish deity, whose minds were not capable of understanding our own reality, subserviently looking to the sky for direction and help. The restoration of our true history liberates us from the imposed and cramping images of our traumatized racial adolescence, away from the internecine conflict and killing, away from the ignorance and superstition that give rise to separation, away from nationalism and regionalism and religious division. The restoration of our true past allows us to be a people "redeemed from time," to reinterpret our present and consciously choose our future.

From where did this history come?
Parched Persian sand is an impartial curator,
A patient and laconic collector
Of fur, feces, kings, or records of the stars,
Indiscriminate, but highly efficient,
Treasurer of the ubiquitous clay archives
Incised with our unthinkable history:
Nondenominational records
Of transcultural gods,
Muscular and imperfect gods,
Known and approachable gods,
Lusting and loving gods,
Goddesses of engineering,
Gods of rocketry and flight,
Goddesses of architecture,
Science, and the birthing
Of our synthetic species,
Multiple mothers of our genetic genesis.

Neil Freer
Neuroglyphs

The Matriculation of Consciousness

We are dealing here with no less that the matriculation of our racial consciousness in the integration of the bicameral mind. This is not a plagiarism of the title of Julian Jaynes' book *The Origins of Consciousness in the Breakdown of the Bicameral Mind* but a slightly satirical play against it since I am concerned here with the next immediate step in our racial psychological evolution. As previously discussed above, I believe that Jaynes is incorrect in his basic assumption that the gods were mythological and his deductions from that premise that humans, previous to 1250 B.C., were hallucinating schizophrenes because they said they heard these (assumed imaginary) gods speaking to them. This racial matriculation to racial maturity and the state of the new human requires our repossessing our planetary history as a first step. If we can now repossess our real history, understand it intellectually from a clear and comprehensive perspective, we also need to deal with it on a psychological level. The racial traumas we can recognize now are the result of our being created as slaves, moved up to limited partnership and then left on our own when the Nefilim phased off the planet. We have gone through nearly three thousand years of grief, denial, guilt, dissociation and attempts to bring them back or at least be doing the right thing in the event that they come back. The godspell syndrome, ultimately, is psychopathological; it has caused deep trauma and scars in our collective consciousness and we need to deal with that as much as an individual needs to resolve and integrate the traumas of childhood and adolescence.

CHAPTER 4

THE GODS UNMASKED

For two thousand years we have reassured each other that the transcultural "gods" of all the ancient civilizations must have been mythical, fictitious, because they were said to be able to fly through space and the atmosphere of the planet, possess awesome atomic-like and laser-like weapons, communicate over long distances, be able to restore the dead to life, live enormously extended lifespans and to create humans. Hasn't anyone noticed that even the capabilities of our current technology from nuclear to genetic has invalidated – vaporized – that entire argument? We either have to come up with something a hell of a lot better or admit that those so-called gods were real.

Frank Clinton
Diary Of A Bewildered Politician

The gods unmasked are found smiling
In our genome's spiral mirror; their history
Traveling the undulating neurolexicons
Of our helical history's precocious repertoire.
Rise, Prometheus, you have stolen
Only your birthright.

Neil Freer
Neuroglyphs

It would seem, on first inspection, that the academic establishment holds a common opinion concerning the nature of the "gods." That is far from true – as it is far from true with regard to the alternative explanation camp. In order to fully appreciate just how strange and contradictory the explanations are concerning the nature of the gods and the source of the worldwide legends about them it is instructive to compare the position of the professional archaeologist with that of Joseph Campbell, the well-known comparative mythologist and with that of Julian Jaynes, a Princeton psychologist who has approached the subject from the point of view of his discipline.

The first point of fundamental agreement among all three is that, certainly, the "gods" of the Sumerians were fictional. This unquestioned postulate seems, at first, to be a part of the conviction that, in general, all "gods" are basically fictions, the products of naive minds attempting to explain awesome natural phenomena, deal with the unknown, etc. For the Western academicians involved with

73

the general subject, it's all mythology – but not quite. The God of their culture is a problem. Religious training, cultural biases, unresolved childhood impressions and personal experiences all contribute to a distinct sense of ambiguity with regard to the reality of the God of their childhood's culture.

The first point of disagreement falls here. Some do and some do not include some variation of the concept of Western culture's God in the category of the fictitious. The rationalization is usually that the god they believe in is some sophisticated cosmic principle that created the universe rather than some obviously primitive anthropomorphic projection. Is the God of the Judaeo-Christian tradition, the God that the proto-Jews made a covenant with different than their cosmic principle? Well. . . things usually trail off into fuzziness here. (So far, try as I might, I have never been able to determine precisely the position of Iraqi archaeologists. Do they hold the gods of their culture's history – it goes back to the Sumerian – to be literally fictions also? I would guess that some do and some don't. . .)

The Jaynes-Campbell Syndrome: Hallucinatory Dances With Archetypes in a Sumerian Garden

By the turn of our century, scholars had become aware of the worldwide similarity of the traditions of remote antiquity recounted by peoples in all regions of the earth dealing with the gods, the formation of the earth, the creation of the human race, the fall or rejection of the human race, a great flood or catastrophe, the wars between the gods, etc. It had become incontrovertibly clear that the basic themes and motifs of human mythology were universal, were similar the world over. One had only to compare such wide flung traditions as the Vedic of ancient India, the Eddic of medieval Iceland, and the Greek to be impressed by the fact that the pantheons of the gods were the same. Comparative studies clearly demonstrated that not only the mythologies and literary forms and languages of the world but also the religions, and even civilizations were startlingly similar. There was already indication of a diffusion of a continuum of culture even across the great oceans of the world that could have carried those mythic themes worldwide.

It had also been established that much of the teaching and traditions and history of the relatively recent Hebrew religion, considered by its adherents to have been revealed directly to the prophets and writers of the Old Testament, actually had been derived from the much earlier and more developed cultures of the Middle East as far back as the Sumerian civilization. These discoveries were a tremendous revelation yet deeply puzzling and disconcerting.

The realization of the universality of the mythic themes and the commonality of the sacred religious traditions once thought peculiar to specific cultures brought with it a broader world perspective. Since these themes were no longer the unquestioned sacred property of any group or culture or religion they then rapidly became fair game for analysis and interpretation on a natural, human basis. This enlightenment brought into question the status of the various religions claiming divine designation as interpretive authority. Because this realization compelled scholars to treat both the "religious" and the "secular" versions of these legends equally and, indeed, as a planet-wide unitary phenomenon, this flood of information triggered tremendous conflicts between authorities, sects and disciplines. Although these radical disagreements presented formidable obstacles to evenhanded integration of the new information, progress, nevertheless, was made.

In addition to these discoveries there had also been major developments in the field of Psychology which were attractive to those studying myth. Freud, Jung, Adler and others had explored the realm of mythology and religion in terms of the human subconscious and had evolved psychological explanations for many of the symbols and themes that were dominant in mythology. This set the stage for an explanation of all of mythology and religion, from primitive or ancient to present, in psychological, psychoanalytic and even psychopathological terms.

Joseph Campbell, Alias "Dances With Archetypes"

Joseph Campbell's work (*The Masks of God* series, first edition, Viking 1959-1976 Penguin) represents, in many ways, the epitome of the scholarly, academic and relatively unchallenged, definition and interpretation of human mythology until the recent past. By the time Campbell had reached scholarly maturity a rapidly multiplying series of discoveries in many fields of study had opened up new perspectives on the past. The end of the nineteenth century had witnessed a tremendous upheaval, still reverberating today, in the areas of comparative philology, oriental and classical studies, archaeology, paleontology, anthropology, ethnology, folklore, art history, psychology and comparative religion. Faced with the significance of this revelation and ferment, Campbell set out to bring this vast information from many disciplines together to write a "natural history of the gods." He conceived of such a work as both a systematic identification and classification as well as the discovery of an evolution of the "visionary world of the gods" that might well exhibit scientific laws, indeed as the beginning of a "science of the roots of revelation." It might be termed a sociobiology of revelation. But, never reaching a point where he could acknowledge the "gods" as actual history, he could only hold that their source had to lie in the psychology of the minds of men, even going so far as to say that there was madness in the claims of the god-kings of Egypt. Seeing no other alternative, he could only recommend a constructive exaltation of what he considered the "mythological" portion of man's nature.

In Campbell's paradigm, then, the gods were psychological projections of the human psyche and that human psyche was substantially the same as our psyche today, indeed had to be in order to create the psychological projections that he understood in the terms of Otto, Freud, Adler and Jung whom he had accepted as authorities, and for him to identify the same mechanisms of what he considered mythologizing in modern man.

Because he accepted the prevalent view that the source of myth was deep in the human psyche, he assumed there must be discoverable structures and dynamics in the psychosomatic system of man that were the source and origin. These assumptions were very critical because the psychological component was seen as primary and any archaeological and ethnological evidence was considered only after the fact and in so far as it could furnish reinforcement or clues.

Once this set of priorities and prejudices has been laid down the logic flowed easily: the sources in the human psyche were presumed to reach all the way back to the most primitive human form, to our very remote past, probably even to the germinal manifestations in more primitive creatures. The Darwinian notion that our evolution had progressed from the very primitive to us in an unbroken continuity reinforced the psychological paradigm. It was these same assumptions of some deep psychological source of that which was presumed unreal myth within the human psyche that eventually had led Freud and Jung, among others, to begin to apply the insights they believed they had gained into the mechanisms of the "neurotic" individual to the data of the ethnologist, the anthropologist and the archaeologist.

It would be expected, on the face of it, that Campbell, therefore, would have reasoned that the human psyche had projected the same mythos, any place in the world, to explain the universality of similar legends and the same pantheon of "gods."

In a very real way, Campbell's conclusions involved him in a self-contradiction: at the very beginning of his work, in the first volume, *Primitive Mythology*, he asked the question, Was the source of the knowledge of the gods and the similar archetypes encountered all over the world to be found in the local projections of deity and personality on the awesome forces of nature, lightning, thunder, great storms, the wind, sun, etc. by the naive human mind of our early ancestors? Amazingly he answered, although local quaint myth may have had its origin in that way, that it was a diffusion of the stunning, sudden high culture from the "little Sumerian mud garden" that was the source of "the whole cultural syndrome that has since constituted the germinal unit of all of the high civilizations of the world." If even the details of the genealogies of the "gods" were the same the world over it must have been a process of diffusion rather than local development. He went as far as to identify Sumer as the primary mythogenic zone, a source of the origination of the universal mythic themes. But having reached that startlingly acute insight, he nevertheless had to go on to explain the gods as mythic archetypes in purely psychological terms – how else can you explain something that you are convinced is unreal?

Campbell's ultimate goal was to develop an encyclopedic "creative mythology" that was based on a mature and sophisticated examination of all this vast material from religion and mythology to art history and literature according to what he understood to be the "laws and hypotheses of the science of the unconscious." If, indeed, the gods were unreal then the work he produced is a magnificent piece of scholarship that accomplished just that. But, because he could not or would not consider that the gods were real and being a person of his time and training (Jesuit, classical, semi-classical), he could only choose between whether any and all "myths" should be rationally dismissed as mere vestiges of primitive ignorance (superstitions) or, on the contrary, interpreted as rendering values beyond the faculty of reason (transcendent symbols). He chose the latter.

We are identified as children (alternatively: a product) of our times most by the criteria we recognize, to what we look for validation. By setting the assumption of unreal myth springing from a psychological source before the archaeological evidence, Campbell did not succeed in reducing the subject to history. He ended using an Eastern spiritual metaphor (he preferred it to the Western) to explain the whole in terms of one of its parts, a sort of Jesuit-Buddhist leading an intellectual celebration of what would still remain, in his own words a "great mystery pageant." Campbell never got free of the envelope of his thought; he, benevolently, considered all the gods of all the traditions equally mythic – except his God.

Julian Jaynes: History as Hallucination

Julian Jaynes, teaching at Princeton University as of this writing, published *The Origins of Consciousness in the Breakdown of the Bicameral Mind* in 1976 and immediately caused a great deal of controversy. Concerning himself as a psychologist with what he understood as the "problem" of human consciousness, in the sense of how it arose, and why, he fixed his erudite focus on the origin of consciousness within the context of the evolutionary process. In contrast to Campbell's acceptance of the Darwinian hypothesis that human consciousness developed from lower forms of consciousness along a continuum of evolutionary development, Jaynes argues that our specifically human consciousness is a learned process. His interpretation of history is that, previous to 3000 years ago humans, if one could call them that by definition, did not have the kind of consciousness we experience now. He contends that civilization is possible without consciousness as we know it. Until the second millennium B.C., he holds, humans did not have any free will or self-consciousness, were not responsible for their acts, and were actually guided by "gods" which were really auditory hallucinations they heard in their heads. He identifies an area of the right side of the brain that corresponds to Wernicke's area in the left side as the probable source locus for these voices of authorization. Previous to the second millennium B.C., he says unequivocally, by our psychiatric standards, all humans were hallucinating schizophrenes. This bicameralism of mind, the pre-conscious human – clearly, to his mind, not human as we are human – driven by auditory hallucinations identified as a "god" or "gods," he considers as genetic evolutionary adaptations which tended to produced stable societies and harmonious relationships. But, he speculates, the bicameral mentality began to break down and give way to a new type (our kind) of consciousness when the pressures of catastrophic events, floods and volcanic eruptions forced humans to confront differences between them. Hostile confrontation abetted the development of the use of deliberate treachery which requires our type of consciousness in his paradigm. The spread of writing, which moved communication and perception outside of the individual, contributed also. As the bicameral mind broke down, the hallucinatory "gods" became silent and gradually people became anxious and lost. This gradually gave rise to religion, divination, and cults. The desperate search for the archaic authorization of the "gods" gave rise to everything from prophets to peyote in his scenario.

He traces the breakdown of the pre-sapiens sapiens mind through the search for authorization down through our day in the gradual degeneration of the religions of the world, the substitution of science as a sort of religion, the turning to the occult, psychotropic drugs, astrology, etc. He is, finally, even able to see that his modeling and explanation of the history of the rise of consciousness is, at the same time, also a part of that very process of development that he is describing.

It is well to note that Jaynes' use of the term "bicameral mind" does not mean the same as my use of the term in this book – it means almost the exact opposite – although the contrast of the two usages is fascinating and significant in itself. Jaynes uses the term to mean the schizophrenic nature, as he interpreted it, of early humans' minds when they were ostensibly hallucinating the voices of "gods." He is describing what he considers a pathological condition at least by modern standards. My use of the term is meant to describe the positive nature of our dual genetic heritage, Homo erectus and Nefilim, both elements of which contribute to and are integrated into, our consciousness.

Campbell and Jaynes Compared

Just how close and how far away from the new paradigm Jaynes and Campbell were is best understood in terms of a comparison of their views. They agreed on many explicit or implicit premises and yet there is startling disagreement on the nature of the human evolutionary process and the specific event that each identifies as primary in our advent as modern humans.

As a main point of difference, Jaynes' understanding of our evolution is not the same as Campbell's. He does not accept the Darwinian hypothesis of the nature of the evolutionary process as a continuum of development from the less complex to the more complex. He finds it highly suspect because the historical facts seemed to indicate some sort of punctuating event rather than a smooth continuum. Campbell saw a more physiological evolutionary process producing our type of consciousness than Jaynes, who understood is as something learned. But, regardless of their different perspective on the mechanics of our evolution, they both understood the mythical as an early and primitive way of thought.

Not only was some very significant specific event indicated by the history (the development of writing in Jaynes' interpretation) at root of the coming into existence of our kind of human, but that event had taken place in a location that both Jaynes and Campbell –and other scholars – are in agreement on: Sumer. We now know that the unique event had occurred far earlier – our genetic invention some 300,000 years ago – and Sumer was only a very recent manifestation of that extraordinary event which, indeed, had taken place in the Middle East. But it is remarkable and significant that, even without working from the premise that the Nefilim were real and had performed such a creation, Campbell and Jaynes very forced by the data to recognize the central role of events in the Middle East.

Both, in accord with other expert opinion, had accepted the generally recognized – and startling – fact which Campbell states in the opening line of his first volume: the cultural history of mankind is a unit. Both were using their knowledge of the major psychological theories of Freud, Jung, Adler and others as a major analytical tool. Yet neither found evidence that the phenomenon of the "gods" was engendered by a common human psychological tendency with parallel local origins. Both agree that the planet-wide similarity of human "myth" is due to a process of worldwide diffusion from a single source. Their judgment in this case is very significant, even surprising.

Jaynes realized, as Campbell did, that there was a process of development in human consciousness which continues to our time and in which we are involved even in these considerations of it. And in that process, he also recognizes, as Campbell did, until the second millennium B.C., the "gods" were the controlling, central, dominant factor.

Jaynes has accepted, apparently without question, the widely held belief that the "gods" of the Sumerians were myth just as Campbell did. That is why, obviously, he finds all humans in ancient times schizophrenic; when they said they spoke with "gods" and received instructions from them, they must have been crazy because the "gods" were unreal.

In retrospect, the work of Campbell and Jaynes was clearly restrained by this unquestioned fundamental assumption that the "gods" were, must have been, unreal. Campbell says that the worldwide phenomenon of the "gods" is not the product of a common neurological system reacting to similar stimulus but of some historical event in Sumer which spread over the world. Jaynes says it was totally a neurological event of a common neurological system reacting to similar stimulus – but somehow centered in Sumer!

An unfortunate flaw arises from the assumption or conviction that the entire panorama of our history is woven of ubiquitously homogeneous myth allows the lumping together of history with nature stories, the admittedly manipulative fantasies of an individual shaman with the proclamation of a responsible king, and advanced psychology and science with superstition. If the "gods" were real then Campbell would not have been able to lump together the real Nefilim with the nature spirits imagined by naive minds in the first attempts at scientific explanation, or the political creations of local opportunistic medicine men, etc. He would have been able to differentiate between "worship" of the Nefilim (working for, serving them), and worship as a later abstraction; attempts to contact the absent gods, the desperate aberrations of regicide and human sacrifice to propitiate the absent gods or bring them back. He might have seen clearly the difference between the manipulation of humans by a Nefilim ruler through a foreman/king and the independent actions of the kings and priests after the Nefilim had left when benevolence often gave way to corruptions of power, advantage taking, and cruelty. Naive science, naive medicine, would have been distinguishable from authentic traditions of healing and even advanced surgery. What the Nefilim taught us or what we overheard about immortality, reincarnation, the nature of the soul, how the Nefilim viewed those things and what they meant to them, would have been easily differentiated from probable hearsay and real myth.

Against the background of the full understanding of our true genetically engineered beginnings, these two scholars represent the near torturous extremes we collectively have gone to, been forced to, to explain, to even make common sense from, and integrate the legacy of our ancient past as the choices of possible explanations have been continually narrowed by more and more thorough investigation, accumulation of evidence and scientific analysis.

If it is Sitchin's master thesis which supplies both the final resolution and context against which we may clearly comprehend the unique nature of our sudden consciousness and the conditions which empower us to work out the ramifications or our unique bicameral nature, then the work of Campbell requires radical revision at the very least; Jaynes' work is clearly false by his own criterion as stated at the end of his first chapter. But credit where credit is due: lesser humans have retreated to the stronghold of institutionalized religious authority, avoided the questions and even buckled in the face of the narrowed set of choices they faced. We, living in the 1990's and early 21st century, may never fully appreciate the great step past the domination of the old authoritarian explanations these two thinkers, though wrong, contributed to our taking. Regardless of the sometimes petulant criticism that they are not specialists, i.e. archeologists, they have made a solid impression on the fields of history, archaeology and sociology.

If Campbell has clearly delineated the startling ramifications of the worldwide character of the great legends, even though he cannot see his way clear to put it in an actual physically historical context, he is preparing the way for that resolution. Campbell also contributed a great deal in that he provided the sufficiently innocuous entree, mythology, to allow scholars of even sectarian affiliation to participate in an investigation otherwise precluded or closed to them. The sectarian theologian can certainly safely read "mythology" without immediate danger to his faith (it's not really "real," after all, is it?) while still privately stopping short of considering his particular version of it also as "myth" or taking it as literal history. And he raised mythology to a new level of respectability as an academic topic.

If Jaynes says that our consciousness is a learned process because of the weight of the evidence from 3000 years ago he is opening up questions that lead directly to the new paradigm. If he uses that consciousness operating in himself, after only 3000 years, to already discern itself as still developing he is pointing directly at the unique nature of our rapid metaevolution. When he expresses doubt about the consensual concept of evolution as it operates in human nature he is shifting focus at least toward the new paradigm's illumination of our development as a special case of evolution, a rapid metamorphosis caused by a synthetic interruption of the natural evolutionary process on this planet.

Campbell and Jaynes are very close to the new paradigm when they identified the ongoing process of consciousness development they recognized in early civilization and continuing on into our day. What they would be describing, if they were working in the framework of the new paradigm, are the effects of the more advanced Nefilim component that drives our precocious metamorphic racial development.

Both, by their insistence on the link between our consciousness and the "gods," even though they were mistaken in the actual details, are only one step removed from the full picture. In their recognition and emphasis of the centrality of the "gods" in human history they mark a stage of development of thought which no longer even questions that fact, no longer even looks for other reasons; the "gods," however one conceives of them, were the major factor in our development as a species. Campbell and Jaynes were still in the direct line of thinking that held that our evolving consciousness had caused the "gods." The new paradigm says that the "gods" had caused our unique type of consciousness and accelerated development, a special case of evolution.

Both made a major contribution in their recognition and emphasis on the fact that religion was an extension of the ancient human-god relationship and, by that emphasis indicated clues to the underlying master-slave godspell.

The ultimate result of these two formidable scholars identifying the unequivocal link between the uniqueness of our consciousness and the actions of the "gods," and our culture with a specific historical event, forces some very hard questions. Was none of this real history? If so which parts? Are we not literally able to differentiate between historical events and fictional?

Jaynes uses, as an important example of what he considers hallucinatory behavior, the bas-relief image of the ruler Hammurabi speaking with his "god" Marduk. Both are depicted as male humanoids, obviously in conversation, Marduk, of slightly larger physical stature sitting on an ornate high-backed chair, Hammurabi standing at ease in front of him. What of the minds of those who recorded the event incised on the clay tablets showing Hammurabi speaking with his "god"? Were they hallucinating the entire scene or only the "god" as Jaynes claims they were? Were the auditory hallucinated voices also accompanied by detailed visual hallucinations? What of their perception of Hammurabi? Was it hallucinated, perhaps, also – or only one of the two figures in the picture? Did they know what the "god" looked like only because Hammurabi told them? Or because they somehow collectively were able to hallucinate the identical images that Hammurabi did? How large was the "god," whether male or female, how dressed, bearded or shaven? If they had to rely only on Hammurabi for a description of the "god" didn't they have some doubts about Hammurabi's mind? Or was the picture we have from the carved image solely the way Hammurabi dictated it to be, exclusively the way he "saw" and "heard" himself when in full hallucinating mode?

Being constrained by their assumption of the unreality of the "gods," Campbell and Jaynes were forced to find psychological/sociological explanations for such astounding behavior. In doing so they identified not the beginning of our type of consciousness but the much later beginning of civilization once we had been upgraded from slaves to partners. By their emphasis on the link between the "gods" and civilization they are only one step removed from Sitchin's scenario.

They also recognized fully that the entire notion of religion, particularly as expressed in the authoritarian institutionalized world religions, was an extension of the ancient human-god relationship and in that realization identified clues to the underlying master-slave godspell. Both point out the obvious and important connection between humans' ancient relationship to the gods and the nature of those religions in our time still claiming to be the sole interpretive authority representing the will and commands of the "gods." They contributed, indirectly, to the full understanding or how arbitrary "religion," as we usually conceive of it, really is. We assume that religion, from the broadest sense of recognition of some incomprehensible power behind the mystery of the universe to the narrowest sectarian ethical prescription, is an inherent, fundamental, essential, natural facet of the phenomenon we know as self-aware consciousness. We assume that since we experience our situation thus, other consciousnesses we would be able to recognize as similar to us would also be "religious." This is even more arbitrary and state-bound an assumption than our projection of a kind of evolutionary development as we experience it here on the Earth on other consciousnesses elsewhere. Given the new paradigm, religion, in any of its connotations would be more reasonably understood as a lingering unique state-bound effect of the original relationship between us and our Nefilim creators. The only place we might expect to find the "godspell effect" might be on another planet (and even the restriction to a planet is a projection) where a species had been brought into existence by genetic engineering or some similar process as we were if the status of the newly formed species was one of subservience. We seldom ever become detached enough from our tradition to wonder why there has been this phenomenon of subservience to very imperfect and limited "gods" much less to wonder why religion at all – or, especially, why an integral part of that phenomenon has been a taboo about questioning it. And, ultimately, even our concept of our own evolution, given the unique acceleration caused by the driving force of our Nefilim genetic component is clearly state-bound and not something we should automatically project on another species of our type.

Campbell's forceful presentation of the compelling evidence from comparative mythology of the world-encompassing continuity of the same mythic content forced recognition that the "religious" and the "secular" elements of the worldwide tradition are the same. So a large change was introduced and accepted by modern scholars in resonance with Campbell's work that, in retrospect, is clearly not in keeping with the previous mind-set of western culture when it was dominated by Christianity. The overwhelming evidence of comparative mythology just turned that around by sheer weight of evidence.

"Mythology," although it is understood as a respectable area of study, as a given, as a fact in our culture, is really no more than an arbitrary interpretation. It turns out, paradoxically, to be our greatest cultural myth. We have obvious, naive, local myth, certainly, which is, as a rule, a first attempt at science, the explanation of an unknown in nature. But, in the light of the cumulative archaeological, astronomical, genetic and historical evidence we are now able to see clearly that the mythic pantheon of "gods" known the world over is not myth but the persistent history of real humanoids. In the recognition of that simple, yet almost unthinkable, fact we attain the resolution of the otherwise almost bizarre enigmas of our racial beginnings – and our own selves. Campbell, in his search for a kind of mythology of mythology, and Jaynes, in his search for some adequate, sane explanation of the beginning of our kind of consciousness stand at the brink of resolution. Campbell found only a broader, self-conscious mythology; we are still only heroes with a thousand hang-ups. Jaynes found only precocious madness. Almost in anguish, in a sense, over the ultimately preposterous facts of our history, yet not being able to see or accept the reality of the Nefilim, both Jaynes and Campbell perceived the nature of the transition required if not the outcome.

The significance of the theses of these two erudite scholars taken together is that, by their time, the weight of evidence for the literal physical, historical existence of the humanoid Anunnaki/Nefilim living and breathing and fighting and loving and performing political machinations as well as huge construction projects among humans for thousands of years had become intense. By their time the choices have been narrowed down to two: it's either history or hallucination; either the humans of those times were really naive schizophrenes who built their entire civilizations, physically and socially, under the direction of delusional neurological auditory and visual hallucinations – or the "gods" were real.

The forms in which Campbell and Jaynes conceived these fundamental problems of our beginnings and our history and their approaches to answering these profound questions differed in style and focus but the similarity of their interpretations and conclusions is startling and highly significant in its coincidence. Insight into the assumptions, prejudices, cultural and academic forces operant in these independent thinkers provides us both a realization of the inescapable focal dominance of the questions concerning the "gods" and our beginnings and, especially, a sense of the extremes to which the western mind-set has driven us in dealing with it.

Campbell, with his characteristic tempered impatience could only ask, Are the modern civilizations to remain spiritually locked from each other in their local notions of the sense of the general tradition; or can we not now break through to some more profoundly based point and counterpoint of human understanding?

Jaynes, with a keen perception of the sweep of our puzzling history, saw the process if not the resolution, thus:

> "This drama, this immense scenario in which humanity has been performing on this planet over the last 4000 years, is clear when we take the large view of the central intellectual tendency of world history. In the second millennium B.C., we stopped hearing the voices of gods. In the first millennium B.C., those of us who still heard the voices, our oracles and prophets, they too died away. In the first millennium A.D., it is their sayings and hearings preserved in sacred texts through which we obeyed our lost divinities. And in the second millennium A.D., these writings lose their authority. . .We, at the end of the second millennium A.D., are

still in a sense deep in this transition to a new mentality . . . And in the last part of "the second millennium A.D., that process is apparently becoming complete . . . We, we fragile human species at the end of the second millennium A.D., we must become our own authorization. And here at the end of the second millennium and about to enter the third, we are surrounded with this problem. It is one that the new millennium will be working out, perhaps slowly, perhaps, swiftly, perhaps even with some further changes in our mentality." (pp. 436-9)

Having attained the genetic enlightenment to realize the nature of the problem, we should strive, in this new millennium, to swiftly work out this process of breaking the godspell, the final totemtaboo, creating further changes in our mentality if required, so that the recognition of the "gods," unmasked, is an act of self-acknowledgment, self-recognition, and liberation of the new human.

CHAPTER 5

THE NEFILIM, NASA AND NIBIRU

. . .no concrete problem is going to be solved as long as the experts of astronomy are too supercilious to touch "mythical" ideas – which are firmly believed to be plain nonsense, of course – as long as historians of religion swear to it that stars and planets were smuggled into originally "healthy" fertility cults and naive fairy tales only "very late" – whence these unhealthy subjects should be neglected by principle – and as long as the philologists imagine that familiarity with grammar replaces that scientific knowledge which they lack, and dislike.

Giorgio de Santillana
& Hertha von Dechend
Hamlet's Mill

If the Nefilim were that advanced 300,000 years ago they could have genetically engineered our species, could have given us a history of our solar system and a description of their home planet. Yet the archaeologist and the astronomer are caught in a recursive pas de deux out of which neither is willing to make the first professional break.

Frank Clinton
Diary Of A Bewildered Politician

The godspell, until now, has caused a mobius loop in the relationship between archaeology and astronomy. It is most evident in the area of archaeoastronomy, the specialization devoted to the study of the astronomy of ancient times. The loop has precluded astronomers from benefiting from the information in the ancient records because of the assumption that it could not be there. It has precluded archaeologists from appreciating the astronomical information translated from the ancient histories for what it really is because of their assumption that it could not be such. Archaeoastronomers have gone to great lengths to discredit any suggestion or claim that the human astronomers of the remote past had any source for their knowledge and skill other than their own minds and physical resources. At the same time they cannot but marvel at the level of sophistication of that knowledge and skill.

We think of ourselves as advanced in the science of solar system astronomy.

We put out investigative probes, one of which has already passed beyond the limits of the solar system's boundaries, and stare in sophisticated wonder at the close-up pictures of our system's most distant planets. We are willing to concede that we are still young in our knowledge of the universe but we think that we have finally understood our local sun's system. We compare ourselves to the astronomers of the last century as they looked on the astronomers of the 16th century. Yet it is so relative: some expert astronomers say the rigorous evidence indicates that there is another planet in our solar system which we have not located visually as yet. It is called Planet X and has been searched for since before Tombaugh discovered Pluto in nineteen thirty – indeed, at first, it was thought that Pluto was Planet X. But Pluto's size ruled that out and the search has continued through to this day. The gravitational effects on the orbital paths of Uranus and Neptune caused by some unseen large body tells some astronomers that it is a Neptune-sized planet that takes at least 800-1000 Earth years to circle the sun once. The search for it is concentrated, because of the clues, in the Southern hemisphere, in the southern skies. Yet the Sumerians have transmitted the precise details of the information about Planet X/Nibiru down to us and we have only to open our minds to that possibility and take advantage of that knowledge. But those parts of that body of knowledge that are advanced beyond even ours today such as knowledge of the size, orbit, periodicity of the planet Nibiru astronomers feel compelled to reject as such because they are convinced that ancient humans could not have possessed it. They even reject the clear, documented fact that the ancients knew of Pluto simply because they say they couldn't have. That is simply preclusive thinking. The humans of ancient times obviously couldn't have known of Pluto, Nibiru – or even the great year of precession, for that matter – by themselves, but they could have been given that information and, quite unequivocally from all the indicators, were given the information.

But the godspell is broken, the door to the wealth of knowledge from the past stands open. It would seem, at first consideration, that the potential for practical benefits opened up by the new paradigm is most immediately and dramatically manifest for the science of archaeo-astronomy, the multi-disciplinary study of astronomy as practiced in ancient times. But, to be precise, the benefits should be far more profoundly beneficial to the science of modern astronomy itself. The reasons are simple and obvious: not only will reexamination of the translated history contribute to discovering the planet Nibiru but actually give astronomers and us the framework of the history of the formation of our solar system from its beginnings. In addition it will allow us to model the entire solar system, including the planet Nibiru, and such modeling by computer should indicate when to expect Nibiru's return to the inner solar system and what effects on the earth to anticipate due to its close passing. This is no matter of passing interest. There may be times when "The Passing," because of differences in the relative positions of the planets with regard to each and the sun, may be far more catastrophic in its effect. This may be very beneficial, perhaps even vital, for our comfort if not survival. Since there is some very obvious potentially profound politics associated with the possible return of the Nefilim, the foreknowledge of such would be valuable.

A summary of the astronomical information gathered from the recovered text, *Enuma elish* ("When in the heights"), now called *The Creation Epic*, is presented by Sitchin as follows: The planet Nibiru is described as a radiant planet, probably indicating a high degree of internally generated heat, somewhat ruddy in color (dark red when first visible to the naked eye) and five magnitudes of the size of the earth. When visible from earth it can be seen at sunrise and disappears from view at sunset (presumably from the typical Sumerian location of 30 degrees north latitude in the area of present day Iraq). The radiant nature of Nibiru emphasized in the ancient records may be a product of high geologic stresses. Astrophysics tells us that a body with an elliptical orbit – Nibiru's is very elongated and comet-like – is constantly tending to resolve to a circular orbit. This causes heat generating stresses more or less proportionate to the degree of ellipticality. This relatively high degree of heat may be a reason why humanoid types very similar to us can survive on that planet even though it moves so far from the heat of the sun in its orbit.

Nibiru was captured into the solar system very early in the existence of the system and was not a "native" planet. It travels in a huge elliptical, comet-like orbit through the region of the asteroid belt and far out past Pluto. It takes 3600 years to complete one orbit. The ancient records say clearly it orbits clockwise, opposite the counterclockwise direction of all the other planets and is tipped at a 45 degree angle to the ecliptic, the plane in which the other planets generally orbit (Pluto is tipped up at 15 degrees from the ecliptic). Nibiru passes through the Great Bear, Orion, the vicinity of Sirius and the constellations of the south. It moves from Taurus to Sagittarius at some point in its incoming progress, coming in from the south in a clockwise direction, apparently makes Venus "brighter" by merging of the two images at one point, "upon its appearance: Mercury; rising thirty degrees of the celestial arc (at an angle of 30 degrees to the imaginary axis of Sun-Earth perigee?) Jupiter." It increases in brilliance as it passes from the location of Jupiter toward the west; when at its closest point to the sun, passes through the point of the original collisional location, the Asteroid belt "in the Zodiac of Cancer." The effect of the passing, the return of Nibiru through the region of the asteroid belt, is described as causing (on Earth) earthquakes, flooding, torrential rains and storms, volcanic eruptions, an actual sonic effect, the sun going down at noon, for one day no light ("neither day nor night"), uncommon freezing and the moon being "as red as blood."

When Nibiru is within the inner solar system there is a period when, from earth, it is visible at sunrise and disappears from view at sunset. There is clearly enough information here for a computer modeling of the orbital path of Nibiru and its interaction with the other planets. It is fascinating to watch the information gradually being gathered by our astronomers concerning our solar system come more and more in line with the ancient records. What do we know now about the existence and nature of planet Nibiru?

Some astronomers, calling it Planet X, do believe it exists strongly enough to devote time and effort and money to a search for it. An excellent article in Astronomy magazine (Aug., 88) summarized the information accumulated in recent times concerning its orbital path, present location, size, and periodicity. The clues to its actual existence were deduced from noticeable, though slight, effects on the path of two outer planets, Uranus and Neptune. The astronomers call these effects of a gravitational pull on these giant planets by some unseen body, tugging them into a slightly wavering path, "residuals" because they are undeniable after rigorously excluding all doubtful observations and measurements. Simply put, they are the deviations between the predicted position of a planet and its observed position. (This is precisely the way Neptune was discovered, by deducing its existence and position from the residuals in the orbit of Uranus.) Because the evidence is strong enough, an observatory in New Zealand was designated to search for this tenth and last planet in our system using a twin 8-inch astrograph telescope on loan from the Naval Observatory in Washington, D.C.. It reported back to NASA where the observational information is analyzed. For reasons given that were a bit confusing to say the least the scope was removed and the search transferred to an observatory in Chile.

To date, although still theoretical, the information is building and indicates that the suspected planet does not fall within the ecliptic (the plane of the earth's orbit) and must be tipped at an angle to it of about 32 degrees. One trip around the sun, in a very elliptical orbit that takes it far out past distant Pluto, is projected from the clues to be somewhere between 350 and 1000 years. The search is now concentrated on the southern skies as its most probably current location. Its size is deduced to be two to five masses of the earth, somewhere close to the size of the earth and Uranus. Its mass and the eccentricity of its orbit are signaled indirectly and curiously by the status of the planet Pluto.

The theory that Pluto was once a satellite of Neptune has been advanced by a number of astronomers since 1934. Since Neptune, whose orbit is extremely normal, nevertheless has moons that show very erratic behavior, astronomers have postulated that a massive planet must have almost collided with it at one time. This near miss is thought to have dislodged Pluto into its own independent planetary orbit and caused the bizarre behaviors of Triton which circles Neptune backwards in a decaying orbit, and Nereid which travels in an elongated orbit. Working from this hypothesis, astronomers Harrington and Van Flandern of the Naval Observatory were able to show that a planet almost colliding with Neptune and causing such disruption would indeed have had to be two to five masses of the earth and lie in an eccentric orbit from all the indicators. Perhaps even more important is the detailed information given us by the Nefilim about the formation of our solar system. It records the formation, in historical sequence, first of Mercury closest to the sun, then Tiamat (a very large proto-planet later split by collision into the asteroids and Earth), then a "pair," Venus and Mars between Mercury and Tiamat, then Saturn and Jupiter even before the "maturing" of Venus and Mars, then another pair, Uranus and Neptune. Pluto was not an independent planet at that time. The ancient records state clearly Pluto was a moon of Saturn, not Neptune, becoming an independent planet only after and as a result of the readjustments due to the capture of Nibiru. Neptune is said to have been responsible for stabilizing the system at that point by some sort of field balancing that prevented the sun from producing more planets.

When Nibiru was captured into the solar system there were nine native planets. Nibiru is described as belching eruptions and emitting vast amounts of radiation at that time. The relative gravitational forces between it and the outer planets as it passed by them caused its path to be bent inward first by Neptune, which caused the side of Nibiru to bulge, then Uranus, which forced Nibiru to emit four satellites (see: Uranus, Andrew P. Ingersoll, Scientific American, Jan 1987). Saturn then was forced to release its moon Pluto while, together with Jupiter, it drew three more satellites from Nibiru. With Nibiru on a collision course with Tiamat, the perturbations of the fields caused Tiamat to produce from itself a total of eleven satellites, ten small ones and a large one known as Kingu. But the two planets did not collide at this first meeting. The satellites of Nibiru, produced in the awesome gravitational interactions during its interjection into the solar system, however, did collide with Tiamat and created some sort of physical cleavage of Tiamat, permitting a huge field discharge between the two planets that left

Tiamat neutralized, scattering Tiamat's ten tiny satellites. On a subsequent orbit Nibiru itself struck Tiamat, splitting it into two parts. The result was the formation of the asteroid belt from debris and the congealing of the remains of the large proto-planet into our Earth, dislocating it into current Earth orbit. These moonlets of both planets would still be members of the asteroid belt although it might be difficult to differentiate them from the pieces of Tiamat and Nibiru that formed the full belt since they were originally formed from Tiamat by gravitational extraction. It would be of great interest to attempt to identify asteroids that have different composition, even spectrally, on the assumption that one type would be pieces of Tiamat and the others pieces of Nibiru. The pieces of Tiamat should have the same composition as the Earth, since the Earth is a congealed remainder of Tiamat. Tiamat's large moon Kingu was exchanged, becoming the Earth's moon.

This catastrophic collisional event and subsequent reforming and congealing of the part of the proto-planet Tiamat to become the present Earth could well be the ultimate cause of our Earth's colossal plate tectonics. Current information says that the plates move apart and then come back together into a large single mass in a rhythmic cycle. It's as if the Earth is still ringing like a bell from the ancient collisional event. This information gives us a possible answer to why less gross material is found in the asteroid belt taken as a whole than expected if a full sized planet meeting the criteria of Bode's law had actually broken up. A comparison of the information contained in the ancient records with the facts and theories advanced about our solar system by modern astronomers obviously shows some startling similarities. The old records say Planet X is about 5 masses of the earth, modern astronomy deduces two to five masses. The old records speak of an orbit highly elliptical and inclined up sharply, 30 plus degrees to the ecliptic, the modern findings the same but with a lesser inclination. Two very striking dissimilarities, viz. the orbital periodicities and the disagreement as to whether Pluto was originally a moon of Neptune or Saturn are, perhaps, the most significant clues to the reality of the matter. The modern astronomer works under the assumption that Planet X orbits, however eccentrically and elliptically, in the same direction as all the other planets. It is on this basis that orbital periodicity (the time once around the sun) is worked out and mass is deduced through the effect on other planets. But the ancient records claim unequivocally, however, that Nibiru, Planet X, orbits in the opposite direction of the other planets. This fact alone, if taken into consideration, should account for the disagreement on periodicity, resolve the enigmas of residuals that still do not allow a long term ephemeris (a chart that predicts the precise location of a planet in the near future) to be constructed for Uranus and Neptune. Since the original capture entry of Planet X was in this opposite direction it might explain why the old records say that Pluto was originally a moon of Saturn rather than of Neptune, as the modern astronomers hypothesize.

The way the gradually accumulating knowledge about our Earth's moon increasingly comes into line with the information in the ancient records is fascinating to watch. Previous to our actually visiting and the obtaining of rock samples from the moon it had been assumed that the moon was inherently "dead." No plate tectonics (the gigantic, very slow, "flow" of sections of surface crust driven by internal heat) were evident and the assumption was that there was probably no internal magnetic iron core. The Soviet Luna 2 probe in 1969 indicated that the moon possessed a detectable residual magnetic field, reinforcing the concept of the moon as internally inactive, much the same throughout its entire structure.

But the information and samples gathered by the astronauts showed otherwise.

Previous to the Apollo missions, scientists held three possible theories concerning the origin and physical nature of our moon. Either the moon was of the same physical constitution and had developed similarly to the earth; or it had been originally part of the earth and had been torn away from it; or it had been captured from inside or outside of the solar system. By the time of Apollo 11 we had recovered and analyzed rocks from the moon which showed them to be completely dissimilar from those of earth (containing high quantities of titanium, no water, and of an age all over 3 billion years old). Apollo 12 through 14 returned the same results but, even after an accumulation of some 245 pounds of rocks, there was still some doubt since all had come from one general location. With Apollo 15 through 17 providing samplings of highland outcropping areas in a different location the conclusion had to be drawn that the material of the moon was so different from the earth that the moon must have been captured by the earth at some point early in the formation of the solar system. When a close examination of the returned samples was carried out, some of the rock samples were found to be magnetized. This has led to a "surprising hypothesis." S.K. Runcorn (The Moon's Ancient Magnetism, S. K. Runcorn, Scientific American, Dec. 1987) and his colleagues along with other groups are of the opinion that the moon at one time may have had its own magnetic field – nearly twice as strong as that of our present day earth. In addition, Runcorn is of the opinion that the indications of polar wandering, the relocation of the north and south poles of the moon at three different times in the early history of the moon, was probably caused by three small satellites of the moon itself eventually breaking up from gravitational forces and crashing into the moon, causing it to adjust its position. The source of these other bodies in the earth moon system is still uncertain. The details of the Nibiru-Tiamat interaction in which the ten tiny satellites of Tiamat were scattered certainly suggests the possibility that three of them could have become moons of Kingu in unstable orbit just as possibly some of the satellites of Nibiru could have, although the tiny ones of Tiamat seem better candidates. The ancient account of the formation and subsequent readjustment of our solar system with the capture of Nibiru is very clear. Kingu, originally a satellite of Tiamat, the planet torn into two parts, originally had all the characteristics and status of a planetary body, a magnetic field, convective churning of its crust and internal parts, an iron core and volcanic activity. It is also clearly stated that, after the catastrophic collisional events, all these characteristics were stripped from it and it shrunk. Kingu became an inert "mass of lifeless clay." And, slowly, as scientists learn that our moon must have been captured by the earth at some point early in the formation of the solar system, at one time must have had its own magnetic field – nearly twice as strong as that of our present day earth – that it had probably acquired or was part of a system that contained small satellites, we see the gradual alignment of the evidence from our most advanced investigations with the ancient accounts.

It seems clear that there is enough factual material in Sitchin's translations that a model of the orbital configuration and path is apparent even without computerized modeling, keeping in mind that the relative location was the Middle East about 30 degrees north. Computer modeling would, however, provide an added dimension allowing running our solar system backwards and forwards in time rapidly to study the details of the past and to anticipate the future. Gerald J. Sussman, Ph.D., of MIT has developed an electronic "orrery," a group of microcomputers, each one dedicated to a single planet, linked together in a parallel processing mode capable of simulating the entire solar system. It can "run" the solar system backwards and forward in time. Using such a system, with an additional computer programmed to represent the planet Nibiru, a simulation of the complete solar system as it is now could be developed. This model run backwards, if sufficiently precise, should eventually arrive at a point close to the time of collisional impact. Further modeling of the proto solar system before impact from the information preserved in the ancient texts should give us the details of that dramatic history. Elaborating the collisional details should provide us with an accurate 3-dimensional model of the direction and trajectory of the ejecta from the impacts and their interaction with and effect on the various planets and satellites. The model, run forward, should indicate when Nibiru will return to the inner solar system the next time, its orbital path and whether its passing will constitute any special danger to the inner planets during its next approach.

Once we have allowed ourselves to recognize that the Nefilim built the original pyramids at Giza as important beacons and markers, perhaps even three-dimensional records of the vital statistics of this planet, we should be able to exploit the information contained therein. There are also specific astronomical alignments built into those pyramids such as the southern sight line of the pyramid projected towards Orion's belt, among others. Since the old records tell us that Nibiru passes through Orion at one point in its orbit it may be possible to use the pyramids to shortcut the search for Nibiru. The astronomical sight lines, calendric elements, mathematical and geometric significance of the Nefilim-built pyramid at Giza and its relationship to the grid system they apparently laid out over the entire earth should be intensely reexamined for the same sort of clues and information. The detailed information the astronomer searching for Planet X/Nibiru seeks may be physically encoded into the pyramid. How this could be accomplished is dealt with in the chapter on the basis of what we have traditionally called the "occult."

Modern science and astronomy have not yet come up with a satisfactory explanation why comets and meteorites contain complex microbiological forms. These intriguing early signs of potential organic life are routinely detected in meteorite materials which have managed to get through the destructive friction of our atmosphere and are found throughout the world. The conditions of the Antarctic are ideal for preserving them in a relatively uncontaminated state and many have been found there. Science has postulated a number of hypotheses for their existence in these stony missiles but the hypotheses remain such. The details of the collisional event given in the ancient texts of planetary collision and fragmentation would clearly point to these comets and meteorites as chunks of planet still retaining traces of early microbiological forms. The future should see a further development of their analysis seeking to determine of which planet, Nibiru or Tiamat, or of which moon of which planet they are fragments. In this context there is a continuity between the collision generated fragments that became asteroids, intermediate bodies such as Chiron, meteorites, comets, those pieces that have splashed into other planets and moons, and the Oort cloud of finer "junk" that may have been dragged along and may still be being dragged along by Nibiru on its outbound path.

The fact stated in the ancient records that Nibiru was not a native planet but was captured into our solar system by the complex gravitational attraction of our system raises questions as to how that could have occurred. Napier and Clube, a pair of astronomers working at the Edinburgh Observatory, have provided an explanation of how that could have happened through their exhaustive study of comets. In their book *The Cosmic Serpent* (Universe Books, 1982), they state that modern astronomy has learned enough about the mechanisms of our galaxy to be able to trace the forms of its spiral arms and their motion as they pinwheel around the central hub (which may contain a black hole, a gravity sink so strong even light cannot escape its pull) of the Milky Way. They have pointed out that the gravitational field of our galaxy controls the motion and orbit of a star like our sun. We can visualize our sun and its planets circling the center of the galaxy, moving through its spiral, pinwheel-like "arms" one after another, crossing them in turn at about 50 million year intervals. They postulate, therefore, that, as the sun moves through the spiral arms which are made up of solid bodies from the size of dust particles to planets, our solar system scoops up whatever is in its path. It acts like a gigantic gravitational vacuum cleaner. The different velocities of different solar systems, determined by a number of variables, can cause them eventually to move farther from each other – or to move closer, even collide, with one another. Although generally the majority of the bodies resident in the spiral arms are of sub-planetary size, it can be seen that our Sun and its planets, sweeping through the spiral arms, could have a close enough encounter with another star's planetary system to capture a planet from an outer orbit (analogous to electron capture from atom to atom in the old Bohr model) – or to attract one that had been ejected from its orbit by an exploding star. To date this seems to be the most plausible scientific theory as to how the planet Nibiru could have been literally captured into our original solar system as the ancient texts clearly state. We should therefore be looking for evidence of three distinct phenomena: the capture of the planet Nibiru and its resultant effect; the capture and presence of sub-planetary bodies (objects less than the size of planets) such as comets, asteroids or meteorites before and after the time of the capture of the planet Nibiru and their subsequent effect on and action within the solar system; and lastly, the bodies constituting the debris of the early catastrophic collisional event.

Central to the ancient cultures is the important event of the return of planet Nibiru to the inner solar system. This meant a return of the Nefilim to the colony planet, a change of personnel, of "shifts" so to speak, a visitation of officials, an accounting perhaps. While the Nefilim were here themselves it was important to those stationed here. It also was important to humans and became more so as the Nefilim phased their active operations off Earth. It may be reasonably assumed, therefore, that the focus of much of the ancient preoccupation with the heavens was the careful keeping track of time to be able to know when the return would come, and systems of observation and tracking to spot the returning planet and keep track of its passing. It is possible, therefore, that when the archaeoastronomers begin to work within the framework of the new paradigm a new chapter will be written. The possibility will be opened up to begin to restudy, reexamine the ancient astronomical sites around the world in order to determine if they contain alignments, sighting points or configurations that are directly related to Nibiru and its orbit, its path through the heavens, its return.

Taking Stonehenge as a single example, it would seem profitable to systematically restudy its organization for possible alignments and sightlines for such clues with emphasis on those elements of this giant astronomical construction which are still enigmatic or may have even been missed or dismissed. It should be obvious that, if indeed the new paradigm is true, then all the information we need to locate, identify, track and computer model the planet Nibiru is available to us if we allow ourselves to recognize it. The most critically important fact to keep in mind is that without reference to the new paradigm, some of the puzzles and mysteries of archaeoastronomy will not even be recognized, much less solved. I submit that the Sumerian account of the formation of our proto solar system, the capture of Nibiru, the subsequent collisional events and readjustments, the 3600 year periodicity of the return of Nibiru through the vicinity of the asteroid belt, is the only scenario that finally explains all the current astronomical enigmas.

We thought we had Planet X when Pluto was discovered. But this has proven to be an error. In January of 1931, when Clyde Tombaugh was given the Jackson-Gwilt medal by England's Astronomical Society for his meticulous work in discovering Pluto, the Society's president noted "There is a further note of fitness in the fact that this medal bears the portrait of Sir William Herschel, who was the first man since prehistoric times to add a new member to the list of known planets." Pre-historic is defined as the time in any society previous to that society keeping written history. The president, I submit, was very wrong. We have a record in a specimen now in the Museum of East Berlin from 4500 years ago that clearly depicts all the planets in our solar system in their relative positions and sizes ringed around the sun – including the planet Nibiru.

Perhaps the Hubble telescope will accelerate the resolution of this frustrating situation. Sitchin has even suggested that Planet X/Nibiru was actually discovered in the early 80's with the Infrared Imaging Telescope (in orbit around the earth) but the information was suppressed because it was recognized that, if Nibiru actually existed and had the characteristics described by the ancient records, it could also be inhabited by the Nefilim – and the profound repercussions would immediately be felt worldwide. The use of the Black Birch, New Zealand observatory began shortly thereafter. After all, Harrington must have realized that the southern skies were the place to look after the IRAS satellite, as early as 1984, returned, among other interesting objects, sightings of a body that would qualify in that area of the skies. Sitchin suggested that it was being used to get visual verification of the IRAS satellite discovery before releasing the information. Be that as it may, it points up the importance of the new paradigm and the profound and far-reaching effect it will have when recognized and accepted. It is clear to me, from many perspectives, including the scientific, that more than sufficient data is in. We not only can restore our real history to ourselves here on the planet but we have a history of our solar system from the time of its early formation.

Harrington, former head of the Naval Observatory, now deceased, after becoming aware of Sitchin's detailed translations of the *Enuma elish* account of the formation and history of our solar system, met with Sitchin at his office and stated in a video-taped interview that the information that he saw developing concerning the tenth planet corresponded so closely with the more detailed information from the ancient documents that he felt it appropriate to invite Sitchin to discuss the matter with him. This is one of the rare scientists who have been secure enough in their position and open-minded enough to recognize the potential advantage of the huge body of data in the ancient records.

On the basis of their performance to date, we should not wait for petrified politicians, supercilious scientists, or archaic academics or look to them for ratification in this matter of the existence of Nibiru and the Nefilim. Certainly we may be faced with very significant events, decisions, choices and politics when Nibiru returns, swinging in through the inner solar system – probably not for another thousand years – or when the Nefilim decide to approach us in whatever mode, or if we send a probe there. And, with that event, if there is sufficient disruption of the inner solar system caused by Nibiru's passing – and there seems to be enough details in the ancient records of such events – to threaten the Earth in whatever way and to whatever degree, we should be prepared.

PART II

THE TRAUMATIC TRANSITION

CHAPTER 6

THE TRAUMATIC TRANSITION

I do not hold the opinion, as Dr. Velikovsky did, that the great catastrophes represent the central traumatic experiences leading to our collective amnesia. The central event was BEING LEFT ALONE by what I have termed the Factual Deities, who had – in full view – been ruling the world ever since humans had become conscious of life. They lived in the collective memory system – language – as Verbal Deities in many philosophical, religious, and political images, compensating the fear of being abandoned in chaos. . . Our era of Verbal Deities should come to an end, once the reconstruction can be brought to the collective minds' consciousness.

M.A. Luckerman, Editor
Catastrophism & Ancient History

. . . surely it is folly to preach to children who will be riding rockets to the moon a morality and cosmology based on concepts of the Good Society and of man's place in nature that were coined before the harnessing of the horse! And the world is now far too small, and men's stake in sanity too great, for any more of those old games. . .

Joseph Campbell
The Masks of God: Primitive Mythology

"Come off it, Prometheus, you fool! You have only stolen your own birthright!"

Petrus Protagoniste
Between a Rock and a Hard Place, Act 2

Our history reveals, painfully, that we have always looked to anyone or anything other than ourselves for authorization. We have always considered ourselves the least qualified to know what is best and right for us. We have long been taught to see ourselves as a danger to ourselves, even as "sinful" and prone to evil. This has been the result of the godspell slave-mentality, that which sees itself as the property of another with no rights or dignity. The godspell subservience is mani-

fest most acutely in the major and minor religions which are codified, institutionalized forms of it sublimated over the long period of our traumatic transition begun when the Nefilim left the Earth. The influence is found in every facet of our
cultures and civilizations, not just in religion. We are so close to many of these
elements that we do not even see them. The use of other humans as soldiers to be
killed or kill in our political wars, the practice of slavery, the institutions of kingship, royalty and social class are but a few of the manifestations of the deep patterns of subservience and domination. But, as surely as we have reached the point
of genetic enlightenment, we inevitably will evolve sociobiologically to the new
human status. Because that status of unassailable integrity will bring with it the
need to deal with the universe directly, the individual will no longer be able to go
through life as a robot living in a cocoon spun of the threads of someone else's
freeze-dried, pre-packaged, off-the-shelf, second hand worldview. We must become our own authorization. The necessary transition we must make, are making,
from the old godspell traditions to a completely new paradigm is as profound a
change as we have made as a species at any time in our 300,000 year history –
probably much greater. At this point it is quite clear that we are backing into it
with our eyes closed in a state of confusion and denial. T. S. Eliot's idea that "
human kind cannot bear very much reality" is, in a profound sense, exactly wrong.

We have an excess capacity of consciousness for our current state of knowledge but we are hampered by a completely inadequate and erroneous world-view
and set of criteria with which to judge its truth. We have too much consciousness
for our current information base and control over physical reality. By control I do
not mean the exercise of power in a destructive way as we have witnessed ourselves destroying our environment driven by disassociation, ignorance, fear or
greed; I mean the harmonious, conscious control of those parts of our nature and
nature around us that are in keeping with the dignity and dimensionality of the
fully human. We have more than enough capability of, and attraction to, a benevolent, planetary, communal unity. But we are constrained by the amazing and
appalling fact that we do not have a consensual definition of what human nature
fundamentally is. We simply don't know, literally, what and who we are talking
about to each other when we discuss our beginnings, our nature, our potential and
our future and it keeps us radically estranged from ourselves and one another.

Although we may often fold under the frustration of the state of imbalance
between our consciousness potential and our knowledge and control, it is frustration that is the cause, not a lack of potential. The godspell mind-set is still the
primary handicap, the result of our being a synthetic species, brought into existence for the explicit purpose of being a subservient slave species for the convenience of another species. We, as a bicameral species, driven from the beginning
in a precociously accelerated form of evolution by our Nefilim genetic component, had the quality of consciousness, from almost our very beginning, for a
level of complexity of knowledge and technological capability far beyond that
actually possessed or allowed us while the Nefilim were present. We have been
handicapped over the course of our development as a synthetic species, because
our subservient position prevented our knowledge acquisition and our inventive-
ness from keeping pace with our psychological development. Until only recently
the lingering subservience has kept us that way even though the Nefilim have
been gone for a long time. As a single clear example, witness the suppression,
over centuries, of science and technology by the Church, a latter day claimant to
absolute, exclusive, proprietary authority to reveal to the rest of humanity what

the will of one of the Nefilim "gods," Yahweh, is. If we are going through trauma in this transition to a new concept of humanity and a new civilization it is due, not so much to the intellectual reinterpreting and reconstructing of the available data and our ideas about our history, but to the psychological readjustment that we must, collectively, inevitably pass through as a result, a strenuous phase of emotional catharsis of this profound, ubiquitous, racial trauma.

In the context of humans understood as the bicameral species Homo erectus Nefilimus, the mutations of our social, political and psychological conceptualizations of the nature of the gods constitute a clinical study in loss, grief, guilt, abreaction, sublimation and compensation due to our having been subordinate to the Nefilim and then left on our own when they phased off the planet. We need to do more than change our ideas. It is critical, in this phase of transition to new human status, to acknowledge, resolve and integrate the traumatic elements of the godspell syndrome. We need to undergo a racial "psychotherapy" to recognize and heal the scars and uncertainties of the last two and one half millennia of dissociative racial behavior to become sufficiently clear about how to create our independent future and determine our evolutionary trajectory. It is only in our day that we have progressed far enough to evaluate and acknowledge our current situation to actually begin the process of integration of the knowledge of our true beginnings and to complete the emotional readjustment required. We need to intelligently and profitably integrate the new paradigm, as individuals and as a species, a race, with the least emotional trauma. A reexamination of the last three thousand years of our history in terms of psychological reaction and attempts at adjustment is most helpful. Approaching our history as a process of psychological analysis provides a context for a collective therapy.

Why Did We Put Ourselves Through All This?

Very fundamental questions must be asked: Why did we transmute the real, flesh and blood "gods" to cosmic God, actually putting ourselves into the godspell mentality and a pervasive racial amnesia? Why should the tradition not have been continuous, the Nefilim recognized for what they were even after they phased off the planet? Why should the knowledge of our real nature, of our unique beginnings and our history not have been preserved in its original form, the nature of humanity recognized in its generic commonality and our precocious development as a species celebrated with pride on a planetary basis? Analyzing our history, using these questions as probes, affords us the greatest insight into what we have put ourselves through and why. There are a number of important factors involved in this process that has extended over millennia.

The triggering conditions for the process was the disorientation and grief that ensued immediately on the departure of the Nefilim. But, at the same time that we attempted to maintain the status quo as we knew it when the Nefilim were here, we also were being driven by the fundamental evolutionary process toward independence and self-definition. The process was muddled and hampered by the opportunism of those human elements who acted to take power and control in the authority vacuum. We gradually began to develop our own science and philosophy and these were both an advantage and a hindrance. The myopic confidence of science, over the centuries, would declare the deeds of the Nefilim fantastic and mythological. Our budding metaphysical thought would supply the absolutes, abstractions, and idealized concepts that contributed to building up a concept of an infinite cosmic being into which we gradually transmuted the very finite and less than cosmic Anunnaki/Nefilim. This metaphysical process of transmutation was fueled by the political monotheism at root of the Hebrew tradition. As the accurate traditions and scientific information we possessed from the Nefilim gradually decayed over the centuries the more we were prone to misinterpret, distort, and overlay fiction.

More unconscious than conscious, our desire to be like the Nefilim prompted us to sublimate the physical immortality that they withheld from us to a spiritual immortality and to denigrate the body and the material realm. Specifically, this pervasive mind-set, so deep in the human psyche, is responsible for the schizophrenic attitude manifesting in the "mind-body problem" in philosophy, psychology and science in general, and acutely in religion. The mistrust of the body, the "flesh," and the emotions and passions, which St. Paul and the ascetics and those of Puritanical persuasion preached so vehemently, based on the "Fall Of Man" misinterpretation of the Old Testament Garden Of Eden history, is at root of the general mistrust of self that holds even modern man in bondage to external authority and rends the integrity of the individual self. The new human will be free of the primitive theological concept of body separate from immortal soul and the entire subject of the definition of the body-mind and mind-brain relationships will have been worked out in terms of our most advanced knowledge and in terms of our unique evolutionary status as a genetically engineered species.

With the elimination of the godspell mentality, any concept of human nature being intrinsically untrustworthy and prone to evil will be eliminated. We will know ourselves as completely independent of external determination and responsible for our own destinies. The fullness of that shift of responsibility and authorization will result in our seeking and attaining an "unassailable personal integrity." This is the key to all the other characteristics of the new human, the human with the potential for an immortality in which to choose and pursue a conscious evolutionary trajectory; it is recognizable even now in a very few who have broken free; its fullness will be the result of planetary genetic enlightenment.

Does this mean that the redefinition of human nature will inevitably be in what are currently considered exclusively "materialistic" terms? Does it mean, in other words, that we are just a biological bundle of bone, flesh, nerve and wetware programming with no "higher" components or dimensionality? I believe that "dimensionality" is the key concept here. I believe that all the traditional ideas of "mind," "spirit," "spirituality," "transcendence" and "soul," once individual godspell religious and theosophic traditions have been set aside, are based on a common denominator of dimensionality. We are evolving toward habitual four-dimensional consciousness; generic transcendental experience is participation at the leading edge of that trajectory. Early indicators of such a generic redefinition may be found in such works as *The Dancing Wu Li Masters* (Gary Zukov) and *The Tao of Physics* (Fritj of Capra) which point out the parallels of Eastern metaphors of transcendence with the metaphors of higher dimensionality of relativistic and quantum physics. Previously, because of the proprietary claim of individual religious sects to the authority to interpret reality to the rest of the species, any activity of the human being that in any way resembled that broad classification called "spiritual" fell, as they see it, under their jurisdiction. Dogmas of the absolutistic religions hold the transcendental, in any form, as being essentially a "spiritual" seeking and/or contact with the Divine. Transcendental experience has always been in that category in the eyes of those institutions. At one extreme there are those that would hold that it is not even attainable unless directly granted as an arbitrary gift by a Divine Being; at the other, more philosophical end of the spectrum, it is considered, still, to be something at least supernatural, if not Divine. Inevitably, therefore, any claim that it, in its essence, is a natural, normal focal activity of every human was vigorously resisted or denied by those theologically based groups or institutions. The matter has been further complicated by the absorption of the notions of the Divine and the putative metaphysical nature of transcendental experience, even in philosophical terms, by latter day Protestant splinter sects and many of the eclectic New Age cults.

The Trigger of the Transition

As the stag pants after the waterbrooks, So pants my mind after you, O gods! My mind thirsts for gods! for living gods! When shall I come face to face with gods?

– Psalm 42

The primary factor driving our millennia-long transition to racial maturity is the effect of the apparent abandonment ("apparent" because the evidence seems to point to their discreet monitoring of the Planet ever since) of the planet and us, the human species, by the Nefilim. The resultant phases of trauma, attempts at propitiation, rationalization, denial, adjustment and, eventually, power struggles and internecine conflict in the authority vacuum resulted in the godspell syndrome. The reason why the Nefilim exodus was such a watershed event may best be understood by considering the human situation previous to that point in time.

From the time, 300,000 years ago, when we were brought into existence until the time of the final phase-off by the Nefilim around 1200-200 B.C., the records we have recovered tell of a relatively stable time for humans. The Nefilim were in control. Humans were their subjects and what was expected, allowed, anticipated, understood, commanded and obeyed was clear. The world-view was common. Where the species had originated and how was known, the definition of what a human is and what a human's purpose was taken for granted. This habitual, stable condition, when things were generally managed by the Nefilim, and humans could depend on it, is the source of the tradition of a Golden Age in our past, a time when things were calm and prosperous and humans could flourish harmoniously. But the "peace" of that golden age was the price of our independence as a species. We were not our own masters for many thousands of years, kept in a relatively suppressed and deprived state, without the freedom to develop and experiment. We existed and worked for the Nefilim. To fully appreciate the degree of subjection we have only to consider a relatively latter day event. When we became precocious and knowledgeable enough to attempt a technological project on our own, (the Tower of Babel account in the Old Testament), the Nefilim stepped in and squelched it. That incident, clearly defining the Nefilim's attitude toward us, is very significant. It was critically important enough to be recorded in the historical documents of all the ancient Middle Eastern civilizations because, among other things, it marked a point where at least some humans, populous and precocious enough to have some limited freedom to develop on their own, were able to construct a rocket vehicle of serious nature enough to disconcert the Nefilim who promptly decreed the division of languages for the sake of effective "crowd control."

When the Nefilim phased off Earth, simply left the colony Planet, the situation radically changed. Our dependence on their dominion is graphically described in the texts recovered that speak of the reaction of humans when the Nefilim

finally left the Earth. Humans were dismayed and distraught that the guiding prin-
cipals of their lives were suddenly absent without notice and they did not know
what to do, literally. We had the potential but not the experience or the knowledge. The impact of inevitable radical transition that the departure of the Nefilim
forced on us is graphically recorded in the depictions of a king, Tukulti. Previous
to the Nefilim departure he is shown calmly standing and conversing face to face
with his Nefilim ruler who is shown seated on a chair on a dais, giving his ser-
vant/foreman instructions. After the Nefilim departure he is shown pointing in
dismay at the empty throne. The accompanying text poignantly expresses the
distress and anguish and depression caused by the Nefilim ruler's absence. There
is the intriguing possibility that there was, at least for a time, a continuing communication at a distance by the Nefilim and at least some humans. Problems of
authentication and authority begin to arise. In Numbers 12:1-2 Miriam, Aaron
and Moses all hear communication from Yahweh and they can't be sure whose
messages are most authentic. This is certainly in contrast with Moses' earlier situation of which it was said that no nabi had been like Moses "whom Yahweh knew
face to face" (Deuteronomy 34:10) – but the communication, in whatever form it
took, was still recognized as being with the god. If, indeed, the Ark of the Covenant was a radio transceiver, as has been demonstrated with a duplication of it by
scientists, then that may have been a type of communication used in the time of
separation and transition.

The records indicate that the departure of the Nefilim was carried out without a great deal of preparation of humans for it and, apparently, with little overt
concern for them. It seems they simply quietly left, metaphorically closing the
laboratory door on the way out. It may be speculated that the reason for this attitude was simply consistency with the Nefilim view. Perhaps they had mined sufficient gold from the Earth to accomplish their mission. Perhaps, consistent with
their view of humans as inferior subordinates, they deemed their actions good
enough or, perhaps, they wished to wean humans from their supervision and deliberately orchestrated their leaving to force humans into the next stage of development.

The latter scenario is intriguing. Did they realize that our rate and degree of
increasing precocity exceeded their expectations and they were moved to allow
us another degree of freedom to develop on our own because we deserved or
required it? How would we have acted as an advanced species toward a species
which we had genetically engineered into being and which had evolved in such a
fashion? We are vicariously working on those questions in our science fiction
right now. I tend to lean toward this latter explanation, all things considered, at
this point. Whichever the Nefilim motive may have been, the effect of departure
was immediate and acute. The records show humans moving quickly from incredulous disbelief to depression, grief, and some anger to self-blame and alienation. Julian Jaynes – although working from the old premise that the gods were
hallucinations – studying the phenomenon from the viewpoint of a psychologist,
noted that two factors, not present in the human situation when the Nefilim were
here, came into existence for the first time when the "gods" withdrew: propitiatory supplication for forgiveness and help (prayer) based on the notion that our
misfortunes were punishments for somehow offending the gods.

When they were here, there were calendrical rituals of a scientific nature.
The calendrical rituals, probably better called procedures, were basically technically astronomical in nature. They were focused on the measurement of the pas-

sage of time, of the year, the precession of the great years, the marking of the precise moment of the equinoxes and the solstices. This technology, familiar to the Nefilim, was taught to a select group of humans (usually called "priests" because we mistakenly associate it with "religion"). This technician class was trained in architecture, astronomy, mathematics and other practical disciplines. Their craft was transmitted by them usually to their offspring. By the time it had come down through many generations, the Nefilim not being here any longer to supervise, the remnants of the technology became part of the religions we now know. The time of the feast of Easter, in the rituals of the Christian Church, as a single example, is still determined according to a modification of the ancient Egyptian calendar, originally given us by the Nefilim. When the Nefilim were here there were supplicatory and adulatory glorifications of the Anunnaki/Nefilim including songs, hymns, anthems and salutations which could be equated with our national and social holiday festivities.

Prayer, as we know it, did not exist previous to the phase-off by the Nefilim. In the new context that would be quite logical: when the Nefilim were here they could be seen, approached, communicated with, petitioned for help and material things. Once they left, prayer, as we think of it, directed to a non-present, nebulous deity, became a new mode of human action stemming from puzzlement, fear, insecurity, need, self-doubt and blame and an attempt to appease and reach reconciliation with the masters who were no longer physically present. And it was not only prayer that came into existence under those new circumstances: over centuries, at the peak of the despair, generated by the lack of response from the departed masters, appeasement and attempts at reconciliation took some peculiar and then some very extreme and frightening forms.

From Service to "Sacrifice"

A widespread custom that developed after the departure of the Nefilim was the use of a statue of the departed Nefilim ruler or rulers as a surrogate. The custom has carried down to the present time in both Eastern and Western religions. With many variations on the theme, very often the statute was bathed, offered food, carried about from location to location, honored especially on days of special celebration and generally revered as if alive. If there was no longer the live Nefilim ruler to prepare a bath for or bathe, or to set out garments for, then a statue would be the next best thing with which to symbolically continue the daily routines. These practices have traditionally been interpreted and described in the archaeological and mythological literature as naive rites to honor mythological deities. Within the new context, if the Nefilim were real persons, a completely new perspective is available to us. The departed Nefilim ruler is honored and served through the statue just as he or she was when present. A very probable psychological explanation of this is the intention of those priests (chief servants) charged with the direct service of the "god," the Nefilim ruler, when present. Their objective would be to maintain their functions and services precisely as before, both with the intention of having all things right in the event that the ruler returned and/or with the intention of enticing him or her back. As time went on and the memory of the real Nefilim rulers faded, it was inevitable that the rituals and rites degraded, more and more superstition crept in and even the understanding of the real nature of the original activities became more and more abstract and gradually lost. Eventually the rites were still performed simply because that was the way it always had been done. Clearly, the rituals of "religion" – as we have been conditioned to think of it – such as sacrifice, oblations, praise, celebration, commemoration of special events, feasts and honoring a cosmic, transcendent deity have their roots in the daily service of the local Nefilim ruler by the humans who were his or her hierarchy of servants both in domestic service and in technical and craft service. "Sacrifice" of animals, prepared food, wine, beverages or precious items and gold, understood traditionally as something given up even with hardship, springs from the routine preparation of the Nefilim ruler's food and drink and daily necessities from a glass of wine or beer to filling a bath. Humans were not giving up something of their own at a cost, a hardship "sacrifice" to themselves when the Nefilim were here. They were presenting to the Nefilim what belonged to the Nefilim and they were taken care of, provided for, as slaves, then as servants, and eventually as limited partners as part of the social structure. During the transition phase, once the Nefilim had left, the character and meaning of service changed to our current concept of sacrifice when the serving of food became abstracted into presenting it to a statue or offering it on an altar (table) of the now absent Nefilim ruler and, eventually, into a propitiatory offering to entice the "god" to return. The records of the ancient civilizations speak of oblations, the ritual pouring out of beverages on an altar or on the ground in honor of the "god" in what we have been taught to think of as religious rites. If the Nefilim ruler was no longer present to drink from the cup which his human servants used to fill for him or her, the ground, symbolically, would have to do. Along with the records of sacrifice (slaughtering) of a "fatted calf" or the preparation of beer or wine for the Nefilim "god," we should always consider the recipes for the preparation of those foods and beverages we have also recovered. The Nefilim were quite interested in their food and had gourmet taste.

The examination of the causal factors we have been considering here provides explanations, especially of some of the more bizarre types of behavior we have collectively exhibited over the millennia coming through the firewalk of transition. The recovered history shows clearly that, not too long after the Nefilim departure, some of the peoples of those ancient Nefilim centers of civilizations, both in the Eastern and Western hemispheres, went through a period where the dominant character of their "religious" rituals focused on the propitiation of the "gods" who were no longer present. This obsequiousness, coupled with an attempt to get them to return, eventually led to the hideous extremes of human sacrifice. By that time the real nature of the "gods" already had been superstitiously melded, at least partially, with the awesome phenomena of nature or simply distorted or perverted or partially forgotten. But the relative degree of comprehension did nothing to lessen the horror of the act. This was not an almost intellectual, inquisitional winnowing of heretical thinkers by execution that would come much later, but a simple attempt to appease the absent masters and entice them to return with the lives of humans. It is difficult to explain in any other way why relatively high and sophisticated civilizations such as the Mayan and the Sumerian went through such a period of degradation . . . Eventually, the situation evolved to a very macabre stage. In an effort to demonstrate our subservience and zeal to make things right if they would just come back, we kept the rules, we maintained the routines of service, but after a long time of disappointment we reached a point of abject, abysmal desperation where we would do anything to get them to come back – and we did. Perhaps remembering the Nefilim males' attraction to our young women, we began to cut their hearts out on top of the empty pyramidal palaces in a collective, craven, pleading shriek to the heavens from whence they had come and gone. But that unspeakable and unappreciated horror could not last: we began to doubt, to entertain frightening cynicism, secret thoughts of independence and why bother. This marked the peak of the traumatic reaction to the departure of the Nefilim in the Eastern and Western hemispheres.

The Power Struggle in the Authority Vacuum

Parallel to this focus upward on the departed Nefilim was the positive thrust toward liberation and independence. Good and righteous kings and servant classes tried to maintain the traditions of benevolence toward the people with integrity. But the inevitable usurpation of authority in the vacuum by self-serving despots, opportunistic officials, and second class elders is written large across those early cultures in letters of misery. Precarious oral traditions of the real gods, the unvarnished history, threading the eye of imperfect intellects, eased furtively into library, then sanctum, and finally into cave, occulted gradually by the now unfettered power of the unscrupulous – and the human imagination. The flesh and blood, real and imperfect gods were replaced gradually by rote remembrances, rituals, cargo-cults and, eventually, their palaces in cathedrals. Stained glass, the color of blood, became the mirror of our guesses. The rifts between humans exponentiated and hardened into the absolute chasms of religion – and fanaticism. Eventually, in the name of a newly transmuted Deity, human eventually came to slaughter human in crusades, jihads, Inquisitions, persecutions. Our real history, remote and obscured, seeming more and more fantastic as the quality of human existence degenerated and our technology decayed, became a dream within a dream, transmuted into a meta-myth, gradually taken over by the academicians and called mythology, taught, in our day, in the college classroom. Transmuted and institutionalized versions with enough power to win the dogmatic battles and suppress competing factions and dissenters became the major religions and enforced their version as reality even politically. The institution of "king" with all its royal trappings, claim to "divine right," authority, exalted status is a continuance, sometimes in form only, of the position of intermediary between a Nefilim ruler and the mass of humanity under his jurisdiction. Even in these latter days, when the Nefilim have been sublimated into a cosmic God, the relationship of our modern kings and queens to religion, as, for example, is the case with the royalty of England, still mirrors the ancient relationship between the Nefilim ruler and HIr foreman. The Pope of Rome has gradually acquired the same kind of status – even the tiara, the tall hat worn by the Pope, resembles the distinctive headdress seen in reliefs and sculptures of the Nefilim rulers and which was described as containing some sort of control unit – perhaps a miniature computerized communications device. A fax from. . . never mind.

The Fundamental Evolutionary Process

Even more fundamental than the drive to assume responsibility for ourselves is the drive of the evolutionary process itself and of which the natural tendency to assume self-reliance is only a symptom. The fundamental evolutionary process, however we characterize it and whatever mechanism we postulate as its engine, has driven us from our beginning toward physical adaptation, greater self-awareness and greater dimensionality of consciousness. This positive drive propelled us toward the precocity that eventually raised us to partnership with the Nefilim and the independence we exhibited even when they were here. I have identified it as a peculiar kind, a "special case" – as the scientist would call it – of evolution since we were literally invented for a purpose in a laboratory rather than blossoming naturally on the evolutionary tree. Fundamental evolution gave rise to the growing sense of independence that resulted in a denial of and dissociation from our subservient beginnings and our Homo erectus heritage. But the natural processes of evolutionary development, active since our inception, have also brought us to the point in our history where we are now beginning to determine how to determine our own evolutionary trajectory. From our beginning as a synthetic species we have been rapidly metamorphosing under the impetus of our unique bicameral genetic makeup and that fundamental engine of change operated and operates whether the Nefilim were present among us or not. It has impelled us toward greater consciousness, independence and understanding. At the most obvious level, physically and mentally, we adapted to climate and conditions in whatever area we found ourselves and the results, in the colors of our skin and in our various physiologies, are evident. But we have been changing in ways that are subtler but far more profound. In the context of the new paradigm, this has occurred more rapidly than would have been anticipated in a "natural" species because of the impetus of our Nefilim gene component. The speed of the changes also makes them easier to identify. The most obvious characteristics of our rapid metamorphosis are a heightening of the dimensional expansion of our thinking, a rapid progression in our technical capabilities and an accelerating understanding and intensification of our self-reflexive consciousness. We are focused here on the manifestation and effect of these elements over the time of the last 3500 years, the period of transmutation and adjustment since the Nefilim phased off the planet. The positive evolutionary pressure to expand and adapt is the most critical factor that has helped us adjust to existence as an independent species. If we have come to a turning point in our species' development where we are ready to emerge from the final stages of racial amnesia and denial it is because we have evolved sufficiently to do so. But, in the course of that difficult transition of adjustment to racial independence and responsibility, the very qualities that would eventually enable us to succeed on our own were the factors in our psyche that contributed to the slow and subtle transmutation, within the collective consciousness, of the imperfect, flesh and blood humanoid Nefilim into a single transcendental being who created the universe.

A Theoretical Comparison with a "Naturally" Evolving Species

An insight into the difference between the relationship to the universe of a "natural" and a "synthetic" species provides us with insight into why we have suffered such trauma. I suggest that a "naturally" evolving species, the product of whatever natural evolutionary process on a planet produced it, would develop gradually in comprehension and technology and uniformly, from a base of a relatively common view of the universe. If we were not genetically engineered into existence for a pragmatic purpose, I suggest we would have evolved also at a pace much slower than we have experienced. A form of humanoid – obviously without a bicameral alien genetic component – perhaps similar to us, in this view, would not have arrived until, perhaps, one to several stages of evolution after Homo erectus if the Nefilim had not interrupted the process. The relationship of the species and the individual with the universe would be direct and integral without the schizoid problems we have experienced because of being forced to deal with the universe through the Nefilim as an intermediate filter.

Consider the difference between even the "sacred" traditions of the religions springing directly from roots in the ancient dominion of the Nefilim, Hebrew, Christian, Moslem, Roman, Greek, Vedic and others and the "religion" of the Indians of North America. The universe and the individual's relationship to it is understood, in the traditions rooted in the Nefilim centers, as explained by the gods or, later, God (the sublimated, metaphysical form) and the individual becomes conceptually distanced from nature and the universe. But in the thinking of those peoples who had been separated from direct domination and contact with the Nefilim for an extended period of time, the human being was understood as an integral part of nature and the universe, one with it.

From a unitary, planetary perspective, however, our situation has been far more akin to that of children who were being given an intense and accelerated education by very sophisticated parents. That is very, very, different from natural evolution as usually conceived and exhibited in the history of hominid species that preceded us. The fact that we have lagged behind our technological progress in our racial psychological development due to the residual trauma and confusion due to separation from the Nefilim masters seems quite clearly to be the reason for our stupid misuse of our natural resources, bad ecological practices, pollution of the environment and insane treatment of other humans. Our racial performance in this regard exhibits all the symptoms of dissociative behavior. Our former dependence on the Nefilim, at the same time their advanced genetic component was driving us rapidly forward, caused us to undergo a heavy trauma. To this point in our evolution we should be indulgent toward ourselves and view that kind of behavior as we would the rapid transitional mistakes made by adolescents on their way to adulthood. From now forward, however, we should recognize the pathological godspell element for what it is and deal with it.

A Comparison of the In-law and Outlaw Transitions

It is fascinating and enlightening to reconsider the situation of those humans and their descendants who had been expelled into the outback, away from the Nefilim city centers and who, therefore, were least controlled, if at all in some cases, by the Nefilim and most free, literally, to think for themselves. It seems that this higher degree of independence made them less vulnerable to the traumatic effects of separation. Many investigators have noted the striking similarity between the customs, art, traditions and beliefs of cultures as widely separated, geographically, as, for example, the early aborigines of Australia and the aborigines of Southern Europe. These cultures were incredibly fine-tuned adaptations to their environment on the obvious level. But these cultures did not originate and develop without "outside" influence either. We have seen that the original humans of these cultures were of the same origin as the "in-law" humans, in touch with the Nefilim tradition to one degree or another. And even their descendants, eventually inhabiting areas apparently very remote from the Nefilim centers, may have remained in touch, occasionally, remotely, with the original traditions and at least knowledge of the Nefilim and the city-centers.

It is a tradition of the Australian aborigine peoples that the "gods" mapped and marked the entire Earth when they were here in the ancient times, the "dream time" as their concept is translated. Their concept of a world grid mapping is remarkable in it's description of the geographical world-grid imposed and used by the Nefilim. Recent studies have also shown that some of the cultures and tribes that were previously considered to be out-back, " out-law" types were not that at all but just the opposite. In the old paradigm the civilizations of the Americas were considered, by the invading Europeans (in the direct line of the old Nefilim "in-law" tradition sublimated into the Judaeo-Christian tradition) to be pagan savages. In reality they were, in Central and South America, also of the direct "in-law" tradition, from city-centers created by the Nefilim at the same time that some of the Middle Eastern Nefilim centers were being developed. It may eventually be demonstrated that there was direct Nefilim intervention in North America also. The level of these American civilizations was highly developed and extraordinary and far beyond anything found in those regions where Nefilim influence and control was minimal. These sophisticated cultures not only had been under the direct aegis of the Nefilim and been taught by them but close examination shows clearly that the populations were, in Central America, composed of Amerind, Negro and light-skinned groups, showing direct transoceanic contact. We will have to systematically go back and reevaluate all the imputed superstitions, "magic" and ooparts found in these civilizations and sort out the truly superstitious from the residual real technology.

We already have a fairly detailed understanding of what groups were under the direct control and tutelage of the Nefilim and which groups were relatively on their own. A preliminary comparison of those two groups seems to indicate that those groups which were in direct contact over their history with the Nefilim were the groups which underwent the most severe trauma when the Nefilim left the planet. This is reasonable. Those groups who had been relatively on their own in the remoter regions of the planet such as, perhaps, Eskimo groups, or those who had penetrated the remote jungles of Africa and South America, would logically have gained a bit of detachment and independence. It is certainly clear that their belief systems seem more in harmony with nature and are psychologically more settled than those groups from the Nefilim centers. It seems consistently true that the Nefilim centers produced the most absolutistic "religions" and the strongest forms of the subservient godspell mentality. In the last three thousand years we have moved from abandonment to disassociation to sublimation to religion to rebellion to recovery. The classic syndrome of the racially dysfunctional family is dispelled. We no longer should be standing subserviently looking to the sky asking Daddy to please come back and tell us what to do.

The Role of Metaphysical Thought in
the Transmutation of "gods" to God

Among the several factors involved in the gradual transmutation of the plural Nefilim into a transcendental God, two are closely related: the effect of the natural proclivity of the human mind to explore the realm of the metaphysical and the gradual loss of the body of higher science that was given us by the Nefilim. Parallel to the gradual dissociation and suppression of the knowledge of our true history and a falling into a sort of racial amnesia and parallel to the development of human science, lies another major factor in our racial transition, the development of a body of metaphysical thought and ontological philosophy. Metaphysical thought means, traditionally, the type of concepts and ideas developed by the human mind when considering the most ultimate notions of which we are capable: higher dimensionality, transcendence, the "spiritual" in its various definitions, higher consciousness. Meta means above or beyond, in this case above and beyond the realm of the Newtonian/Cartesian physical world at least as we habitually perceive it. Ontological usually refers to the considerations involved in the study of being as such as distinguished from transient existence in any form. It is important to note that the basic ability to think in abstractions is essential (a pun as you will see) to both these areas of thought. An abstraction is a concept which is formed by drawing out of a class of individual items (rocks, dogs, automobiles, humans, etc.) the essence of those individuals that make them what they are. Put as simply as possible, any time we refer to "rock" or "dog" or "human" in general, without any reference to any individual of those types, and really are thinking of "rockness" or "dogness" as a general thing, we are thinking in the abstract.

An important question arises which must be dealt with at this point: Did we, as a species, possess the same intellectual capabilities from our beginning as we do now? We might interpret the words of the Sumerian texts that speak of the earliest humans as eating the grasses and drinking from the ditch as an indication of a state of such primitive consciousness and intellectual capabilities that we were not capable of abstract thought at that time. But it may also be argued that that is simply an indication of a state of social deprivation. Our position of subservience may have forced us into such a basic, seemingly primitive existence. It may be recalled that there is an explicit statement in the ancient records that Adaba (the first perfect human specimen) was taken into the heavens and probably to Nibiru by the Nefilim. If we were developed enough to be considered for inclusion in Nefilim society that might be a strong positive indicator of our capability for advanced thinking even then. Of course it may also be argued that Adaba may have been taken up more like a zoo pet or a curiosity. But, since the story indicates that physical immortality was bestowed on Adaba, the former scenario seems far more probable. He was not just a scientific wonder or curiosity to be inspected, but worthy of being granted the Nefilim's type of existence. Whether we possessed the capabilities of intellect and consciousness that we now exhibit from our beginnings or it evolved quickly over time, the potential, logically and as manifest in the ancient accounts, appears to have been latent. In fact, the potential may have even been a good deal more than originally intended by the Nefilim. They had carried out the genetic manipulation between the genes of Homo erectus and their genes with the intent of increasing the mental capabilities of Homo erectus sufficiently to make the new creature capable of mining tasks that were of moderate complexity. It seems clear that they got a bit more than they bargained for as their genetic component began to accelerate us into a precocity that disconcerted and annoyed them over time. The annoyance became so acute over time that they finally decided to simply wipe us out as a species. That's a high level of annoyance. . . All indications from the ancient records reinforce the fact that, at least, the potential human capacity to think the same thoughts we are capable of today was ours from the beginning. But it is also quite clear that, for many thousands of years, the Nefilim did not encourage, stimulate, or teach us, as a slave species, to indulge in that area of thought. Because the potential was there, it is reasonable to believe that some humans did think such abstract and even profound thoughts but the context and opportunity to develop it into a mature body of collective philosophy was simply not available to us. Once, however, the Nefilim had left and we literally had to fend – and think – for ourselves, we see a gradual rise in independent philosophizing about the nature of reality in general and a gradual increase of self-reflexive analysis of our own thinking itself. That philosophizing took a number of forms. We began – or, at least, some humans with that bent – to think about the ultimate causes of things and the universe in general. We gave more and more thought to the concepts of absolutes. Slowly, over the ages, we developed a body of metaphysical concepts that became an integral part of our language, thinking and traditions. Absolutes such as infinity and eternity or idealized notions such as truth, beauty and goodness sprang from our ability to form an idea of the essence of something. Condensing the long process into a sentence, we see that, many centuries ago, the concept of Being as such became a familiar term in the philosopher's vocabulary and, eventually, the formal focus of the area of Philosophy called Ontology. This ability to mentally abstract the essence of things and to employ abstract concepts is clearly a positive and valuable ability but, in the transition, sometimes contributed to confusion. Two trends in our collective consciousness merged. As the memory of direct contact with the Nefilim faded into the dimness of the past and the clear understanding of our beginnings began to become confused through increasing ignorance and dissociation, our independent thinking about reality was providing us with a rich lexicon of metaphysical concepts. Our evolving consciousness found fertile ways to sublimate the concept of "god" to universal God in our expanding metaphysical considerations. As the "gods" took on more and more of a mythic character in cultures such as those of Greece, Egypt and India, metaphysical concepts began to be applied to the nature of the singular "god" of the Hebrew tradition, the tradition at root of the Judaeo-Christian foundation of Western culture. In the minds of humans within that tradition, the flesh and blood Nefilim ruler of the Hebrew tribe was transformed into an ineffable transcendental singularity. It is not a difficult step from all-powerful to Omnipotent, from the Watchers in orbit to the Omniscient, from creator of humanity to Creator of reality and then to Creator of the Universe, from the cause of humanity's existence to the Uncaused Cause, from the wise and technically advanced to Infinite Wisdom, from ruler gods to God.

An Historical Overview of the Transition Phase

In the broadest historical perspective, from the time of the Nefilim exodus from the Planet to our time, several major phases of transitional gambits may be distinguished. These were attempts by humans, usually led by the advanced thinking of powerful intellects, to set down new definitions of what a human being is and what kind of society would be most conducive to harmonious human living. These major philosophical shifts were no less than attempts to redefine human reality.

The Age of Right Action

The history of our species on this planet, since the time of the final phase-off of our Nefilim masters, has seen one attempt after another to do away with the memory of the gods or to free ourselves of the domination of religious forms. By the sixth century B.C. an extraordinary constellation of intellects, Buddha, Pythagoras, Ashoka, Confucius, were promoting various solutions as to how humans should define themselves and live in harmony with each other. Buddha taught a doctrine of transcendental detachment, Ashoka a way of right personal and social action, Confucius a set of natural moral principles and responsibility, Pythagoras yet another doctrine. It is significant that all of these doctrines were not focused on the "gods" as the ultimate authorities and givers of law; their focus was on a reasoned, even secular order. This worldwide phenomenon marks the first major intellectual plateau we reached, as a race, in an attempt to collectively adjust to the absence of the Nefilim and to take over for ourselves. It was not as if these leaders had dreamed up some theory in isolation from their contemporary civilizations. Ashoka, as a single example, finally came to his vision of a better way after killing, it was said, one hundred thousand people in his conquests. The transitional nature of this major shift is painfully clear in that the measures used to bring them about or enforce them were many times harsh, even draconian by our standards, as slavery was still allowed and social inequalities were tolerated or taken for granted. But it was a start toward relatively enlightened human self-regulation by human standards after the uncontrolled, almost hysterical, power struggles of kings, tyrants, heroes and ruthless opportunists in the authority vacuum.

The Age of Denial of Self: Asceticism

The second major worldwide movement toward human self-regulation was asceticism, expressed through the phenomenon of monasticism. In its generic form, monasticism was a withdrawal from the preoccupations, attachments and cares of the material and the everyday world to focus on the contemplative activities of the higher faculties of human nature. It was, as pointed out by Gerald Heard in *The Five Ages of Man*, a reaction against the rampant egotism of the hero by a denial of the ego. The methods used to achieve this focus varied from East to West. Fasting was a favored technique for consciousness change and meditation was its companion in the Middle East and the West. Yoga techniques, psychoactive chemicals, meditation and trance became prevalent in the East. In the West asceticism was integrated into the godspell religions. In the East asceticism was more independent, its goal being transcendental states of being as such, rather than being seen as associated only with a Supreme God, as in the West.

Regardless of the metaphors used by either the Western or the Eastern traditions, the fundamental concept was to attain higher transcendental states though denial and discipline of the body, which was considered inferior to the "soul" or the "spirit." The mastery over self and one's fate and destiny gained through these disciplines is key to understanding how this phenomenon fit into the long process of traumatic transition. The immortality that we lacked was sought in a transcendental state; the body which, subconsciously, was identified with our Homo erectus component and resented, was "mortified" (subdued, deadened). In the final analysis, asceticism was another attempt to control, take over our own self-direction and gave rise to a cultural form that lasted for centuries in both its strict monastic form as well as its modified versions meant for the layperson.

The Hebrews: The Tribe of Traumatic Transition

Because the Hebrew polity and religion arose and developed in the latter time of transition when the Nefilim were beginning to phase off the planet, it affords an excellent example of the actual dynamics of interaction of the various factors operant in the transmutation. The Hebrew tribe uses as its founding date 3760 B.C.. Moses received the Ten Commandments around 1500-1000 B.C.. Solomon, his son, built the Temple at Jerusalem dedicated to Yahweh, beginning in 960 B.C. Approximately three hundred years later, Josiah found it necessary to revive the worship of Yahweh in Jerusalem. These significant dates bracket the time between 1300-1200 B.C. when the Nefilim seem to have begun phasing off the Earth. Moses received the Commandments close to this pivotal period as a representative of the Hebrew peoples. He is known in the Old Testament as the last of his kind to have been known by Yahweh "face to face." Yahweh is not said to have actually lived in the Temple at Jerusalem built between 960 to 925 B.C.; it was only dedicated to him. Some three to four hundred years later the clear memory of Yahweh had already faded so much as to have to be revived and "worship" of him restored by Josiah. No matter how one views it, the Judaeo-Christian ethos is a major subliminal and pervasive influence on our thinking always – whether we subscribe to its religious authority or not, simply because we are part of Western culture – and especially when considering the nature of the gods and the transmutations our thought has gone through with regard to them. Because the Hebrew tradition holds this influential position in our thought it is most critical that we understand the factors operating in our current transition. In examining it in this way we are able to see all the complex factors we have described, so far, operating dynamically and interactively in the process of traumatic transition. The same traumatic transition was undergone by the other cultures rooted in the Nefilim centers around the world, from Sumeria to the Americas but we have a more complete record of the Hebrew history. We might have had an equally complete history of the Mayan civilization if it were not for the ruthlessly systematic destruction of their records by the Spanish.

When one consults an encyclopedia of religions and studies the fundamentals of the Hebrew religion as expounded by scholars, one soon realizes that there is no single authority recognized by all Jews, as with the Pope within Catholicism by comparison, who or which decides the truth or falseness of opinions or interpretations or doctrines. The Orthodox Jew has one interpretation of the Torah, the moderate Jew another, the liberal Jew another, including interpretations of how the Jewish tribe came to be, the way monotheism came to be a focal principle, what is the nature of God, etc. Rather than enter into a convoluted analysis and discussion of all of those pros and cons, argued even between Jews, I simply present a synopsis of the history of the Hebrew tribe and experience according to the interpretation of, and within the context of, the new paradigm.

If, indeed, the word Hebrew clearly derives from the Akkadian word khabiru, meaning vagrants, refugees, then the proto-Hebrews derived from the amorphous flow of humanity, desert dwellers or the homeless masses living on the periphery of the city-states, escapees from wars or persecution or invasions. If, indeed, the interpretation of some modern Jewish historians can be taken as accurate, some of them entered into a contract with a personage who was the most powerful in the area for the sake of survival and advantage. The original Covenant, in this interpretation, was a contractual agreement by which they bound themselves to serve this person in return for his patronage and protection. The scholarly conflict between those who would hold for a "spontaneous monotheism" and those who would see a gradual development from polytheism to monotheism can be resolved in the context of the new paradigm. Yahweh, the one with whom the proto-Hebrews made their contract, was a Nefilim among Nefilim. There were many Nefilim and they controlled the planet and therefore, if we call them "gods," that was polytheism. Yahweh was in charge of one area, of one tribe, and the exclusive administrator of that territory. Therefore, since Yahweh made it very clear that his subjects were not to put "strange gods before him," that was monotheism. The transmutation of this original situation into what is considered the religion of the Jews today, is the real "mythization of history" to take Martin Buber's phrase from another perspective. It is easily seen that both the "spontaneous" and the "gradual" interpretations of the development of monotheism are partially correct. The Hebrew tradition centers around the concept of being the chosen tribe of a singular deity who is supposed to transcend all deities. This concept of monotheism holds that there is only one transcendental God. But this political monotheism did not form in the Hebrew tradition until around 500-600 B.C.. I have called the first element the "political" monotheism of the Hebrew tribe rather than religious monotheism in order to jar fixed concepts hardened over time and to overcome the difficulty encountered when dealing with this phenomenon. Because of the unique and sacrosanct position it occupies in the development of the Judaeo-Christian ethos, which is at base of Western culture, there is a problem in even dealing with the monotheism of the Jewish religion, much less questioning it. If one examines the various books on mythology or histories of mythology in our libraries one finds that the deities of the religions of the world are listed as being mythological but the deity of the Judaeo-Christian religion is not listed as such – without explanation or apology. We have discussed the example of Joseph Campbell, the scholar of comparative mythology, who, benevolently and impartially, treats all gods of all systems as myths – except his God. His God is the unquestioned God who is taken as given in the title of his *series The Masks of God*; all the mythology that he presents as such is simply, in his mind, a proliferation of symbols and metaphors, masks behind which his transcendental God plays. Joseph Campbell was Jesuit trained; his religious, cultural and personal biases are clearly evident. These are but two examples of the mind-set of Western culture in this matter. I have discussed, above, how the capacity of the human mind to formulate abstract and metaphysical ideas facilitated – actually still facilitates – the transformative sublimation of the flesh and blood Nefilim into the notion of a transcendental Being. Within the Hebrew context there is an additional and related factor that operated in much the same way, the Cabalistic tradition.

The Cabalistic tradition is said to have arisen around the time of the beginnings of Christianity but the fundamental doctrines in their essential form seem to date from much earlier times. We shall discuss the import of this tradition for our science in a later chapter and here note its general contribution to the phase of transition. Although for many centuries, due to the loss of its true meaning by the Rabbis, it has been interpreted as myth, metaphor, poetry, magic, mystery, allegory, the Cabalistic tradition has been rediscovered (Stan Tenen: Meru Foundation) to be a coherent body of high science. That high science, it appears, was based on a unified field concept of the universe that was taught to us by the Nefilim. The universe, understood as a cosmic field working according to an all-pervading law couched in the metaphor of self-reference, seems to have been viewed as a transcendental Unity. Previous to the departure of the Nefilim it was understood correctly by those humans who were taught it as science and as a view of the universe. It may also have been correctly viewed as the closest thing to "religion," in our terms, that the Nefilim possessed. That is to say that the Nefilim had a respectful philosophical-scientific view of reality but they do not appear to have practiced anything that we would call worshipful religion of a transcendental Being. Their "rituals," as described in the ancient texts, were political or calendrical celebrations, as when Nibiru, their home planet, came into view and passed through the inner solar system, sometimes marking an important visit from their ruler to the colony planet, Earth. But, after the Nefilim had departed and the scientific tradition gradually decayed and became misunderstood, the conceptual framework lent itself easily to a shift to the realm of superstition and the identification of the scientific metaphors of self-reference and unity with the departed Nefilim. This is particularly obvious with regard to the Hebrews who, charged with protecting the doctrines, realized their importance but gradually lost their meaning as the scientific concept of an indescribable singularity gradually was merged with the sublimated ruler of the Hebrews, Yahweh. This phenomenon, in some respects, is almost identical with the way ontological and transcendental abstract concepts of philosophy have been gradually transferred to the Nefilim as discussed above. This is not to say that the real nature of the high scientific tradition was entirely lost, at least parts of it were understood by teachers down through the ages if only as a profound and mysterious body of knowledge, the key to which had been mislaid. It's importance was never lost sight of, at least within the Hebrew tradition, even though its real content became almost entirely obscured.

The Cosmofication of Yahweh

But the more the Nefilim, and, specifically, Yahweh, the Nefilim ruler of the Hebrews, were sublimated into a transcendental God, the less possibility there was of understanding the Cabalistic tradition as a body of advanced science taught us by a humanoid race. This was especially so because any consideration of plural "gods" – even though mentioned in the Torah (Anakim, Nefilim, Elohim) – was anathema in a monotheistic context as polytheistic paganism. Reciprocally, the more that tradition was lost or misunderstood, the more mysterious it became, the more potential there was for interpreting it as mysticism and transferring the gradually more abstract and, assumed, transcendental concepts to the sublimated Yahweh now seen as a cosmic God.

In the largest perspective, the Hebrew tribe epitomizes the traumatic transitional phase. Their Nefilim ruler, Yahweh, was jealous by his own admission and dominated his slave-subjects absolutely. He regulated their every action with minute laws from diet to dying, insisting on complete subservience. When he left the planet, along with the other Nefilim, his dependent subjects were devastated. They lamented, grieved, were confused and argued among themselves as to who was still receiving the commands from Yahweh. They learned to pray. Over the slow centuries, some began to "prophecy," some to wander into the desert seeking communion with their departed Nefilim ruler, some to consult omens and oracles. Their laments and self-blame, their loss and their desolation ring pitifully through the high emotion of the Psalms. Their humiliation burns in the book of Job. Gradually, as the memory of the flesh and blood Yahweh fades, the ancients seek moral wisdom rather than the living ruler, the concepts of the high science learned from the Nefilim are minted into metaphysical concepts and slowly transferred to the notion of a deity now considered to be transcendental. The confusion, born of the transmutation of grief and dependency and subservience, is complete. Atonement (at-one-ment) must be sought over and over and with the displeased deity. Even in our times written messages and pleas are inserted in the wailing wall of the olden Temple in Jerusalem. (There is a company that will even do it for you if you mail or fax your note to them.) "Religion" is born. Theology takes the stage to deliver its lonely soliloquy. The godspell mentality is full-blown. This process of transmutation, begun with the Hebrews and other ancient civilizations, grew in scope and seriousness the more that time passed. By the time we witness the advent of the Christian and Moslem doctrines, the gods, the Nefilim in general, had been turned into myths. The lofty abstractions, lent by metaphysics, were transferred to the now monotheistic deity of the Judaeo-Christian ethos who was now understood as the Infinite, the Ultimate Good, Infinite Reason – among other absolutes. The gradual transmutation from flesh and blood humanoid aliens to the Supreme Being in the minds of humans was complete. In a slow and amazing process the imperfect humanoid Nefilim individuals had been transmuted into a single infinite being who created the universe. The theologian, the student of the nature of God, came to be when the philosopher turned his natural metaphysical bent, as a religious believer, to the logical consequences of the projected nature of a transcendental God – who, most amazingly, still spoke in the words of an ordinary, "jealous" Nefilim ruler through the history in the Old Testament.

What of the Original Tradition?

If the pain and confusion and despair of separation from the Nefilim masters was the fundamental reason for the ubiquitous and profound psychological sublimation of the Nefilim into "God," was the knowledge of them lost completely?

Has there been a continuous thread of tradition from the earliest times that the Anunnaki were real flesh and blood humanoids? If so did it go "underground" because it became unpopular or did it exist in the open? If it has always existed, where is it found today? The answer to that question is too obvious: the entire vast body of history (oral, inscribed, carved, sculpted, and written), recovered over the last one hundred and fifty years from the sites of those first sudden civilizations all bear witness, take as given, speak always in such an unambiguous way, that the Anunnaki / Neter / Anakim / Nefilim were precisely that: real, flesh and blood, physically present humanoids. It is not a tenuous frayed thread of tradition, it is a veritable river of knowledge and information. The entire stream of tradition should be seen as the answer to the question. The ancients were not in doubt. It has only been our interpretation and retroprojection that has even caused the question to be asked in the first place. Did the tradition go "underground" for any reason? It went literally and physically underground by being buried by the sands over time. And it went underground, metaphorically, to survive. The latter-day, proprietary, sublimated monotheistic "religions," having come to ascendancy and exercising power and control over the intellectual development of Western culture, simply ignored the past to the degree that the ancient civilizations and what they said were largely forgotten. What written and oral traditions survived were consigned to the categories either of myth or paganism, remaining so until renegade thinkers like Schliemann followed the remnants of written accounts and the old maps and began to dig it up again. Where can the record of this generic, pristine tradition be found today? In our museums, libraries, collections and in the documents recovered from the once forgotten sites of the ancient civilizations, written on clay tablets and papyrus and stone and carved in heroic proportions on the walls of the ancient cities and palaces, recorded in the words of those ancient peoples who knew the gods among them as Moses knew the Nefilim ruler, Yahweh, "face to face."

We have examined the contributions to our racial transition of the psycho-evolutionary changes the human race has gone through as we gained independence, precocity, experience and maturity as the Nefilim became more and more remote in time; the expanding metaphysical speculation of the precocious human mind and the eventual transference of those idealized concepts to a gradually cosmified Nefilim; the evolution of the political monotheism of the Jewish tribe; the residual deposit of high science taught humans by the Nefilim which was easily mistaken, as time passed, ignorance increased, and the keys lost, as magic, a mystical (mysterious) metaphysical doctrine, poetry or myth; the psychological pressure of our yearning for immortality contributing to our sublimation of the heavens to Heaven and the Nefilim to God; the institutionalizing of the ancient godspell mentality into "religion." We have reviewed the long agony of uncertainty and subservience that has caused us so much racial trauma. But, just as the pain, wrenching dislocation and traumatic sublimation have caused us as a race so much separation and grief over the last three millennia, so should the time of resolution be cause for celebration and joy.

Where are We in the Process Now?

Where in the process of transition to the new paradigm do we stand as a race and as individuals right now? Breaking the godspell involves two enormous shifts, one of ideas and the other a profound psychological readjustment. There is a spectrum of attitudes toward the new paradigm, from incredulous and belligerent rejection to advanced understanding and enthusiastic acceptance.

The State of the Religions of the World

Those who adhere to specific religious doctrines and rely on their clergy for evaluation and judgment are at a standstill; the reactions of officials of the various denominations vary from the extreme of liberal consideration of the new paradigm to the extreme of radical rejection as some sort of atheistic or sinful doctrine. A great deal of private, internal conflict is generated within the religious institutions with regard to the new paradigm due to the intramural variations in position with regard to how much respect scientific evidence is to be accorded. The churches are losing members and clergy by attrition, disinterest, suspicion and the development of a growing awareness of something being very wrong. A small percentage is actually breaking the godspell, most people simply recognize that something is "not right" but cannot articulate it. We may ask why humans have always found religion oppressive, why they have always tried to get out from under its onerous yoke. Should not something that is supposed to be such a glorious and joyous accord with the very nature of reality itself be blissfully embraced? We might ask the even more fundamental question, Would theology and religion, as we know it, have taken the forms it has, or even come into existence if the godspell syndrome had not evolved out of the ancient situation?

The Academic Community

The academic community is caught in an embarrassing position with regard to the new paradigm because it does not simply correct an existing set of data or improve a postulate or represent a new plateau of knowledge within the old context. If accepted, it replaces the entire structural base of religion, philosophy and parts of science. It is a direct contradiction of a fundamental assumption of all of Western thought which has been based on the postulate that the gods were mythological. It is only now that "mainstream" academicians are willing to allow themselves, for fear of peer pressure, ridicule and tenure tetanus, to even look at the evidence for our genetic beginnings, much less to consider the ramifications. Undoubtedly, breaking the godspell will be most difficult and radically revolutionary for the religions of the world. But the academic establishment has a problem with it also.

Traditionally, in Western culture, history is taken as history up to the point where what is recorded cannot be accepted for whatever reason. The Sumerians and the ancient civilizations were real, their history as they recorded it has been verified through archaeological research, they had advanced learning institutions, wrote bi- and tri-lingual dictionaries, kept libraries, and were to be marveled at for their sudden full blown efflorescence. But, when the Sumerians said the Anunnaki/Nefilim were real, scholars rationalized the texts to mean something different because that just couldn't have been so – in their minds they simply could not accept it. There is a direct and clear parallel between the relationship of the oral tradition and the Torah and mythology in its relationship to history as it is understood in the academic traditions of today. Both have developed as adjuncts to the text primarily for the sake of explaining what the parts of the text "really" mean when what is stated in the text cannot be understood or accepted, literally, as it stands. Mythology as a concept, a field of study, is the academic oral tradition that puts itself forth as the essential interpretation of what their texts really meant. And it has been heresy to go up against that tradition. The cities of ancient civilizations were considered mythological until they were literally dug out of the sands that had buried them. When the cities of those first civilizations were discovered it reversed the belief that the peoples of those societies had never existed. And when the reality of those ancient humans and their amazingly sophisticated cultures was beyond doubt, the history of their civilizations, their deeds, politics, science, languages and beliefs all had to be reconsidered as real. But, at that point, something happened. Scholars gradually came to agree on all those things – except one; they balked at the gods, the superior beings that all the ancient societies said were the creators of the human species, had taught us civilization and technology and, indeed, were the transcultural focus of all those ancient societies. The academic community is the last to get the message to accept the evidence. That is significant in itself.

The Scientific Community

The scientific community, by and large, would rather it all just went away, would rather be able to preclude it from consideration, consign it to the lunatic fringe with sarcasm and ridicule. If we wait until enough dry and clinically scientific evidence has been amassed to make even the scientific establishment secure enough to consider the topic, we will have put ourselves through enormous unnecessary trauma. It is the rare scientist, indeed, who will even consider the evidence. The scientific method, as we have it now, is the current form of an approach to the understanding of the universe that arose, in large part, as a reaction to the old godspell theological and philosophical dogmatism that controlled Western culture for more than a thousand years. It was, in a very real sense, a manifestation of reaction, even rebellion. It is well to keep that perspective, particularly in this time of transition. The absolutistic, authoritarian doctrines, philosophies and viewpoints kept us separate, divided, with radically opposed definitions of human nature and its origin, its place in the universe, its purpose and end. It was because of the stultifying restrictions of that terrible separation that we have developed the scientific approach if it is examined in its sociological aspect. Philosophers and even scientists have criticized it as being inadequate, because of its postulates and narrowness of focus. Opposition has always been present from the theologians who claim it must submit to what they consider the higher authority of Revelation. But, from a sociological perspective, it acts as a sort of second class common denominator, a sort of generic methodology and criteria base that cuts, more successfully than any other methodology so far, across the boundaries of all disciplines. It has been science, after all, that has brought about the change from the view of the earth-centered solar system to the heliocentric one, vindicated Galileo, Giordano Bruno, and opened up the processes of human thinking to allow us to reach a point where we can grasp and accept the new paradigm. So far it is reacting to its authority being challenged, its dogma being questioned as theology has reacted previously, but science is faced with a profound shift of consciousness again with the coming of the new paradigm. It no longer will exhibit its reactionary character, no longer be needed to act as common denominator because the absolutistic theological authoritarianism will be gone. If we have a common, consensual understanding and definition of what a human being really is on a planetary basis, the invisible walls of those two thousand year old divisions will vanish. Science in a unified planetary society will be quite different. It will no longer have to be fallen back on in frustration and despair of ever agreeing on a fuller and more satisfactory criterion of truth. So much of scientific effort has been devoted not so much to pure science as to the struggle against the robot dogmatism of the godspell mentality. All of human thinking should be freed of that burden and the future embraced in common.

The Role of the Development of
Human Science in the Transition

In the process of transition we have been both aided and abetted by our technological progress. At the same time that it has increased our knowledge and independence it has, curiously, in these latter days, worked against our re-recognition of our true history by viewing it as mythological. It was the growing understanding of the world through our science that gave rise to the skepticism that reinforced the denial of our history as real. Our natural proclivity to seek explanation for physical phenomena has certainly increased our independence through a greater ability to explain the universe in general without reference to the "gods" or God. Our potential curiosity and desire for control were no doubt with us from the beginning but the complete domination by the Nefilim at first and their ongoing control, even after we were accepted as limited partners, kept us in check. The science we learned was that which was taught us or which we copied or, perhaps, stole from the Nefilim. In an encapsulated view, once we were on our own, in an attempt to find out precisely how certain things worked so that we could control or duplicate them for our practical advantage, we began to seek the "laws" by which they worked. Even though our first groping for explanations of how and why things happened the way they did appear much like superstition and our dealing with nature like magic, the motivation was clearly toward intellectual comprehension. In the earliest cases we had so little to go on and were so fundamentally ignorant of even the most basic principles, the explanations we developed sound, to us today, still like the most primitive superstition. But superstition is almost always quickly given up when we find a more precise and intelligible explanation. Humans do not seek superstition in preference to hard knowledge unless there is an individual emotional need that overwhelms reason. The classic textbook explanation of the cause and source of mythology is that the human mind, faced with the powerful and awesome forces of nature and the beneficial and disastrous effects spawned by them, tended to project godlike characteristics such as intelligence, personality, benevolence or evil on them. Superstition, as witnessed among humans anywhere in the world where sufficient knowledge of physics is lacking, is a simple fact of human existence in any age. Superstition, inevitably, gives way to more accurate science when available due to the nature of the human mind. Our naive attempts to "do science" have gradually become more refined and accurate, become more and more rigorous as we increased our knowledge and skill. At the same time that our gain in technological and scientific knowledge afforded us a higher standard of living, our increased insight into and understanding of the nature of the universe gave us greater confidence in our knowledge of how things really work. With the technology taught us or copied from the Nefilim mostly lost to us over time, we gradually began to develop our own science. Even that relatively primitive knowledge made it plain to us that the technologies that legend taught us were possessed by the Nefilim were far beyond us to the extent that they seemed literally fantastic. And the more we learned, the more fantastic those legends seemed to us. It was not difficult to eventually relegate the legendary technologies of the "gods" to the classification of myth, fantasy and quaint imaging by naive and probably simpler minds in the ancient past. If the technologies of the "gods" were fantastic, it was a short step to considering the "gods" so also. It is significant that science, developed to be independent of the theological context, has played directly into the hands of the organized religions on this point. If the "gods" were fantastic, mythic and unreal projections of our naive ancestors, we could see ourselves as more intelligent and sophisticated than those early humans. Our science has been influenced by that seductive appeal to our uncertainty and fragile status during the transition. But "gods," in whatever form, are the polytheistic, pagan, evil, anathema deities, acknowledged and worshipped by "heathens," condemned by the monotheistic theologies. Science, in reactively classifying the "gods" of the ancient civilization, the Nefilim, as mythological, reinforces the attitude of the monotheistic religions in a major way. It removes the possibility of the theologians having to confront the dawning fact that the God of their devotion is a psychological sublimation of those ancient "gods" into a transcendental metaphysical abstraction. And that they are out of a job. But, by this time, we have gained sufficient genetic enlightenment to see a possible resolution to the historical questions and conflicts. That allows us to at least begin to consider the new paradigm and to move beyond religion.

Getting Rid of the Post-Traumatic
Stress Syndrome: The Godspell Broken

From our historical point of retrospection it seems simply incredible that thinking humans should, could, at any time, subscribe to the concept of a cosmic Absolute as a male – or female – entity with humanoid characteristics. But that is the undeniable, simple fact. In these latter times, however, millions actually believe in some kind of muddled anthropomorphic projection and, even for those who reject it entirely, the fabric of our civilizations and cultural mind-sets are profoundly effected by that syndrome. Over time, millions have died because of it, the Babel factor is still potently divisive, we are still fearful of the displeasure of a jealous Father in the sky, we still interpret goodness in terms of subservience. The garden variety of atheism may be understood as a sort of racially adolescent rebellion against such a crass anthropomorphic notion. It does not matter whether the latest sophistication would have it that these concepts are potent psychological archetypes, or the crude notions of literal fundamentalism, they effect us directly or indirectly in a profound way. A survey taken in the United States within the last few years shows forty percent of the adult population believes that God made the earth and man and the flora and the fauna within the last ten thousand years. The tenacity of the belief in a sublimated concept of the "gods" turned into a cosmic God signifies the severity of the trauma due to the depth of dependency on the Nefilim. As a result of the processes of the development of racial and individual independence, we now stand at the threshold of a new dimension of human consciousness. Suffice it to say here that the redefinition will, obviously, be a protracted and interesting process to say the least. It will be one of disagreement, controversy, and constructive debate; it won't be boring. The critically different factor will be that we are all fundamentally talking about the same thing, a genetically engineered species now coming to the point of independence and self-awareness driving it into directing its own evolutionary progress. The debate will, therefore, be about the details, not the essentials. It will be about the potential this creature, this human, possesses, what this amazing entity can know, discover, achieve, what dimensions this intriguing organism can perceive and operate within. It will be about how to redefine the potentials the species already exhibits and project the future possibilities of realities the species can create. This relativistic renaissance will be the most difficult we have ever undertaken as a race. The redefinition of ourselves will be done by ourselves and, therefore, the degree of self-reference will be intense; we will be looking to ourselves as source for a correct redefinition at the same time as we are redefining ourselves. We, literally, will be redetermining the criterion to which we refer in judging the correctness of our redefinition at the same time we are looking to it as a criterion. That will be a great stress to some, a key to the future for others.

It is not enough to just get free, to break the godspell racially and in one's own life. The intense intellectual experience of genetic enlightenment, although certainly exhilarating, is brief; one realizes the truths of our beginnings and real history, senses the degree of freedom and potential and quickly appreciates the ramifications. But the actual absorption of those elements into one's life has an even more profound long term effect. The essential focus is to recognize the traumatic nature of the racial transition from godspell to god games that has spanned nearly three millennia and to identify the elements of that trauma that one has acquired from one's ancestors and to deal with them. The nature and character of our individual therapeutic processes is far more than the experience of intellectual liberation from the godspell. It should be the total integration of both the intellectual and emotional realizations and experiences of a new dimension of consciousness. To put it in terms of some of our most classic metaphors, Prometheus can get off his rock and reach genetic satori. Buddha can get out from under his tree and reach genetic enlightenment. Job can get off his dung heap and complete his est training. Jesus can come down off his cross and finish his degree in political science. But what will they do for an encore? We will begin to act as the self-validating "gods" that we are. Once having gotten used to that level of self-esteem, we will begin to argue about the direction humans should take in their own evolution, finally see the humor in that and begin to find a common ground in supporting each other in our own individual evolutionary trajectories in the radical freedom of determining our futures.

CHAPTER 7

BEYOND RELIGION

/

"What we must do must come from ourselves. . . we fragile human species at the end of the second millennium A.D., we must become our own authorization. And here at the end of the second millennium and about to enter the third, we are surrounded with this problem. It is one the new millennium will be working out, perhaps slowly, perhaps swiftly, perhaps even with some further changes in our mentality."

Julian Jaynes
The Origin of Consciousness in the
Breakdown of the Bicameral Mind

Imagine, even no religion,
They say I'm a dreamer,
But I'm not the only one. . .

John Lennon
Imagine

The godspell, a dream within a dream,
The meta-myth of our myths,
The Babel factor exponentiated
Into a war of absolutes,
The certain indicators of uncertainty.

Neil Freer
Neuroglyphs

We will soon be faced, unavoidably, and negatively for some, as a liberation and challenge for others, with determining the most fundamental orientation of our lives to the universe as individuals and as a race, as the new paradigm takes hold. There are and will be a myriad of attempts to adjust the facts, to construct new contexts of semi-secular belief systems as substitutes for the old, to build "natural" or strictly "scientific" or generic systems of neuveau species of pantheism, worship of the Universe, extensions of channeling of "higher" entities, private cults, schools of enlightenment, philosophical systems, revisiting of the "occult" traditions and dozens of others. It's already happening. But that will be a passing phase of transition. We need, therefore, to discuss how it will be truly beyond religion as we know it. But, before we do, there are questions, the answers to which will, perhaps, give us some perspective on such a novel topic: What really is religion as we know it and what is its ultimate source? Who really was/is the God or Gods of the Judaeo-Christian-Islamic religion, the fundamental formative engine of our cultural heritage?

Who was Yahweh, The God of the Judaeo-Christian Tradition?

Sitchin begins the Endpaper of *Divine Encounters* with this question, "So who was Yahweh?" The inherent implications of the question itself contain a great deal of information. By the fact that the question is about the "God" of the Judaeo-Christian religious ethos and not about the "god" of some other cultural tradition, we know that it is the basis of Western culture with which we must be concerned. By the fact that there is a question we know that there is a puzzle, some uncertainty. Could it be possible a deity of this importance and power not be completely and unambiguously identifiable? Why should there be this problem in the first place? To answer this question, Sitchin proceeds in his usual thorough and thoughtful way. His first step is to compare the personality and historical actions of Yahweh to those of a number of known Anunnaki/Nefilim personalities in an attempt to determine if Yahweh was actually one of them. He systematically considers Enlil, Enki, Marduk, Ninurta, and many others including Anu, the chief Anunnaki who resided mostly on Nibiru, and eliminates the possibility that Yahweh was, indeed, one of them but known by another, possibly cultural or regional, name. Having done so he moves directly to the Old Testament records for an answer to his question.

Although it is. almost amazing and certainly significant that he even considers this possibility in the first place, it is disconcerting that he never considers the possibility that Yahweh was only another, distinct Anunnaki/Nefilim individual. In the event that Yahweh was really only a new Anunnaki personality coming on the Middle Eastern scene at a relatively recent date, actually when the records seem to indicate that the Anunnaki were beginning generally to phase off the planet, there would be an immediate problem. The possibility that Yahweh was simply a lower echelon Anunnaki official assigned to handle details of the phase-off transition, or as a minor official in one geographic location would contradict or at least denigrate the claims for his glory and greatness made by the biblical authors.

Sitchin's next methodical step, then, is to identify the source and nature of the Old Testament itself: " . . . texts found in the library of the Assyrian king Ashurbanipal in Nineveh . . . recorded a tale of creation that matches, in some parts word for word, the tale of Genesis. George Smith . . . published, in 1876, *The Chaldean Genesis*; it conclusively established that there indeed existed an Akkadian text of the Genesis tale, written in the Old Babylonian dialect, that preceded the biblical text by at least a thousand years. Excavations between 1902 and 1914 uncovered tablets with the Assyrian version of the creation epic in which the name of Ashur, the Assyrian national god, was substituted for that of the

Babylonian Marduk." (*Genesis Revisited*, pp.41-2).

What was this "tale of creation" that was the original source of the Babylonian, Assyrian and then the Hebrew version? Sitchin defines it unequivocally as the body of knowledge, "a true cosmogony . . . taking us to the beginning of time" that we know through the recovered Babylonian document, *Enuma elish*, The Epic Of Creation. " . . . the most hallowed historical-religious epic of Babylon . . . Intended to propagate the supremacy of Marduk, the Babylonian version made him the hero of the tale of Creation. This, however, was not always so. There is enough evidence to show that the Babylonian version of the epic was a *masterful religious-political forgery* of earlier Sumerian versions in which Anu, Enlil, and Ninurta were the heroes."[italics mine]

"But why not," asks Sitchin, "take the epic at face value, as nothing more nor less than the statement of cosmological facts as known to the Sumerians, as told them by the Nefilim? Using such a bold and novel approach, we find that the "Epic Of Creation" perfectly explains the events that probably took place in our solar system." The source, "Sumerian knowledge," "Sumerian science" is clearly defined as the detailed teachings humans received from the Anunnaki/Nefilim concerning the formation of our solar system beginning at the time when the Sun existed along with Mercury and Tiamat (the planet later shattered into the asteroid belt and the remainder of which was displaced to form the Earth in its current orbit). It is to be noted that it does not deal with the beginning of the universe or of space/time; it deals with the formation only of our solar system. The question immediately prompted is, How could the scientific knowledge of the physical events of the formation of our solar system, in which the Anunnaki home planet, Nibiru, played a major role, become somehow transmuted into the Genesis story? The key to all of this are the words "masterful religious-political forgery."

The Babylonian Version of the Forgery

What was the nature of the forgery and where was it initiated? "There [Babylon] the original Sumerian Epic Of Creation was translated and revised so that Marduk, the Babylonian national god, was assigned a celestial counterpart. By renaming the Annunaki home planet, Nibiru, "Marduk" in the Babylonian versions of the creation story, the Babylonians usurped for Marduk the attributes of a supreme "God of Heaven and Earth." This version – the most intact one found so far – is known as *Enuma elish* ("When in the heights"), taken from its opening words." (*Genesis Revisited*, p. 41) There was a clear recognition, on the part of the Babylonians, that Marduk was an Anunnaki/Nefilim "god," "physically present" among them and also that he was associated with planet Nibiru in order that the attributes of that major planet could be associated with him.

The Assyrians Version of the Forgery

The Assyrian forgery, in their turn, simply replaced Marduk with Ashur, another Anunnaki/Nefilim individual, most probably "a resurgent Enlil," (*Divine Encounters*, p. 367) for the same purposes and with the same intentions. They forged Ashur's name on the Sumerian astrophysics of the formation of our solar system, to elevate and glorify Ashur as their national "god."

The Hebrew Version of the Forgery

What was the Hebrew version of the forgery? Sitchin says, " . . . what the editors and compilers of the Book of Genesis had done was no different from what the Babylonians had done: using the only scientific source of their time, . . . [they] also took the Epic Of Creation, shortened and edited it, and made it the foundation of a national religion glorifying Yahweh." In doing so, "Following the teachings of their Sumerian forefather, Abraham of Ur, the ancient Hebrews also associated their supreme deity with the supreme planet." Furthermore, "Condensing the many gods into a single Supreme Deity, the biblical tale is but an edited version of the Sumerian reports of the discussions in the Assembly of the Gods." (*The 12th Planet*, p. 338)

One may reasonably question, in light of the clear evidence of the forgery, why it would be necessary to carry the inquiry any further. Sitchin does so because this even more elaborate and sophisticated, third or even fourth generation of "masterful religious-political forgery" executed by the Hebrew revisionists for obvious partisan reasons incorporates claims of unique, even universal status for Yahweh that must be examined and evaluated. In their elaboration of the forgery, as Sitchin points out in *Genesis Revisited*, "The Hebrews followed suit but, preaching monotheism and recognizing – based on scientific knowledge – the universality of God, ingeniously solved the problem of duality [god/associated planet] and of the multitude of Anunnaki deities involved in the events on Earth by concocting a singular-but-plural entity, not an El (the Hebrew equivalent of Ilu) but Elohim – a Creator who is plural (literally "Gods"), yet One." This was, as Sitchin himself says, a "departure from the Babylonian and Assyrian religious viewpoint" (page 44) – which was itself a departure from the Sumerian viewpoint: if the Hebrews preached monotheism and the "universality of God" but the Sumerians, Babylonians, Assyrians and the Akkadians before them did not then what the Hebrews did was a further extension and elaboration on the basic political forgery.

But, if the Hebrews based the doctrine of monotheism (Yahweh as the only god to be worshipped) and the recognition of "the universality of God" on "Sumerian scientific knowledge" why do we not find any indication of that doctrine or recognition among the Sumerians, who were closest to the Nefilim source of that scientific knowledge, or among the Babylonians, Assyrians and Akkadians who came after the Sumerians by hundreds or thousands of years? Sitchin's understanding of this departure from the Babylonian and Assyrian religious viewpoint is that it "can be explained *only* [italics mine] by a realization that the Hebrews were aware that the deity who could speak to Abraham and Moses and the celestial Lord who the Sumerians called [planet] Nibiru were not one and the same scientifically, although all were part of a universal, everlasting and omnipresent God – Elohim – in whose grand design for the universe the path of each planet is its predetermined 'destiny,' and what the Anunnaki had done on Earth was likewise a predetermined mission. Thus was the handiwork of a universal God manifest in Heaven and on Earth." (*Genesis Revisited*, pp. 45-6) "These profound perceptions, which lie at the core of the biblical adoption of the creation story, *Enuma elish*, could be arrived at *only* [emphasis mine] by bringing together religion and science while retaining, in the narrative and sequences of events, the scientific basis." (*Genesis Revisited*, pp. 46)

There is much to be sorted out here. It is necessary to ascertain precisely what the "science" base is as distinguished from the "religious" element that the Hebrews injected. It is necessary to understand precisely what Sitchin means by the "universality of God" as it relates to the concocted singular/plural Elohim and to Yahweh. We need to understand the basis of the doctrine of monotheism and, ultimately, why the Hebrews had to elaborate the forgery even further than any previous nation.

The "Science" as Basis of the Forgery

We have seen that Sitchin says that what the science part of what the Hebrew authors of Genesis did was adopt the Sumerian sourced *Enuma elish* detailed astrophysical account of the formation of our solar system and the major role their planet, Olam/Nibiru played in it, "no different" than the Babylonians did in their "masterful religious-political forgery of earlier Sumerian versions, in which Anu, Enlil, and Ninurta were the heroes."

The "Religious" Basis of the Forgery

What was the religious part of what the Hebrew authors of Genesis did? The essence of it, according to Sitchin, is the recognition of the "universality of God." What does Sitchin claim was the source for the Hebrews recognition of "the universality of God"? It was "based on Sumerian scientific knowledge" (*Genesis Revisited*, p. 44) There seems to be a profound contradiction involved here: if the only way to understand what the Hebrews did is to see that they had to combine the separate elements of religion and science but the religious element came from the science, what is happening here?

A First Clue

Notice what different reactions the reader would have if Sitchin had said the "universality of the Elohim" or the "universality of an Ultimate Principle" or the "universality of Yahweh" rather than the "universality of God."

What if Sitchin had written instead "the universality of Yahweh"? There would have been no ambiguity, no question, no problem, only the need to justify that claim. If he had written the "universality of the Elohim," meaning the humanoid Anunnaki/Nefilim, it would require the acceptance of the Nefilim as somehow at least present throughout the universe from the beginning of the universe, and the historical facts negate this. This would have been the equivalent of the Sumerians claiming that Anu was everlasting and universal or the Babylonians claiming that Marduk – who almost perished for his outrageous local behavior if he had not been rescued from the Giza pyramid in which he was imprisoned in time, as eternal, or the Assyrians had attempted to portray Ashur as infinite Spirit. If he had written the "universality of an Ultimate Principle" then the reference would be much cleaner emotionally and certainly clearer. It would invoke in the mind of the Western reader all those numinous metaphysical attributes of an Eternal Spirit or Infinite Mind found in modern day religious doctrines, many of which no longer even refer to or even know much of the biblical Yahweh as such. But is that truly what the Hebrews' "profound perception" was? So we have to determine what Sitchin means by the Hebrews' perception of "universality of God." He says that the Hebrews "were aware that the deity who could speak to Abraham and Moses and the Lord whom the Sumerians called Nibiru were not the same. . ." and perceived that Yahweh, the Anunnaki, the planet Nibiru .". . .all were part of a universal, everlasting and omnipresent God – Elohim – in whose grand design for the universe the path of each planet is its predetermined 'destiny,' and what the Anunnaki had done on Earth was likewise a predetermined mission. Thus was the handiwork of a universal God manifest in Heaven and on Earth." (*Genesis Revisited*, pp. 45-6) This statement gives the key to Sitchin's fundamental conception of the matter and it is this precise point with which I must respectfully take issue. In order to advance this explanation one must presuppose that there is a "universal, everlasting and omnipresent God". Now, either that word "God" represents an entity entirely separate from Marduk, Ashur, Yahweh, the Annunaki, the planet Nibiru, El, Elohim, discoverable by the human mind without any contact with the ancient records or the Old Testament, or it represents a concept present in Western culture derived directly from the Old Testament as a translation of a specific word.

Is This Dualism?

There is no indication that there were two Gods that the Hebrews recognized or worshipped, in this case Yahweh as one and Elohim as the other: indeed they preached monotheism. And Sitchin has already made it clear that Elohim is a concoction of expediency.

Does the Tetragrammaton Really Spell "Pantheism"?

From another unavoidable perspective, Sitchin's explication may be easily understood as an unequivocal statement of classic pantheism: All is God or Everything is God, everything is God, God is equated with the Universe or the Universe of universes, etc. In this meaning the physical planet Nibiru, the Anunnaki, the individual, Yahweh, as some sort of local avatar manifestation perhaps – and everything else – would be simply manifestations of Elohim the Everything. But, although his words strongly invoke this concept, Sitchin makes it very clear that is not what he means: "Although the biblical admonition against the worship of pagan images gave rise to the notion that the Hebrew God had neither image nor likeness, not only the Genesis tale but other biblical reports attest to the contrary. The God of the ancient Hebrews could be seen face-to-face, could be wrestled with, could be heard and spoken to; he had a head and feet, hands and fingers, and a waist. The biblical God and his emissaries looked like men and acted like men – because men were created to look like and act like the gods." (*The 12th Planet*, p. 338) But this still leaves another question: If, unequivocally, the humanoid alien Yahweh was the God that the Hebrews worshipped, was the word Elohim a philosophical conceptualization of ultimate reality, i.e., a metaphysical concept forming only an intellectual context for the overall forgery?

Is Elohim a Metaphysical Overlay?

If the "profound perception" of "the universality of God" is not dualistic nor is pantheism then what is it? Is it a metaphysical concept, arrived at by Hebrew philosophers, overlaying a sort of transcendental semi-mystical cosmic, religious aura – and justification – on the political forgery of Yahweh's glorification? Did the Hebrews arrive at this notion as their thinking evolved independently of any consideration of the knowledge they derived from the *Enuma elish* tradition? Had the Hebrew metaphysics progressed to the point where they had developed the notion of an Ultimate Principle, perhaps an infinite Mind or Eternal Spirit, either stimulated by the mind expanding planetary panorama of the astrophysics of *Enuma elish* or on their own, had overlaid those abstractions on the person of Yahweh? Sitchin makes no such claim for Hebrew speculative philosophical thought. He goes to great length to leave no doubt that it was all astrophysics, not metaphysics, all science that was the source, stating emphatically, ". . .as I have repeatedly done in my writings, . . .the information of how things began – including how Man himself was created – indeed did not come from the memory of the Assyrians or Babylonians or Sumerians but from the knowledge and science of the Anunnaki/ Nefilim." (*Genesis Revisited*, p. 46)

We are compelled, then, to ask again, if Sitchin recognizes the fundamental nature of the Hebrew concoction as a patent forgery and the God of the Hebrews as a humanoid, why does he feel it worth pursuing the subject further? What unique element makes the forgery so attractive? It is apparently because of the unique claims made in the biblical texts for and about Yahweh. Apparently holding in abeyance judgment as to whether such a revisionist forgery can be taken as a truthful and unbiased source for the sake of elucidating precisely what the Hebrews did and thought in those ancient times, Sitchin quotes the text which state or suggest that Yahweh was not just the god of the Hebrews but also the god of the Anunnaki/Nefilim. This would certainly be a revelation if taken at face value and was a true statement. Having established that possibility he then quotes text that suggest, at least as one possible meaning, that, although the Anunnaki/Nefilim possessed extreme longevity by our standards, they were not immortal – but Yahweh was. Referring to a passage in Psalm 82, "I have said, ye are Elohim, all of you sons of the Most High; But ye shall die as men do, like any prince ye shall fail" to demonstrate his point, he also claims this passage suggests that Yahweh created the Anunnaki/Nefilim. He does not consider the possibility that Yahweh's statement in the Anunnaki assembly could be understood as Yahweh, if also an Anunnaki, rhetorically reminding his kind that they would all eventually die, including himself. Again, if "Most High" is a title frequently used for "God," is he also acknowledging the Anunnaki as his sons?

Having suggested these extraordinary characteristics of Yahweh, Sitchin then considers the "puzzle that has baffled generations of biblical scholars" (*Divine Encounters*, p. 375), why the biblical authors did not begin the first word of the Torah with Aleph, the first letter of the Hebrew alphabet which signifies beginning. He notes that the canonization of the biblical texts took place while the Hebrews were in captivity and probably was not begun with Aleph for fear of offending their captors' god. The Aleph beginning would make the text read "The Father-of-Beginning created the Elohim. . ." rather than the actual text as it has been transmitted through the ages, "In the beginning God created the Heaven and the Earth." But, he states, he believes that it should have begun with Aleph and, if that were, .". . .all at once Science and Religion, Physics and Metaphysics, converge into one single answer that conforms to the credo of Jewish monotheism: `I am Yahweh, there is none beside me!' It is a credo that carried the Prophet, and us with them, from the arena of gods to the God who embraces the universe." (*Divine Encounters*, p. 376.) But, regardless of the reinforcement this desired beginning would give to the arguments for Yahweh's unique status, the fact of the matter is that the Hebrew biblical authors did not begin the beginning with an Aleph even though, once out of captivity they could have made that fundamental alteration – if, indeed, they ever really intended it in the first place. Sitchin's conviction that it "is not to be doubted" that at one time the text did begin with Aleph allows him to finally conclude Yahweh was an alien, humanoid, "COSMIC VOYAGER," whose domain extends beyond our Solar system even to the stars, indeed the Creator of the Universe. This being so, he concludes, we are Angels of Yahweh and carrying out his universal plan. Sitchin has come full circle, if in quite an unorthodox manner, to the rather orthodox worldview of the believing Jew.

I believe that Sitchin has indeed portrayed precisely what the Hebrew authors *did* in this matter, fundamental to the basis of Western culture and religion, by putting it in the context of our aboriginal relationship with the Anunnaki/Nefilim.

I disagree only – although this is pivotal and crucial – with his extension of the Hebrew authors' intention from the clear agenda to exalt Yahweh as the singular god among the gods as their ruler to claiming that they *intended* Elohim to mean an infinite cosmic God. I see our current positions and postures, relative to the Anunnaki and Yahweh from a different perspective. Let us assume for the sake of the discussion that we set aside the facts that, as Sitchin himself has pointed out, many deeds attributed unabashedly to Yahweh were actually the deeds of a variety of Nefilim individuals; that obvious lapses were committed by the revisionists wherein the plural "us" in the Babel story, the creation of humans, etc. was carried over from the Sumerian where the subject was meant to be the singular Yahweh; that the compression of the multitude of the Anunnaki into a single deity forced the authors to concoct an ambiguous term, Elohim, to cover all the awkward possibilities; that, ultimately, the elaborate forgery was transparently imperfect. Let us also set aside the fact that the only information we have concerning Yahweh comes solely from the partisan biblical texts. Further let us assume, at least for the sake of the discussion, that Sitchin's arguments are sound enough to conclude that Yahweh, a ruthless, peevish, male chauvinistic, megalomaniacal alien humanoid who traveled in a space ship, who could simply kill two of his servant priest for burning the wrong incense, turn his minions on an offending general's army and slay 186,000 in a night, allowed his subjects to attribute to him the atomic vaporization of two cities of humans that he didn't actually do, was never mentioned by the Anunnaki or outside of the Bible, was sometimes unknown even to neighboring rulers, often ignored by his own slaves, was/is really the Creator of our Universe. Let us even grant that it is at least conceivable that an advanced humanoid intelligence, could know enough about the meta-universe scientifically, to be able to the manipulate the physical laws to initiate the beginning of a new sub-universe. Let us, for the sake of this part of the discussion, allow even this to Yahweh.

Even if these cosmic claims for Yahweh were all true, and I don't think they can be fully substantiated if, indeed, they are just extensions of the overall forgery, I would categorically and absolutely refuse to submit to Yahweh's dominion even under threat of his notorious wrath. How can one respect, much less work for, any entity who acts in ways that do not even measure up to the imperfect ideals of the highest development of an ordinary human, who does not act in ways that are even up to the relatively primitive standards of international law or the Constitution? If Yahweh was really the God whom the Anunnaki worshipped, revered or even worked for, would they not have communicated that central fact of their lives to humans some time over the thousands of years they trained and taught us? If they wanted us to learn Yahweh's brand of monotheism because it was theirs why no mention or training of humans along these lines even by Enki or Enlil who was known for his strictness and righteousness? Isn't it quite clear that those gods who were "abominations" in Yahweh's opinion like Inanna and Marduk and Ba'al should be considered great "sinners" in such a context? If, as Sitchin has pointed out, they were, with the rest of the Annunaki, lumped under the term, Elohim, how could that be? If Yahweh was one of the Twelve of the Assembly of The Gods why do we not ever hear of him or his associated number in all of the Sumerian and other records? If Yahweh is truly the top management of this particular universe then I'm frankly disappointed. If this is typical behavior of supposedly superior beings in general then they have lost their ratings as far as I'm concerned. The worship of such a deity, inculcated into the human psyche by our subjection to the Anunnaki/Nefilim over thousands of years, is precisely the subservient godspell attitude of which we must be free to find our unique racial and individual identities.

With regard to the Judaeo-Christian tradition and the Islamic tradition considered here as example, the points of the argument over whose God is the only one are still being delivered by car bomb and missile. We need to break this godspell and move beyond religion as we know it.

The Godspell Revisited

Father Charles Moore reports that when he asked Mr. Sitchin the question "Is the God whom we worship the God whom the gods worship?" Mr. Sitchin ". . .answered, after a moment's thought: 'Yes, exactly'." (Sitchin, *Of Heaven and Earth*, 1996, p. 34) Father Moore is a Catholic priest. Mr. Sitchin is a believing Jew. Father Moore concludes from this that "In other words God and the gods are not mutually contradictory, but are related in what we might call agency. Agency, that is, in regard to us, with God as principal and gods as agents. More or less like angels, and perhaps devils." (p.34) This is an excellent example of a mind-set that one encounters halfway to freedom beyond religion: it may well be called the Big-G, little-g syndrome.

The Big-G, Little g Syndrome

There is a common rationalization employed by those who are free and open-minded enough to accept the new paradigm but, because they have had a strong affiliation with an organized religion, find real difficulty in completely reversing their position through embarrassment, some uncertainty, fear of stepping into a perceived vacuum, alienation from family or friends, etc. When confronted with the real nature of the "gods" and the fact that, over time we have transmuted them generally and one specifically, Yahweh, in the Judaeo-Christian Western context, and another, specifically, El/Nannar/Sin, in the Islamic context, psychologically into a cosmic Being, God or Allah, many have adopted the position embodied in the typical statement, "I resolved the problem when I realized the difference between "god" with a small "g" and "God" with a capital "G." By this they mean the difference between a "god" or "gods" as humanoid alien and a cosmic and usually infinitely transcendent "God" – usually the "God" that the individual confronted with the problem had been taught to worship in the religious context in which she or he was brought up. . . It matters little whether "the problem" arises for the individual in the form of science versus religion, the contradictions between history and religious doctrine, or the conflict directly between religious doctrines. This rationalization is a very common, safe way for those who have been ministers, priests, teachers, officials or adherents of a religion to adjust the knowledge of the new paradigm to their belief system: they can remain worshippers of a cosmic God and still accept the historical facts about the "gods." It is fairly easy, then, having taken that stance, for one to answer straightforward questions such as: If the "gods" literally created us in the laboratory then God did not? with Well, Genesis is somewhat symbolic: the "gods" did but as "agents" of God. Or, If the word Elohim clearly was a plural-as-singular term for the Annunaki as given away, as example, by the "us" in the Babel narrative, then how can you claim that it really refers to a cosmic creator God? with, Well, it was the intention of the Hebrew authors of the Torah.

But the inescapable, awkward and problematic fact is that that history, in Sitchin's own interpretation, shows clearly that the entire concept of a

projected cosmic deity as embodied in the teachings of the world religions, God with a Big G, has its source in the Nefilim individuals who were the "gods with a small g", the plurality of which the Levite authors of the Torah concocted the plural-as-singular Elohim to express as a monotheistic appellation for Yahweh, the essence of the "religious-political forgery" to elevate Yahweh to cosmic status. Regardless of how much we shift focus to some concept of a metaphysical, infinitely unthinkable God abstracted and sublimated from that concocted, forged term, Elohim, as the theologians have done, the humanoid Yahweh is the god of the Judaeo-Christian tradition.

Although one can sympathize with the awkwardness of the situation and the sometimes painful adjustments forced by the new paradigm, the rationalization may buy some time but it simply does not hold water. The Big-G, little-g syndrome is a stall tactic halfway out of the godspell.

The View On the Way to Freedom

The politics of our unique racial history may be characterized as a rapid evolution from slave to serf, to saviors to self. We were literally invented for the Nefilim's pragmatic purposes as gold-mining and agricultural slave animals, became their serfs and limited partners, with kings as local go-between foremen, after they almost wiped us out in the Flood. Once they had phased off the planet and left us on our own, we began to look to charismatic saviors, political and eventually "religious," to lead us in the same submissive way we had formerly looked to our Nefilim masters. I feel fortunate to be living in and contributing to an age when we are emerging from racial adolescence. We come alive in a laboratory, the mutant fusion of an alien race with slightly more advanced knowledge and science than ours is today, with an indigenous species with intelligence requiring genetic manipulation to bring it up to adequate for basic gold mining operations. We are slaves of these far superior masters, looking up, innocent, naked and history-less, in awe at their power, knowledge, history and amazing activities. As we develop over time, more precociously than they perhaps anticipated, some of us, at least, are recognized for a growing potential, elevated in status and function, sometimes even taken as sexual partners. Occasionally the offspring of such a marriage, like the king, Gilgamesh, who knew his mother to be pure Nefilim and his father human, demanded the rights of the Nefilim. Gilgamesh literally demanded immortality on a purely legal basis as a demi-god, a half-god. Surplus of us are pushed into the outback and develop our own adaptive native cultures from scratch. The out-backers retain some of the old memories from the masters' centers and hand them down as venerable, "sacred." And our own "religions," naive proto-scientific explanations of the awesome forces and mysteries of nature evolve over time, mix with the venerable traditions and legends. This "out-law" culture spreads over the globe carrying with it the ancient history of our beginnings and the watershed events (the Orinoco Indians knew the Flood story. The American Indian cultures had 42 versions of it) and the veneration of the Nefilim Goddess Mother, Ninhursag, the Nefilim geneticist who literally made us. Gradually over time the two traditions con-fuse.

But the "in-law" culture, still in direct contact with the Nefilim, know the masters to be flesh and blood humanoids and definitely in control. Kings are put in charge when the populations get large and the candidates for this foreman position are often very capable but aware of the danger of mistakes. The rules are simple: do exactly as you are told as a foreman or you die, or maybe be lucky and just get exiled. The Sumerian word for worship meant "work for," or serve. And, from the beginning that we did, including acting our servant parts in the palace, serving the masters the coq au vin, the beer, the wines they loved, the fatted calf, the bull and grains.

When our females became attractive to the Nefilim males and they began to collect harems of these desirable female slaves and beget children by them, we became too much of a nuisance-threat and the Nefilim attempted to wipe out all of us in the Flood and afterward, rethought their decision. Having decided to keep us in rebuilding their centers after the devastating Flood, they taught us "crash courses" in civilization and technology as necessary for specific functions, farming, husbandry, textile production, food preparation, beer and wine production, mining, construction, slave raiding, scribing, keeping the calendar, knowing how to tell when the home planet, Nibiru, would again pass within the inner solar system and the Nefilim return.

The Beginning of the Traumatic Transition to Independence

And then they phased off the colony planet. Pretty much just left without closing the laboratory door, apparently beginning around 1250 B.C.. The foreman-kings are suddenly depicted in the stone carvings standing where they used to stand when listening to instructions from the master, pointing to the master's now empty chair in utter dismay. The laments are still engraved in the clay tablets: "What do I do when my master is no longer here to instruct me. . .what shall I tell the people." We went into grief, despair, denial. We blamed ourselves and looked to the sky for their return. The good kings did their best, the leaders sometimes were told to go up the mountain to get some instructions long distance from space or make a wooden box lined with metal, just so, to act as a receiver. "Now hear this." Finally we were alone and in confusion, beginning to fight over who still knew what the master really said, really meant, what we really should be doing if he did show up. Service at their table transmuted into ritual sacrifice of food; attendance at their baths turned into bathing and clothing of surrogate statues of them; the routine services gradually turned into cargo-cult rituals and their palaces became empty temples. And "religion" became.

And the less than good kings began to take advantage, began to swagger. Sometimes they got away with it on their own. Sometimes the people, in desperation, raised their king to a symbolic god. And the god-king and the seed of the notion of the divine right of kings became. The chief servants went along with it because it was to their advantage to become known as priests or to preserve their jobs and status. And those who had been taught, seeing that the advanced knowledge of technologies, science and the arts, learned as part of their function, of writing, mathematics, astronomy, science, metallurgy, and the fine crafts in general was being lost, set an agenda to preserve it. In the face of misunderstanding and threat they disguised it, withdrew it, hid it. And the "occult" became.

Slowly a classic disassociational process developed due to separation in time and we began to sublimate the flesh and blood Nefilim into cosmic absolutes, and their personalities into mythic archetypes. Looking back over the history of our species, the traumatic transition we have gone through might well be characterized as the creation of the concept of a cosmic God by us through a series of psychological mutational phases. Eventually, we simply began to forget. And religion and theology became.

The transition from racial amnesia to racial maturity has been a very traumatic passage. Breaking the godspell has seen us go through the stages of abandonment, to disassociation, to sublimation, to religion, to rebellion and now to recovery. It is the classic syndrome of the dysfunctional family on a planetary scale. Are we not haughty Egyptians, to whom the sky gods were everything, our rulers weakening and abandoned, now fearing that we had done something very wrong, searching the stars for some sign of their awesome craft? Are we not Hebrews, the chosen of our Nefilim master, Yahweh, preservers and transmitters of the advanced technological knowledge entrusted to us, becoming dismayed that the static is getting so bad that the words of the retreating Yahweh from the Ark loudspeaker can hardly be distinguished and Moses, the last to see him face to face, is aging and even the nabi, the seers, argue about what he said? Are we not Inanna's women of Mohenjo-Daro, for long sure of our dignity and her patronage, suddenly abandoned by our goddess queen? Are we not enlightened and democratic Greek citizens, logical people who could hardly accept the impossible deeds of the now-remote gods as anything but uncertain fantasies? Are we not Romans, sophisticated and urbane, masters of our destiny and the known world, who would give anything but lip service to the ancient deities, now almost abstractions on our walls? Are we not Medieval Christian theologians already abstracting and sublimating the ruthless, peevish, jealous humanoid Nefilim Yahweh into an infinite, omnipotent, omniscient, cosmic being beyond our capabilities of thought, who holds the universe in being just by thinking about it?

Where was toil and danger? Down there slaving in the hot gold mines. And hell became. Where is the god and the good things that come from the god? Up there in the sky. And heaven became. Are we not the young Catholic priest telling his docile parishioners not to read the Old Testament and leave the interpretation of Scripture to the clergy, trying privately in anguish to reconcile history with his faith? Are we not the fundamentalist bible school teacher watching herself mold the minds of children to a tradition of unquestioned docility to a God, her own doubts about whom she dare not allow herself to think?

And so, down to our day, incredibly, we have remained still Babel-factored for good crowd control, broken into tribes, each proprietarily telling the other that ours is the only accurate tradition of what the god intended, what rules to follow, what we should be doing to demonstrate that we are still loyal and docile servants. Sometimes we just kill each other over it. And persecutions, crusades, jihads, Inquisitions, the saved and the damned, the martyr, the infidel, the saint, the Protestant, the fundamentalist, the atheist and evil empires, became.

A very serious question is: Do other planets in other galaxies have the same overarching phenomenon of alien humanoid master/creators-> gods-> god-over-gods-> God-> GOD-> CREATOR OF UNIVERSES transformational sequence in their history because they were genetically engineered also? The convolution, sublimation, reverse anthropomorphizing (anyone have a better term for that? maybe anthropodeification?), intellectualizing, abstraction, that the notion of deity has gone through over the last 3000 years is, to be wryly positive about it, a tribute to human genius. . .

Atheism Revisited

To clear up an important point: it might well be assumed, in light of what I have said so far about institutionalized religion, that I am preaching atheism. That is not inherent in the case I have been making. The tendency of humans towards the transcendental has formed a body of speculative information regarding an ultimate cosmic principle that is, at very least, worth investigating. It is my own opinion that our attaining to an unified field theory, a theory of everything, will throw some light on that topic. So it is worth reiterating here a point that I have emphasized before: breaking the godspell does not mean atheism as in any of the ways that we have defined and thought of that phenomenon especially in Western culture. It does mean that we must reevaluate and discard institutional religions. It does mean that the theologies of the world religions are false. It does mean that the God, Yahweh, of the Judaeo-Christian-Islamic tradition was an alien humanoid, not some infinite cosmic God. But it does not mean, intrinsically, that there cannot be some unthinkable principle that causes the Universe and is capable of creating an infinite number of universes and playing games with them. Whether there is a principle of such status is a question that is not necessary or probably even possible to attempt to answer definitely here. I recognize that that fact may, at first, seem contradictory or at least confusing to the reader. But atheism, at least in its usual form, is a denial of and/or protest against a deity whose characteristics are all too human, who deals in an arbitrary and peevish way with "his" subjects, who often seems unconcerned with the well-being and fortunes, even the lives of his most loyal subjects and is said to punish with gruesome torment, serious infractions of his rules for all eternity. That kind of image derives from the anthropomorphic characteristics of the God of the Old Testament and the New within the context of Western culture. The rejection of that "God" by atheism is justified and correct in historical and archaeological perspective. Garden variety atheism can now be understood as an early sign of racial adolescent rebellion and questioning of the authority of the obviously all too humanoid characteristics of the particular local Nefilim "god" or "gods." It is simply a long overdue correction of some local solar system politics, relatively rather pedestrian in cosmic perspective. In the West it takes the obvious form of rejection of Yahweh who was imperfect, limited and, by his own admission, jealous (paraphrasing the first Commandment: I am the Lord, your god, and you shall not go over the border and work for my brothers and cousins. . .) and clearly male chauvinist and disposed to violence in dealing with his human slaves if the high-tech tricks of magic and high wattage loudspeakers to which he resorted to intimidate them didn't work.

The new paradigm, once the godspell is dispelled, simply frees us to go one on one with the universe and to seek directly whatever unthinkable or thinkable ultimate principle is behind it. Probably everyone, at some time in their lives, has speculated privately what it would be like to be an "atheist," an "agnostic," an "unbeliever." But those are all basically negative positions, or defensive attitudes. In the new paradigm the atheist who, formerly, denied the existence of a Yahweh-like god can only now deny and the agnostic only express uncertainty about, the existence of some transcendental principle as philosophical convictions. The classic conflict between the believer and the atheist or the agnostic about the existence of the Judaeo-Christian-Islamic – or any – God will be no more. Unbeliever, a title usually bestowed by believers, will no longer have meaning. Because there will be no more institutional or private religions demanding acts of faith in an unseen god. How soon? Probably not very soon at all. It will take generations, it seems, to effect the transition so deep is the change required. Think about what it would take just to dismantle the Catholic Church as an institution, much less as a belief system. The Pope is not going to accept early retirement easily even if he does like to ski. By the definition of "atheist" we have used here, the new paradigm is not atheistic. But what of the position of the atheist that not only is the notion of an anthropomorphic god false, but that there is no ultimate Principle behind the Universe? The question of a transcendental principle, discoverable or not discoverable, comprehensible or non-comprehensible by us, regardless of its ultimate answer, is different and not resolved, inherently, by the new paradigm. The new paradigm resolves only the paradoxically rather pedestrian intra-solar system politics of the real nature of the Nefilim and their relationship to us as our genetic creators. Although it does shed light on the real nature of the theological doctrines we have developed as sublimations of the old master-slave relationship with the Nefilim and even why we have an evolutionary trajectory toward transcendence, it does not deal with the existence or non-existence of some ultimate cosmic principle. It does open the door to a one-on-one relationship of the individual with the universe in a new dimension of freedom.

The New Paradigm and the Religions of the World

It is not simply that religious institutions and individuals committed to their particular religion feel so fundamentally threatened by the new paradigm because it clearly demands a complete turnabout on their part, the godspell mentality is embedded deeply in human nature by this time – so much so that E.O Wilson considers religious proclivities to spring from a neurological source. As individuals we are practically imprinted at birth with this subservient view of the universe in whatever context, religious or secular, it happens to infect us through our parents. Our family or social context does not have to be explicitly affiliated with a specific religion: the godspell is an attitude, an orientation to reality and the universe peculiar to our planet. And neurological imprints, as we have come to learn, are normally difficult to access and to change. We walk around as if everything was just all right and human existence was a richly pleasurable event. But the problem that few seem to want to face is that we are fundamentally unhappy because we do not consider the Universe ours. The conditioning we receive from the time we are born or before is that it is someone else's universe, God's in most cases, and we are here by permission and we had better act just so or we will not only be censored but, in the minds of some, punished. We think that we can demand nothing from the universe which seems impersonal and often threatening because we have no concept of our having a right to anything in this universe, not existence, life, material survival, happiness or information about how the universe really is and works.

Even those with no religious convictions at all are effected by the implicit underpinnings of the residual belief in some kind of Supreme Landlord. This is the lingering effect of the godspell in its most subtle and pervasive form, the deepest cause of the unhappiness so widespread in the human planetary community. It not only separates us from one another but it reduces us to cringing impotence, feeling alienated from a universe of which we are an integral part. No one even articulates it thus anymore, but the certain implicit concept is that the gods own it all and we had better watch out and not overstep our position as non-gods. Nietzsche, intuitively, called it "slave morality." We can call it the godspell because we now have the information at hand to identify it and its effect and throw off its planetary tyranny once and for all. It will require us to literally start all over again as a planetary society and as individuals.

The godspell mentality is, unquestionably in my mind, the most fundamental flaw in our planet's culture. It is the most pervasive, most fundamentally deep element at the root of war, conflict, division, and misunderstanding among humans. It is such, precisely, because it is essentially a complete falsehood and the use of "spell" in the word godspell is nearly literal. It is an ideology based on millenniums of fear, deep conditioning, further and further separation from the physical reality of our inventors and the truth about our invention, and sublimations on top of sublimations of the Nefilim into a Supreme Being. It is an elaborate, false, hollow intellectual construction that still forms the skeletal framework of our planetary cultures, manifesting as religion and theology, and reinforced by strong emotional evocation. It cannot be proven reasonably and, therefore, there is "faith," the belief in that which cannot be seen, usually justified on the basis that that which is believed in is transcendental and/or ineffable.

The most charitable judgment on the phenomenon of the godspell mentality is that it was probably inevitable, given the Nefilim departing the planet, depriving us of the direction and teaching we had known under their domination, thereby forcing us to do for ourselves. We have been explaining, adjusting, arguing among ourselves and killing each over who knows what the god(s) really want from us and what we are supposed to be doing as humans for some three thousand years now and the process has always been stultifying, very counterproductive and horribly tragic in the vast number of humans who have lost their lives in the name of one god or another.

But now we are beginning a profound transition to a new time, a new era in human history. In that time we will finally come to live as humans beyond any form of religion as we know it today. We are already into the transition phase. We must make the elements of that transition clear, understood as a characteristic of our evolutionary direction, so we may facilitate it as consciously as possible.

The Final Stage of Transition

We have discussed in detail the nearly three thousand years of traumatic changes we, as a new species, have gone through dealing with being left on our own by the Nefilim and having to find our own planetary way. We are now reaching a plateau of racial maturity which marks the end of that transition. We are entering the end-game phase. As with most profound revolutions in our world views this phase will be marked by confusion, turmoil, desperate attempts at return to the old ways and conflict, all of which, at times, will become very intense as the pressure for change steadily increases and the reality of the new era becomes clearer and clearer. But this will be counterbalanced and overshadowed by the promise of relief from divisive strife, ignorance, absolutist doctrines and, even more, the vision of human existence with a known history, a full understanding of our true bicameral nature, and the dawning freedom to determine our own evolutionary trajectory as individuals and as a race.

Currently, the processes of transition are accelerating. The weight of evidence for our real beginnings and the real nature of the "gods" has already begun to effect institutional religions severely; the attrition rate can only be held in check if the masses are held in ignorance or intimidated. The conservative, reactionary, fundamentalist elements, embarrassing and troubling to those of a more sensitive bent, dominate more and more as the threat of total dissolution grows. Because it is, literally, our cultural base to which we all, so far, have been conditioned, the removal of that godspell skeletal structure is and will be resisted fiercely with instinctual survival reactions by religious institutions and individuals. But this is their final stand. We are about to move beyond religion as we know it.

CHAPTER 8

BEYOND THE OLD NEW AGE

The New Age vision is of a new planetary civilization coming about through the gradual dissemination of expanded consciousness bringing about the heightened ideals of brotherhood, tolerance, benevolence, ecological balance and a kind of natural spirituality.

The new paradigm reaches even deeper. It begins with a redefinition of human nature, establishing the commonality of human beings on the basis of a generic genetic base. All of the generally held ideals of the New Age are then seen to be manifestations of the potential brought about by the new paradigm. The New Age sees a "new" human, enlightened by assimilation of high spiritual ideals but the real nature of our racial genesis, the radical redefinition of what human nature is, is not in focus. The result can only be the intellectual rehashing of a generic kinder and gentler neo-Scholastic theology, recast in modern psychological terms such as found in Capra's *The Turning Point*. The New Age vision relies on an individual psychological conversion rather than the unifying power of our generic genetic commonality to bring on the new planetary civilization. The result can only be a "third culture" rather than a profound planetary transformation.

A reciprocal process has always been involved in our torturous coming to racial genetic enlightenment: the more we learn the freer and more confident we feel to learn and accept more, to shed the totems and taboos of the ancient godspell. At this point in time we quite clearly are reaching the point of radical paradigm shift. When we get this close to a paradigm shift changeover point everybody can feel it coming in their bones, in their genes, can almost taste it even though they may not be able to articulate it. So the predictions and the guesses and the prognostications about what the new paradigm really is are rife and various.

Some say, variously, the major changes will be focused around the shift into the precessional age of Aquarius; the paradigm shift is already on us and it means that the "vibrational level" of the entire planet will increase to a higher frequency generating a higher human consciousness; the actual crystalline form of the earth is shifting to a more complex form; the magnetic field of the earth is diminishing and magnetic pole reversals will trigger it; it will happen finally when aliens land and go public and, hopefully, offer us the solutions to all our problems; it's coming will be a monumental intellectual one as we decode the prophecies of the ancients; we will shift from the modern interpretation of the ancient world to a recognition of the identity, nature and advancement of the ancient civilization that gave rise to Sumer, Egypt, India, the Mayan empire and other western ancient empires.

Terence McKenna, having programmed the cycles of the I Ching into the computer (McKenna, 1994) can see a startling slide into novelty coming rapidly as we approach the year 2012 and sometimes says the paradigm shift will be a transcendental dimensional shift, that the world as we know it will end and sometimes that he is not really sure. He, in one sense, is using the I Ching as a doomsday machine.

Some, like Marilyn Ferguson, (*The Aquarian Conspiracy*, 1980) say that the big change is manifesting in a rising groundswell to a new level of human benevolence, kindness and love leading to a planetary order of peace. Some say that the paradigm shift will be more political, the manifestation of a new world order leading to harmony and peace.

It takes little reflection to notice that the primary characteristic that dominates all these interpretations of these prophetic perceptions may be called, generically, a profound racial consciousness expansion. I suggest that what we are witnessing and experiencing, in the largest perspective, is nothing less than the dawning of a planetary, racial, genetic enlightenment, that these gods, we and our children, wear designer genes, that we can and should claim our planetary birthright, restore our true collective history in the final dispelling of the haze of racial amnesia. We can integrate our indigenous Homo erectus heritage and become one with the earth again at the same time as we integrate our off-planet Nefilim heritage and move off into space.

As we step out of racial adolescence, struggling through the awkward but inevitable separation from the parent-gods, now aware that our mythology is our greatest myth, we will realize that the myriad predictions, prophecies and pop-eyed pronouncements of turmoil, danger and glory are substantially correct projections of the difficulties involved. The glimpses of glory are intuitive projections of the facets of the character and personality of the new planetary human and the new civilization already brimming over the horizon. The key to graceful passage is an unassailable integrity, both racial and personal, springing from the genetic enlightenment that erases the painful scar of the subservient, godspell slave brand from our personal genetic web. We need, respectfully but firmly, to finally wash the ancient dyes of subjection now sublimated to a focus on a pitifully anthropomorphic projection in the sky from the tapestries of our cultures. We are not dealing with just a change from the patriarchal to the matriarchal, from male dominance to female or partnerships of equality, from unconscious to conscious archetypes, from materialistic science to scientific magic, from West to East, from establishment to New Age thinking, from retrograde nationalism to some new world order. All of those elements are just symptoms of a far deeper revolution.

We are dealing here with nothing less than a profound turning point in the history of this planet. We are dealing with the concept of a new humanity, a new human planetary civilization. The new paradigm, I suggest, is precisely the unexpected but actual, overarching vision that will bring about a new age that will be the fulfillment of the New Age vision and yet far more in that its affords the common ground to subsume all intellectual, scientific, philosophic and New Age world-views and complete them in a transcendent unity. The Aquarian conspiracy is turned into a planetary celebration.

There is a certain resistance to the new paradigm in the New Age community because those who are deeply committed to the new age vision intuitively or intellectually recognize that, if the "gods" were real historical humanoids then the fundamental premises and criteria of the New Age philosophy must be radically reexamined. If the new paradigm is correct, then the usual interpretations of the Cabala, the mystery schools, the occult, the Eastern and Western metaphysical doctrines and the Goddess movement, that the "gods" were "divine," or manifestations of transcendent being, or consciousness with a capital C, or Jungian archetypes springing from the collective unconscious, or interdimensional entities or mythic beings is simply as partially incorrect as institutional religion, Creationism or evolutionary theory. There is the unfortunate possibility that, if the New Age community holds on to the mythic interpretations of the gods as Jungian archetypes according to Joseph Campbell's modern mythos of myth and to invest the Goddess archetype with religious significance, the new paradigm, which is the only one which can subsume and explicate all the previous partial paradigms and resolve the differences between them, will be excluded from the voices heard on the edge of tomorrow. In that event the true fulfillment of the New Age vision will not be realized and the New Age community tomorrow will only be a better version of today.

It is a beautiful paradox that the full fruition of the New Age expectations will be found through the science that has been generally rejected as too materialistic and that the confirmation of the vision will come through an enlightenment which will eventually be verifiable through genetic analysis.

Halfway There

The godspell in psychology is elegantly expressed by Marvin Spiegelman in his book *Buddhism And Jungian Psychology*: the premise that the gods were myth is at the very heart of Jungian psychology; myth is the very stuff of which the analysis of dreams and fantasy is made. But the new paradigm shows that the gods who initiated the religious concepts and attitudes in humans, of myth and all of the symbols and imagery associated with them are real. The entire substructure of classic myth is the foundation of the form, context, and content of the major and minor religions of the world. The conflicts and guilt experienced by those who become involved with and attempt to reconcile the practice of psychoanalysis with religion are the result of the concepts of the numinous, the desire to be ecumenical. The choice is not simply between scientism and religion. The problem is not solved by accepting as a given the old godspell premise that the divine transcends us all and that all religions are simply just different paths to the same goal – especially without taking into consideration the nature of the practices and the doctrines of the various religions.

The nature of religion can now be seen in a new psychological light. The so called mythic elements of human nature can not be lumped together as such under the unspoken assumption that they are all generated by the human psyche and therefore just facets of the same underlying drive. Some of the facts are history. And, if the history is true, then the genetic component is the key to the images, the dreams, the fantasies. And the images and fantasies and dreams take on a new and more meaningful significance and should be dealt with as such without some lame framework of reference that does not redefine and reevaluate the primitive, outmoded religious context.

Breaking the Goddessspell

The Goddess movement, which has subscribed to a new concept of the status of women, looking to the ancient world and its legends for reinforcement, is a primary subject for which the ramifications of the new paradigm holds profound significance. Well known authors and academicians representing this focus are Merlin Stone (*When God Was a Woman*), Maria Gimbutas (on the Goddess movement), and Riane Eisler (*The Chalice and the Blade*), among others. Their interpretations and opinions of the archaeological and historical data carry a great deal of weight because of their depth of research and scholarly approach. The reason for the attractiveness of their thesis is that it throws into perspective the neglected, even denied, status and role of women in history and points to a time in the past when there was equality of the sexes and peace among humans, looked to as a model for our time and our future. But many in the Goddess movement view the new paradigm with suspicion because they feel that it diminishes what Maria Gimbutas has interpreted as the "supernatural" element in the worship of the Goddess. On the basis of the truth, whatever it turns out to be, sets you truly free, I suggest closer study shows that the two paradigms mutually reinforce and the new paradigm ultimately affords an even more profound meaning to the Goddess concept.

In espousing the historical interpretation of the Goddess phenomenon in the semi-religious form currently popular, those involved in the movement have also accepted the implied Darwinian evolutionary theory as a basic assumption. The development of the Goddess cult over the ages is set in the context of our racial evolvement from the most primitive recognition of woman as life-bearer, through the hunter-gatherer phase to an agrarian phase to civilized centers of culture to us. In developing the theory of Goddess worship the various scholars involved in the field have all taken current evolutionary theory as a basic assumption in the explanation of the evolvement of the Goddess cult, from the most primitive recognition of woman as life-bearer, through the hunter-gatherer phase to civilized centers of culture to us.

But evolutionary theory, whether Darwinian, Lamarkian or some variant of these is still in question, in fact rejected by a number of thinkers in our time. If either the fundamental concept of evolution from less to more complex was questioned or the fundamental concept of the new paradigm that we, specifically, are a genetically engineered species rather than any kind of product of natural selection or learned process, some interpretations of the female cult focus prevalent in the Goddess studies field would be thrown into question. A critical question to be addressed in this regard is whether there is hard evidence for what is interpreted as a cult of the female previous to our genetic genesis 250,000-300,000 years ago.

Maria Gimbutas held that, so far, we can't see evidence for the goddess cult much more than perhaps 25,000 years ago – although she "guessed" that it might go back as far as the consensual date of the beginning of the Paleolithic era some 500,000 years ago. I see no contradiction in any form of the new paradigm by the evidence so far adduced in that regard.

The new paradigm provides an overarching context in which to attain a much more accurate explanation of several major features that are involved in the Goddess religion theory: the mythological interpretation of the evidence; the interpretation of the evidence as indication of a supernatural religion; the lumping of all particular goddesses into a single amorphous Goddess; the interpretation of the cause of the apparent "golden age" of peace among humans; the nature of the catastrophic revolution that altered the social structure to male dominant. Because it is doubly dependent on Darwinian evolutionary theory and mythological interpretation, those involved in this phase of the New Age development often find it most difficult to accept the new paradigm. But they are so close, sometimes without realizing it. In the context of the new paradigm, it is no longer necessary to attempt to deduce what human men and women thought in those ancient times, no longer necessary to attribute religious, cult or superstitious status to the bits and pieces of the phenomena we have rediscovered on the strength of our inter-pretations and guesses, no longer necessary to impute some psychological mind-set based on our own judgment to those distant ancestors when the history is before us. If the new paradigm is correct then two major erroneous conclusions drawn by those developing the Goddess theory may be corrected.

From Goddess Archetype to Real Goddesses

Because the consensus of the scholars studying the Goddess phenomenon hold to the evolutionary context, it is easy and natural to project a mythological character on the Goddess in her various forms since the implicit assumption is that the earlier we find humans in history the more naive and primitive they would be and therefore all the more inclined to mythologize. It is with regard to who Inanna — and the other female "gods" — actually were that is the basis of the second serious error of the Goddess theory. The general interpretation of the en-tire scope of the traditions of the "gods" by those developing the theory is based on the consensual academic interpretation of the "gods" as mythological, unreal, projections of the naive psyches of human beings. By extension of this interpre-tation of the Goddess phenomenon as mythological, it becomes easy to conceive of the goddess phenomenon as a single continuum of gradual development cen-tered on just one Goddess, from the earliest crude Paleolithic figurines through the millennia to the very detailed images of specific goddesses in the middle east and Crete. An end result of these interpretations as mythic projection of the primi-tive, naive human psyche in an historical continuum is that the individual god-desses recognized over the millennia as distinct beings, often in very different contexts and times, often with very idiosyncratic differences between them, some-times as different spouses of different male "gods", are freely and generally taken as simply different cultural names or aspects of the Goddess interpreted as an archetype. I believe that we can already see sufficient evidential correlation to say confidently that the basis of the goddess mother cult in the earliest times was a recognition and veneration of the Nefilim female geneticist, Ninhursag, who lit-erally created us in the laboratory. Both the "in-law" and the "out-law" cultures would remember her equally. Gradually, over the millennia, as we became either the subjects of or at least very aware of the various Nefilim females of higher and lower social and functional rank, they also were venerated on a more local basis.

A problem lies in the fact that, taking the gods as myth has allowed the vari-ous female Nefilim with their disparate personal characteristics and foibles to be lumped together as the amorphous Goddess archetype, an innocent but signifi-cant error. Because of this amorphous melding into the goddess archetype, the less than ideally feminine characteristics of some of the Nefilim women tend to be glossed over. Innana provides a good example. She, as a member of the ad-ministrative council of twelve which oversaw the entire planet and a pilot and far ranging traveller, was very widely known to humans. Although idealized as a mythic goddess, as a real flesh and blood person she was capable of demanding the death of human men quite ruthlessly when they refused to become her lovers. Her political ambitions caused her to manipulate in such heavy-handed ways that current slang would term her a "bitch on wheels". Yet she is also the one who probably gave us tantra yoga and the ideal of the fully independent, self-realized woman.

The recognition of the Nefilim as real allows us to see that those "goddesses" were actually individuals of an advanced culture and why, though they were humanoid, humans saw them as superior or at least in a position of superior authority. Some were highly respected and loved by humans, others less so. Some were of very high rank and authority, some were lower echelon technicians, specialists and officials. But they were Nefilim and thereby set apart from humans.

Understanding the veneration of our Nefilim female rulers as both compulsory and conditioned by the relative attractiveness and benevolence of their individual personalities throws light on two other facets of human existence when the Nefilim were here on the planet. Humans, in any given area, did what they were told by their local Nefilim ruler. If there indeed was a golden age of sexual equality it existed because both male and female humans were in an equally servile position: you don't get to dominate as females or as males when you are being dominated by a powerful authority. The time when male dominant groups began to overrun the scene, described as catastrophically disruptive of peaceful human society by the goddess researchers, coincides with the time when the Nefilim phased off the planet. When the cats away...

But there were male dominance problems before the Nefilim left on a local basis. The most significant example is the male dominant characteristic of the Hebrew culture. Yahweh, ruler of the territory in which kabiru, out-backers in trouble, had entered into a contract with him to be his servants in return for his protection were indoctrinated in a social structure in which women were considered pretty much as furniture. If you are going to set your banner in front of your human armies and play GI-Joe with them for slave raiding from your cousins and to assert yourself as a Nefilim over all Nefilim you are going to give preference to the male grunts.

By contrast, the women in Innana's territory in the Indian center she demanded her Nefilim male counterparts put her in charge of, were esteemed, given equal privilege both politically and sexually. But, as has been pointed out by goddess researchers, they were sometimes killed for their independence if they were unfortunate enough to enter Yahweh's territory.

A Slightly Tarnished Golden Age

There is nostalgia in the Goddess movement for the persistent notion of a past time, a hazy "golden age" when there was equality of the sexes, no conflict of major proportions, a time of grace and harmony generally among humans. This fortunate state is usually attributed to the fact that the influence of the female gender was sufficient and beneficial to contribute a modifying and mellowing influence over the minds and hearts of humans. In the context of the new paradigm, with no denigration of womens' influence, the status was such in those times because the Nefilim were absolutely in charge and humans and human kings (the Nefilim's human foremen) had no choice but to do exactly as they were told. You simply do not get to exert dominance as such, either male or female, under those conditions. This condition would have been most probable when we were just developing and stabilizing as a species. Once we had become developed and precocious enough to even become partners with the Nefilim obvious differences in different localities became apparent. Innana treated women preferentially in her locality. Yahweh treated men preferentially in his area. And the attitudes of these Nefilim officials was mirrored in the relationship between the sexes, either by imitation or by mandate in their respective areas.

When Left On Our Own

When the Nefilim finally phased off the planet, left us on our own, the power struggles began. If anything, Nefilim society itself, from the accounts, tended to be male dominant, although the equality between the sexes was obviously recognized. Inanna, although not originally of the ruling group, was a Nefilim female who was intelligent, versatile and ambitious. She, by her demand, was actually given ruling status and put in charge of the third major Nefilim center founded in northern modern day India. Her character formed the character of the human society of that area and the civilization that arose there — a radical contrast to the character of her compatriot Nefilim, Yahweh, whose character created and influenced the male chauvinist, women-as-furniture milieu of his Hebrew slaves.

Due to the Goddess movement and the New Age in general having accepted these two partially incorrect theories from the scientific establishment, the error is further compounded because the romanticized mythic interpretation of a Joseph Campbell becomes seductive. The new paradigm does not denigrate the scholarly, independent and impressive work of the goddess researchers. On the contrary it affords a context in which an even more robust and dignified appreciation of our true history and womens' role in it. Breaking the ancient mythic goddesspell is necessary to arrive at the more profound realization that Goddesshood is not just a high metaphor; it is literally genetic.

Male dominance, even though it has a deep base in the various godspell religions focused on Yahweh who taught his male subjects to treat women as inferior, cannot survive genetic enlightenment. But we won't have to wait for those religions to disappear totally — the process is already well underway through attrition and suspicion - - for it to diminish. Even though the word "obey" has been eliminated from the marriage vows only recently and the Pope is not going to accept early retirement easily, it is going to happen faster than anticipated.

When DNA speaks, everybody listens. The godspell does not yield easily in any form. The worst thing we could do is to turn the new paradigm into some new form of "religion". In the context of the new paradigm, the deep and beautiful intuition of the "goddess" nature in all women is more correct than even realized: it is the upwelling of our genetic, half Homo Erectus, half Nefilim heritage telling us that we are indeed more than we have been programmed to believe under the godspell restrictions to subservience. And we shall be, female and male, our own goddesses and gods playing our own god games in an equality truer and more profound in that it springs from the ultimate knowledge of our true beginnings and nature and potential, from the very depths of our genetic code.

The Awakening of the New Age Mind

The essence of the New Age movement as it developed in the nineteen sixties was consciousness expansion inspired by the psychedelic experience. (It was not, as the establishment documentaries would portray it, the anti-war protest phenomenon which was only a single result among many of that heightened consciousness.) Those undergoing profound experiences of many different states of human consciousness quickly recognized that they were spoken of in various metaphors by the so-called occult disciplines and the Eastern philosophies. But, almost uniformly, the occult sciences ran true to their name: they were obtuse and difficult, spoke obliquely and often gave evidence of speaking about things profound but, by this time, obfuscated or simply forgotten or lost. After all, the word "occult" means hidden and they were such for good cause: they had gone into hiding to avoid persecution during the rise of hostile and authoritarian religions. But the New Age caught the implications of valuable and profound information and inferred, sometimes, more than was even implied. What was decipherable often proved quite valuable and this reinforced the quest for the essential secrets. The end result is displayed in the New Age book and craft store from the familiar scent of incense, to the Eastern mystical literature, the books on alchemy, the use of crystals, Tarot cards, yoga, tantra, consciousness raising techniques, new physics, alternative medicine, mystery schools, meditative and ecstatic music, to healing and healthy dieting.

The Occult Revisited

The new paradigm opens up the doors on the hidden vaults of the occult. It is now clear that these venerable traditions, long recognized as valuable and advanced but obscured, are the residual remnants of high technologies originally taught to us by the Nefilim either for convenience so that we could do the manufacture involved to serve them or for our convenience then and later. The critical factor involved is that the new paradigm, when accepted, allows us to recognize that these bodies of knowledge and sciences and the technologies springing from them could have existed in the past. It is clear that humans were not advanced enough in those early times to have invented these sciences and techniques themselves but they obviously have had the potential to learn and operate them once they were taught. And the new paradigm gives us a clear directive as to where and how to re-search for the true essential data of those venerable technologies which we were taught and trained to administer, operate, produce or practice. There are striking examples of the rediscovery of the real nature of lost sciences and techniques already clearly before us. Stan Tenen's rediscovery of the nature of the Hebrew alphabet system, the decipherment of the Mayan system of astronomy and its purposes by Cotterell, David Hudson's rediscovery of the alchemists' grail, the monoatomic form of gold, the rediscovery by Chatelain, Munck, Torun and Hoagland of the ley line physics on this planet and Mars are the more spectacular examples of the uncovering of the advanced technology already possessed by the Nefilim millennia ago and now being reclaimed by us. There is still more to be recognized and examined and assimilated. The next decade will see an acceleration of that process and great benefit from it because it will shortcut some of our own investigations. We will examine and evaluate the Nefilim legacy in detail in a later chapter.

The New Age and the Nature of Consciousness

Many in the New Age movement, having rejected the shallow pietism and constricting dimensionality of western Christian theology, constraining rationality and materialistic science, espouse as far more adequate, sophisticated, and enlightened some version of Eastern metaphysics and religion. They either are not aware of the new paradigm or, aware of it, ignore its implications. Some of the New Age persuasion are secretly in embarrassed conflict and denial faced with the possibility that those doctrines also are the latter day sublimations of things taught by, plagiarized from, copied with less than perfect understanding from the teachings we received in ancient times from the Nefilim in charge or territories in the Far East. Some, as with the rest of the population, are simply at a standstill because of the puzzle of how to chose a criterion by which to judge what is correct in all this. The rebellion, in the sixties, to narrow and authoritarian mind-sets and a dry rational approach to reality often led to a rejection of all things academic and "intellectual." More often than not this was justified and necessary in the short term. In the long term it tended to cut one off from the essential flow of new information especially if a strong set against "head trips" caused one to reject new ideas out of hand and to rely too heavily on only feeling and intuition. Paradoxically, those who did so fell behind by losing touch with authentic new ideas and insights. After all, the concept of the New Age and consciousness expansion were "new ideas" when they engendered a tidal wave that would change our minds forever. The rejection of the narrowly logical, the questioning of the perceived materialism of science, the avoidance of "head trips" that ignored or neglected the body and the physical, and the insistence on holistic approach to the universe were and are correct. But the attraction to a Campbell style view of the potency of myth combined with a tincture of the old godspell religions still allowed and allows those so disposed to accept the metaphors of "spirit" and "spirituality," "transcendence" and "God" in a sort of amorphous amalgam of re-hashed Eastern and Western mysticism and "religion." One result of this eclectic melding of what is perceived, implicitly as a rule, as the "best" of both traditions is the rise of the semi-theological, psychological dynamics practiced by New Age – sometimes called alternative method – therapists.

Stanislav Grof: Beyond the Brain

The paradigm of the human condition presented by Stanislav Grof provides an excellent example of how all the, by now, classic New Age alternative worldview factors are amalgamated into a new synthesis. This is not to criticize Grof's world view or his work because, indeed, his thought is penetrating and his therapeutic techniques very effective. I reference him and his work here only because it is so completely representative. The precision and clarity of his thought and exposition, most importantly, elucidates how the recognition of the new paradigm completes and enriches the highest ideals and vision of the New Age. Grof's family background was atheistic, he was attracted to Marxism, and formed by the materialistic orientation of medical school. Although he became interested in Freud because of what he could do to religion through the notions of totem and taboo and obsessive compulsive neurosis, etc. he eventually, through his own experiences, realized Freud had missed the boat by associating religion with ritual. Through the use of LSD as a therapeutic technique he became aware of the metaphysical and philosophical significance of non-ordinary states of consciousness. He realized that Western thinking was dominated by Cartesian, Newtonian thinking. Some areas of science such as quantum physics, he came to see, create and work from an advanced concept of the universe. But Psychology, Medicine, Psychiatry and Philosophy do not as yet. These traditional disciplines tend to negate observations of theories that do not comply. What he has termed Holo-transpersonal experiences do not happen in their universe, leading to a blocking of communication. All non-ordinary states of awareness are understood in their context as psychopathology. Traditional science is culture bound, cogni-centric, and ignores data from non-ordinary states, pathologizing the best parts of human history. He became intrigued by the difference in experiences, data, insights, and how one dances between the two realities. He noted that shamans are experienced dancer and can be very comfortable in both. A turning point for him was a reluctant interview with the guru Muktananda which turned out to be an extraordinary non-ordinary consciousness experience for him personally. He began to investigate the various yogas and Buddhist map of reality. He recognized that there was a very developed tradition of non-ordinary consciousness extant in the far East and became aware of kasmir shaivism through two intensive seminars and noted the similarity between that metaphor and the one which he had experienced through the LSD work. His conclusions about non-ordinary states tend to be couched in the mixed terms of advanced science, shaivism, and psychotherapy against the background of a cosmology that is predominantly derived from a sophisticated shaivism. This cosmic view holds that the One becomes many by self-division in a transcendent self-delusion and "forgetting." Shaivism holds that it is a waste of time collecting information because what you are experiencing is part of the story and, as Bohm has claimed, it is an illusion that science is telling you anything about the universe. He accepts Campbell's view that the experiences of deities, gods and goddesses that some people report are culturally based and the archetypes are forms, creations on another plane, is attractive to Grof. Ultimately, one must turn inward, because things out there are not causally related, just a higher order screen play. The solution, Grof holds, is found in cybernetics, systems theory, the work of Bateson and Sheldrake. As noted above, Grof's philosophical stance and practical work is an excellent and comprehensive example of the New Age synthesis: he combines the concepts and data of our most advanced science, the constructive use of the psychedelic chemicals, the positive aspects of non-ordinary states with the concepts of the ancient Eastern traditions and applies the amalgam to therapeutic techniques. In the process he demonstrates an anti-scientific bias while drawing on the best of science, an anti-intellectual bias while drawing on the highest intellectual concepts of which he is aware. He acknowledges the "problem" of consciousness while claiming the solution is to see that all is consciousness. He rejects the Western notion of God and acknowledges the experience of gods and goddesses as culturally specific while taking for granted the experience of a cosmic God. Grof's position is, typically, strongly eclectic. He takes his subjective judgment and that of each person as the ultimate criterion of truth and what is good for him and each person. He differentiates between institutional and ritualistic religions and the direct personal experience of a supreme power in approximately the same context as Joseph Campbell. His psychotherapeutic techniques are based predominantly on this philosophy and its logic. Anything wrong or false about such a philosophical synthesis and its application to psychotherapy? It seems to be efficacious for those who relate to it, at least as far as it goes. But there is a further step required in the recognition and integration of the new paradigm of our synthetic beginnings and bicameral nature which leads to the enlightening fact that the philosophy/religions of both East and West are derivatives of two factors, our subservient relationship to the Nefilim and philosophical principles they taught us. It is necessary to move beyond a rehashing of both Eastern and Western mysticism and their sublimation of the ancient Nefilim gods into cosmic principles. This theological orientation, both East and West, now transmuted and several times removed from the humanoid gods keeps us separated through the proprietary claims to absolute authorization by the "god" or "gods" as to what human nature is, is meant for, what is right and wrong and what our subservient relationship to those gods must be. Until we can get out from under that worldview and recognize our generic humanity we shall remain trapped in isolated groups at odds with each other and compelled to look to some authority or a consensual opinion for a clue to reality with no real freedom to trust ourselves and to honor the freedom of others. No full "spiritual" or transcendent personal or racial evolution will be attained until we gain that freedom. Breaking the subservient godspell frees us to go one on one with the universe. In that expanded freedom we can recognize the traditions of philosophical principles and the applied techniques of consciousness manipulation for what they are and examine them closely to determine how intact they are, if we are interpreting them correctly, if they are actually good for us and how we may transcend them by developing systems that are related to our unique nature and situation. We need to determine if they were generally Nefilim principles transmitted to humans – or casually picked up from the Nefilim by us – or principles meant specifically for the human situation. After all, as advanced as some of those principles may have been when transmitted to us they may have been primarily Nefilim-specific principles and, since we are half Nefilim, perhaps only half correct for us. Ultimately we will have to determine the unique principles that are best for ourselves.

The Next Step

It is obvious that we are already at a major turning point. It's not an intellectualized, romanticized kind of turning point that Capra would have in his neo-neo-scholasticized New Age theology. It's subtle, the information content does not seem to have taken any quantum type jump, we haven't physically metamorphosed noticeably, there has not been any particularly different, sensational overt fourth or fifth kind happening, our emotional set with regard to the phenomenon has not shifted radically on a general basis. Yet, if for no other reason than the marinade, made up of various parts of background neurological processing, mutual reinforcement, best evidence and modulated collective and individual experience, along with other trace elements, has brought us together and to this strange point where the sense is that we are entering, are already accelerating, into the next phase. A bit inchoate yet but here it comes. If I had to say – and that seems to be my subjective genetic imperative for which it seems irrelevant to apologize: don't blame me, it's my tight genes – I would describe the phase as of two major parts. We are now about to step out of a 3000 year period of dysfunctional racial adolescence marked by the classic godspell symptoms of grief, fear, denial, disassociation, craven propitiatory gambits, bravado, subservience, cargo-cultism and sublimation since the Nefilim phased off the planet. Daddy, won't you please come home. We are already phasing out of institutional religion and beginning to phase out of the methadone metaphors of Eastern karma-dharma-diddle. We are generally about as gigglefritzed, (David Pursglove's coin: a word deserving Nobelation) currently, as a gorilla turned out into a new high tech cage in our ad hoc exploration/definition of the strawberry fields of the relativistic forever plenum of our new freedom where one has to mint metaphor in real time in unfamiliar 4-D metre to get some sort of self-reflexive sonar bounce off one's chest just to keep making sure that one is still the anticipatory inescapability one thought was unquestioned – at least until fatigue sets in enough so that we have to ease up. The other part is that it seems that the time has come, in the rhythm of consciousness, or whatever MickeyMousemetaphor we want to apply, for alien contact. Either by alien decision, our demonstration of readiness, or both, here it comes. Regardless of the improbableness, imperfections, strengths, genius or limitations of those involved, the modality developed, whatever, here it comes. The aliens could be the Nefilim themselves or they could be real strangers from beyond our solar system. All signs say to me that both encounters are already happening. I believe that the Nefilim are monitoring us, probably have been since their phasing off the planet and other alien species have already made contact.

Contact

Seems obvious yet almost unthinkable, comfortable and almost anticlimactic for some of us, a bit scary for others. We won't be ready to gracefully make contact unless we clean up the first part and overcome the Babel factors that keep us separate. My genetic imperative tends to keep me preoccupied with the first part, to clean up our act in detail and get ready. But I also look forward to contact; I've got a lot of questions to ask – once we have determined that the Nefilim's intentions are honorable and respectful. What's required immediately? We need to recognize that the immediate ramifications are that we are a bicameral (two part) species, a species with two racial gene codes melded together, with a bicameral mind. That has some profound significance for many facets of our current problems and conflicts. The integration, in our self understanding, of our indigenous racial heritage with our Nefilim heritage shows us a way to get comfortable with the seemingly unmatched parts of our nature. The indigenous half of our nature will put us in tune with this planet, its ecology and its rhythms. Our planet becomes a fine place to live and a fine place to visit and we will honor its spirit and turn it into a beautiful park again, cleaning its air and its wounds. Our Nefilim ancestry will continue to enlarge our vision, prompt us to seek expansion first among our planets and then the stars.

The Contribution of the New Age and What's to Come

From the positive perspective, all New Age interpretations are partially correct in that they tend to recognize, directly through history or indirectly through the philosophic traditions, the central and dominant character of the "gods" in the inception and development of the human species. But they should be corrected to fit the full historical facts: the "gods" were real, historical, flesh and blood. They created humans for their pragmatic needs, eventually accepting us as limited partners, teaching us civilization and technology and their version of the philosophy of the nature of the universe and the nature of humanoid creatures such as they and we. With the withdrawal of their support and through the severe effects of disorientation and readjustment due to their departure we sublimated those limited "gods" and "goddesses" ("gods," more often than not because male humans tended to take over the power) into supernatural and mythic beings. The new paradigm, I suggest, is precisely the unexpected but actual, overarching vision that will bring about a new age that will be the fulfillment of the New Age vision and yet far more in that it affords the common ground to subsume all intellectual, scientific, philosophic and New Age world-views and complete them in a transcendent unity. The Aquarian conspiracy is turned into a planetary celebration.

CHAPTER 9

Adventures In The
Criteria Vacuum

Is it possible that we are totally predetermined to determine our own determinism? How absolutely certain can one be that there are no absolutes? By what criterion does one choose the criterion by which one chooses the criterion by which one chooses the criterion by which one. . . . There clearly has to be something very lacking! We can see the deficiencies, so it must mean that we have to upgrade our language, our logic, our philosophy, our science in order to completely and satisfactorily express *what our consciousness already knows.*

Touchstone

Philosophy, as we traditionally conceive of it, linear, syllogistic, and binary, is an inadequate antique intellectual politic. At the same time that we will have to come to live with the realization that the objective order of the universe is that it is probably fundamentally, relativistically subjective, we will have to adjust to living in a universe known through a Law Of Everything, a universe that we may well be able to completely predict.

Frank Clinton
Diary of A Bewildered Politician

To teach men how to live without certainty and yet without being paralyzed by hesitation, is perhaps the chief thing philosophy can still do.

Bertrand Russell

When all the old godspell theologies are gone like dissipated gray clouds, with them will have gone the traditional theo-philosophic basis for the concepts of a divinely mandated and sustained objective order of reality and the absolutistic doctrines it engenders. Once one realizes that the most profound of the ancient cosmological documents of both the East and the West are not mythic sagas but very advanced astrophysical and scientific accounts of significant events in the development of our solar system, or parable-type teachings of right conduct taught us by the Nefilim, one's perspective is changed forever. Being able to see clearly that our concepts of god, gods, God, and Avatars, are a direct product of our original relationship to our humanoid creators and sublimations thereof, we can move beyond the criteria of reality of both the crucified West and the karma-dharma-diddle of the East. What does one do when the reality of a liberation into a vast personal and racial freedom suddenly dawns and we find ourselves not only beyond religion, but beyond the new age vision and even philosophy as we have known it?

Theology itself, now seen to be an elaborate intellectual forgery generated by the godspell ethos, is defunct. Once one realizes that the reasonable philosophic and scientific methodologies which we have relied on in the West have been based on and developed either as a secular version of theology or a reinforcement of it (Scholasticism, Thomism) or in reaction to it's characteristic absolutism, one's world will never be the same.

The Judaeo-Christian ethos produced a view of reality that took the Bible (at least as officially codified by the Catholic Church at the Council of Trent in 407 A.D.) as the unquestioned word of God. The gradually refined concept of reality that emerged from this attitude was that this God was an Infinite Intelligence who created the universe – or anything that pleased Him – just by an act of Infinite Thought. Reality was then defined as the precise way things were as they were held in existence by the supporting thought of that God. Reality was defined, therefore, as "objective," i.e., things were and are and will be real only in so far as they correspond to and are held in existence by the thought of God independent of any other intellect. Truth was the precise correspondence of any other intellect, the human intellect in this case, to the thought of God about any particular thing.

Reason, taken as a God created and given faculty, was deemed as the proper and adequate tool for discovering truth up to, but not including, Revelation. Revelation was defined as a special category of knowledge about parts of reality which reason could not attain to and which had to be taught, revealed to the human mind by God directly. Traditionally the Triune, Three-Person nature of God and other special items have been considered Revelation. It is rather easy to see that this pervasive tradition holds for an "objective" order of reality, knowable by reason, existing "outside" of and independent of the human or any kind of mind. It followed, quite easily, therefore, that anything that did not correspond, in the opinion of the Church authorities, with their understanding of the nature of that reality was obviously at least unreal, if not evil. Galileo and Giordano Bruno found that out with varying degrees of discomfort, house arrest for the former and burning at the stake for the latter, to be specific.

In retrospect we can see clearly that, philosophically, our dialectical logicizing is both a product and a weapon of that cramping theological paradigm. De Bono has pointed out that our logic is a Scholastic bastardization of the dialectic process deliberately modified to be maximally effective in skewering heretics in their testimony. Once one realizes that the philosophies of the east are not so much philosophies as we know systems of syllogistic thought in the West, but actually psychologies dealing systematically with states of consciousness and how to identify, attain and navigate them, their status as criteria bases begins to fail. This does not mean that there is nothing of value to use as a touchstone for our own independent thinking, it simply removes the aura of divine or superior authority from these ancient sources. Any criterion of truth and reality can only be such if it is unquestionably fundamental, analogous to the way the forces identified by physics (the weak, strong, electromagnetic, and gravity) are irreducibly fundamental in the physical realm. Yet, for some time and now, coming to acute focus in our day, every criterion of truth and reality we have ever concocted or relied on is seen as inadequate. At the center of this silent, swirling storm of chaotic relativity, there is, nevertheless, an almost indefinable point at which we remain essentially intact, able to observe calmly these antique criteria of truth as they falter and blink out like neglected ancient lamps.

The Historic Relationship of
Philosophy to Theology in the West

It is well to fully realize how fundamentally our philosophizing, particularly in the West, has been conditioned and determined by the godspell ethos. Theology has evolved based on a concept of a static universe, one held in existence by the mind of a God ceaselessly "thinking" it to be, and to be in a certain way. The Scholastic logic following from that kind of postulate sees reality as objective, being as it is without any dependence on any consciousness' perception of it – except that of a God – and man and man's capabilities as part of it, static also. Perfection, truth, sanity and beauty are measured by the degree in which anything corresponds with the way God sees them. When Christianity held sway in its full power in the university and society, the philosopher, unless a rebel, understood that creative thought, regardless of how serious and logically correct, was always subordinate to the criteria of theological dogma. At the height of theological domination, if even the most profoundly logical and true conclusion or insight did not agree with theological dogma, it was branded as false or, worse, heresy. One has only to contemplate the plight of Galileo to see how the system worked. Even the dissenting philosopher who rebelled against such stricture was most often working in reaction to the pervading theological world-view. Reason, in this context, is understood as a tool, bestowed on man by God, totally adequate to determine what reality is. Reason, therefore, a static tool in a static universe, becomes a rote kind of bondage, and philosophy only a mechanical exercise, too often only an adjunct to theology and religious dogma and too often subverted into a means of reinforcing theology.

Philosophy "Liberated" From Theology in the West

Regardless of how secular some modern philosophy may seem, the framework of the entire spectrum of philosophy as we know it in Western culture has been determined originally by the theological concepts and strictures underlying our cultural heritage. Even the structure of logic as inherited from the Greek philosophers has been altered to fit the Scholastic mold. For centuries, because a gradual split has developed between philosophy, religion and science we have had three different bases of criteria for truth: philosophy, religion, and science. But not everyone realizes that we have developed those different approaches to reality because each of them have been unacceptable to some portion of the population at some time.

Religion would subjugate philosophy as purely natural and not privy to the higher "truth" of which religion would contend it alone was the proprietor and custodian. Science rejected the static authoritarianism of religious dogma. The scientist insisted that proof must come through controlled experimentation or at least physically verifiable evidence and, therefore, also rejected the "pure reasoning" of philosophy. In effect, science struck out on its own, propounding the dogma that dogma must be rejected. And philosophy has always held itself at least secretly superior to both theology and science since they both must employ the reasoning process to attain their goals, and the assumptions both make to even begin are philosophical assumptions. So, for all practical purposes, we have been working in a criteria vacuum for quite some time. What is fascinating is that, although it might be assumed, because the disagreements were so fundamental, conflict would arise over what reality is, the focus of the primary conflict has actually been over secondary level questions such as what truth is and how we can know it, and whether we can know that we know it. This might seem surprising, at first, given the radical differences between these three major approaches to the universe. But, as example, both the theologian and the scientist, as well as some philosophers, begin with the assumption that reality, most fundamentally, is objective. By objective is meant that the universe has a certain lawful nature and exists in a certain way in and by itself whether there is any consciousness to perceive it or not. The scientific method assumes an objective universe because to discover laws of the universe it must postulate a lawful universe to begin with. The opposite of this view, held by some, says that the universe only is as we perceive it; some going so far as to say that reality does not exist until consciousness, ours in this case, literally creates it.

But, in this time of transition out of racial adolescence many still try to hang on to some sophomoric security God figure, some reasonable philosophic platitudes, some reassuring, concrete, fundamental scientific data as criteria of truth and reality. But none of the above are adequate, comfortable or even acceptable when we dare to really think about them. All have come under scrutiny and questioning over time. Each modality, adequate to the task which it was designed to handle, is now seen to be a special pragmatic tool developed for a specific type of information processing within a certain section of the spectrum of human consciousness. But, even more profoundly and immediately than our gradually accumulated philosophical sophistication, the new paradigm, by its intrinsic character and implications, calls into question the very criteria on which we would have formerly relied to determine its validity. This can cause a deep epistemological uncertainty. (Epistemology is a section of Philosophy that deals with questions such as What is truth? How do we know truth? I use the term epistemological throughout this work for precision: there is no better word for the critical, focal questions that must be faced and answered by the individual as SHe transitions to full new human status.)

The Transition Phase

Once truly freed from the godspell and recognizing the consensual definition of what a human is, the epistemological questions that have been warped and thwarted by the pre-logical absolutes of religion and reactionary science will simply vaporize. We will be able to see the entire process so far in terms of our special case of evolution, and our epistemological thinking in terms of that as a subset. Only then will we have cleared our vision and our collective psyche sufficiently to begin a positive construction of a consensual world view based on a common criterion of truth.

As the traditional, theological structural support for the notion of a divine objective order is removed, the central topic of objective versus subjective reality gets thrown open for reexamination. Even today, as those theo-cultural (the rules and customs springing from the old godspell theologies and religious traditions embedded in and forming our cultures) mandates erode slowly away, the ongoing debate is becoming more intense. The effect of the removal of theological authority and its philosophical derivatives is to remove the framework of absolutistic objective reality and to create an intellectual and psychological partial vacuum. The theo-cultural criteria, formerly used to distinguish real from unreal, true from false, good from bad, are already severely diminished if not disenfranchised. The loosening of structure is and will be perceived as a danger and problem by some, an increase of freedom by others, a difficult dance by philosophers, and a challenging and exciting expansion by still others.

The initial reaction of some to the notion that there might not be any objective order to the universe is very negative. The universe can suddenly seem cold, impersonal, and disinterested as the old parameters by which the individual identifies HIr usual self are seen as relative, even arbitrary, and one's fate as not completely under one's control. The sense of self and its relationship to the universe is totally in question when we no longer can operate according to tradition, custom, taboo, or imitation. Many will falter at that point. Fear often reduces the individual to survival reactions. We look sideways at each other to get peer reinforcement or opinion but, when we permit ourselves to honestly consider, the universe begins to look like an immeasurable, uncaring, impersonal partial vacuum that doesn't even recognize us. Initially, we find ourselves rapidly losing the last handholds of criteria by which to judge the validity of anything.

The Status Quo

The status quo then becomes: no objective order held exquisitely real in the mind of a Dominican God beyond thought; disappointment by the inadequacy of clunky classic logic and painfully aware of the ragged edges of reason's parameters; bruised by the rigid coordinate bars of the antique Cartesian/Newtonian prison; subject to seizures of epistemological catatonia when the last fix of Consensis ' wears off; frustrated with being forced to work with second-hand metaphors under grade school rules. Few, so far, travel comfortably in this vacuum where the antique criteria of Theology, Philosophy and Science rattle about, where the criteria by which we choose our criteria by which we choose our criteria withers on the stalk due to prolonged in-breeding, where we become paranoid or desperate, rubbed neurologically raw by the anticipatory inescapability of subjective self-reference, where we are juveniles in the management of radical self-determination.

The Conservative Reaction

Many maintain some sense of reality only by ignoring or blocking the topic altogether; otherwise they feel they would be inadequate to handle the ramifications and become frighteningly disoriented. Our consensual reality is wearing very thin and tenuous but it still allows an individual of this type to go about the ordinary business of living with sufficient support and tolerance as to seem "normal." The vehement reaction to any suggestion that the objective order of the universe should be questioned or reexamined is usually due to the fear that it would lead immediately and inevitably to social chaos. It is worth examining a cardinal example of this type of thinking: the thesis of Professor Allan Bloom (*The Closing Of the American Mind*, Simon and Schuster, 1987). It is fair to take Allan Bloom as a spokesman for the defenders of the classical concept of the rationally founded university and society of which it is a mirror. It is simply missing the point to say, as Martha Nussbaum and others have, that Bloom ignores history, women, diversity of learners and true philosophy; those are only symptoms of the essential flaw. The real clues to the nature of the fundamental problem are written large across his erudite complaint.

The clue to the rejection of the future and, therefore, the powerful precursors to it operant in the present university: Bloom rejects, abhors, indeed is revulsed by, the phenomenon of the 1960's. Bloom's abhorrence of the offshoot symptoms of that phenomenon tends to distract his readers from the fact that he does not only abhor but will not even discuss the essence of the sixties.

The fundamental nature of the revolution of the sixties was that of internal personal freedom, the ability to change one's own behavior radically by oneself, to reprogram, to literally reimprint the neurological system, to attain the heights of what had been called the "religious experience" at will. Part of that profound experiential exploration and education showed the individual precisely how SHe constructed HIr own personal reality in three dimensions and extended that to the world "outside."

The revolution of the sixties as portrayed by the press and media as a political one marked by war protests, rebellion against parental and social authority, eccentric dress and dropping out from the establishment is simply an inaccurate reporting of relatively superficial effects and results rather than the essential engine that drove it. Since powerful agents of personal evolution such as LSD were the primary tools, the prejudice against such drugs precluded them from being recognized for the positive contribution they made. Ultimately the fear of change itself and the loss of power and control by authorities in all parts of the social arena from parents to teachers to psychiatrists to politicians to the military and the police would only allow the media to portray the sixties as a negative, threatening phenomenon. But Bloom's Jesus is Plato and his dogma comes from a series of European philosopher popes; he sees the 60's as a negative threat, identifying the character of that era with its social effects such as war protest, civil disobedience, and the socio-political manifestations of change in attitudes toward sexuality, the family, government. As a result he could only conceive of the alternative to the classical reasoned ethos as being an uncontrollable anarchy of subjectivity. It is well to note that a scholar as keen as Bloom is very sensitive to a challenge to the reasoning process itself because he is well aware that the validity of reason also rests on the assumption of an objective lawful order of nature. Correct logic, correct reasoning, presupposes that rules of reasoning apply and rules presuppose a lawful order from which they arise. For this reason (pun unavoidable), too often reason, a function of the peculiar way human nature is, is set up as the criterion by which the validity or even the reality of the other functions of human consciousness are judged and evaluated. Reason has always found intuition naive, the transcendental incomprehensible, imagination suspect and ecstasy unruly, disjointed and even debauched. It makes reasonable philosophers very uneasy when you even question their favorite tool.

Bloom's reasoning is that the rejection of any form of objective order inexorably leads to each individual holding for his or her own notion of right and wrong. With no "objective" order and reasoned morality to refer to, each individual would hold that his or her ideas of right and wrong were as true and valid as anyone else's and there could be no effective law or order and society, as we know it, would degenerate rapidly into chaos.

His perceptions of the weakness and hypocrisy evident in the political and academic scene mask the fact that he will not even consider that human consciousness can be more than just rational, that it can and shall transcend linear logical reasoning. He allows himself to tune only to the narrow band of reason, would preserve the classical status quo because the only alternative he sees is moral barbarism (while stating that the newer crops of students are more honest than previous generations!), and sees as "the greatest endeavors" a series of Roman clichés: "victory in a just war, consummated love, artistic creation, religious devotion and discovery of the truth." Bloom has totally rejected Huxley's warning not to prematurely or narrow-mindedly close the doors of perception. His fear, mirroring the survival fear of many, makes any concept of a relativistic universe an unacceptable threat. It epitomizes all variations of the conservative mentality. Ultimately, any epistemology which questions the concept of an objective order is rejected by those of this orientation, paradoxically, not on the basis of some reasoned scientific or philosophical argument; it is rejected from pragmatic fear of the projected social consequences.

The cramping, stultifying fear that prevents many from accepting this degree of freedom may often be managed and overcome by recognizing one of its real causes: unaccustomed responsibility. We may protest that we feel inadequate to deal with the universe as nearly unknown, we may direct our frustration at those who have not prepared us for such freedom, etc. but, inevitably, we must face the fact that it is the responsibility for ourselves, utter and unavoidable, that we must take that shakes us so profoundly. This will no doubt become less of a problem as generations become accustomed to such radical freedom as the normal and essential state of human nature. Meanwhile the profound conditioning of the godspell subtly influences us to accept living and dying under the imperatives of processes we do not control, programs us to look to authority, the group, institutions, religions, governments, even other individuals for guidance and direction and a definition of self, to deny or suppress reactions to those factors which make us recoil, sometimes in secret horror.

It is a sad moment when we realize that there will be some who will refuse the full dimensions of freedom and allow themselves to be hopelessly swallowed by what they interpret as a power greater than they.

The Fullness of Freedom

But, for those who can react positively, what may be seen as an eerie vacuum of unseen boundaries, impersonal forces and cold fate may, alternately, be appreciated as a plenum in which the dimensions of existential freedom are profound, affording the utmost expansion and potential; where that exponentiated awesome potential of freedom is recognized as the most valuable and intelligent feature of existence. Here the universe is an invitation to a dance in which the individual can, indeed is expected, to reciprocally participate in the creation of an on-going reality.

Let us consider the fullest dimensions of freedom of which we are capable of conceiving. Let us start with the most radical approach: For the sake of the discussion I will postulate, flatly, that there is no reason that I cannot demand of the universe literally anything that I wish. The typical objection that I am, therefore, acting in a megalomaniacal mode, is based on a deep and subtle effect of the godspell: the deeply ingrained attitude that we are pitifully limited creatures with no right to demand anything of the universe, in fact it is prideful and dangerous to do so. It is "against God." This is simply an extension of the rules of the master-slave relationship we were invented under – and it was the better part of discretion and common sense to watch what one said or did as a human under those circumstances of subjection. Involved here is a concept and experience of the universe as being impersonal, unaware of us and our consciousness, positively inflexible, a process that is unfathomable, something to be guarded against and fought or given into. As if we were somehow detached, at least in part, from it. But we are a part of that universe, an integral, vital part of it. Simply put, with the removal of the old theological notion of an unchanging divinely held reality goes the implicit "personality" of an ancient, authoritative, dominant, cranky, male deity whom we may not even have accepted but the archetype of which is implicitly and deeply embedded in even the so-called secular ethos of Western culture.

I demand from the universe the greatest degree of freedom of which I can conceive. I demand that the universe teach me about even greater degrees of freedom than I can currently conceive. I demand answers from the universe that will satisfy my capacity to understand anything that I wish to know. I demand the fullest degree of pleasure and happiness of which I am capable. I demand the freedom to demand. I demand the ability to play any game that I can conceive of an intelligence (still an anthropomorphic projection) playing at the "level" of creating universes, interacting with other intelligences of the same or other kinds at that level in whatever modes of interaction I prefer or wish to create.

If the universe is intelligence, or infinite, or possesses or is any other characteristic of which I can conceive, I demand to be fully capable to be that also. (Although it runs counter to some interpretations of the "Thou art that" dictum of Eastern mysticism, I tend to think that what I am speaking of here is the perception which gave rise to that statement.) If I demand to be the universe, has not the universe given rise to an element of itself with the potential to demand that? Ultimately, how could a universe preclude my attaining any element or degree of freedom I am capable of understanding or anticipating without treating me as if I did not have a right to do so? If it can and does restrict itself (me and you) in such a way then I demand the right to protest or remove that restriction. But, it will be inevitably said, "You are arrogant and stupid and naive! How can you even seriously say such nonsense when you cannot cure yourself of some diseases, cannot stop aging, cannot prevent death, cannot control some favorable or unfavorable circumstances immediately around you?" The point is that I can conceive of doing so, my consciousness can intuitively encompass the thought of at least several universes at once, I can conceive of at least a limited version (elegant oxymoron) of infinite freedom, I can be dissatisfied with the limitations that my intelligence and consciousness experience, I am capable of feeling free to consider myself an equal to any intelligence, "infinite" or otherwise, of which I can conceive. Since I am part of the universe, a "product" of the universe so to speak, then the universe itself is, at least in part (me), this way. And what I can conceive I can work toward and sometimes achieve.

It is difficult to grasp, at this beginning phase, that such latitude is the minimum required for the degree of freedom of self-determination of which we are capable. But this imputed void, perceived, at first, as a frightening relativity of echoes, is the plenum of our transcendence. Not the "void as purpose," neither animism nor a culture-spawned cult, it simply is the adequate venue for expansion of our biologically determined philotropism, and our biologically determined determination to determine our own determinism.

When it comes to answers, our children deserve much better. With the acceptance of the new paradigm we will have a planetarily common definition of human nature. With the godspell broken, and a consensual definition of the human attained, we will be free to explore the essence of truth rather than be constrained by arguments about who possesses the truth. We will operate from the deep potential of human nature rather than be trapped at the shallow level of arguments about what a human being is in the first place. We will finally emerge into a time beyond the standstill of dispute between factions, each of which claims authorization by "god"(!), as sole proprietorship and interpretive oracle of the ultimate truth.

The Philosophical Status Quo

Our current philosophical situation still exhibits many of the characteristics of the classic theo-cultural situation even though philosophy purports to be substantially secularized and disassociated from theology. The philosophical arena is still dominated by an argument between those who hold for an objective order and those who hold for a subjective interpretation of reality. This central debate, in its many forms, and the questioning of philosophy and the scientific method in general which has been going on for some time, will be exacerbated by the acceptance of the new paradigm. But it is this central debate which furnishes the key clue to the next step in philosophy because it illuminates the barrier beyond which we must pass to arrive at a new criterion of truth.

Beyond Static Reason: A Reasonable Criterion Shift

Inescapably, as we are thrown back on ourselves as the ultimate criterion of reality, it is how we develop an understanding of how to use our own judgment – and how to trust it – as criterion of truth that is so critical to graceful existence as we go forward into the future.

How shall we view the shift to our subjective judgment as the fundamental criterion of truth? Even as we recognize the fact that philosophy, as we have known it, is no longer adequate to our comprehension and communication, we should recognize that we need to determine the next higher power of reasoning. A fine analogy for what we must do is found in modern physics. The development of relativistic physics and, eventually, quantum mechanics showed that Newtonian mechanics, the laws of Newton, were actually erroneous – but accurate, usable and practical at least at the large scale level in which we apply them; they are a subset of the more precise and comprehensive Einsteinian relativistic and quantum mechanics. So should we view our classical philosophy and logic, as a subset of a larger, more comprehensive four-dimensional logic and broader spectrum philosophy, actually erroneous in their incompleteness and inadequacy but usable and practical at least at the gross three-dimensional level at which we have applied them. This may sound like a strange concept since it implies that we may arrive at truth using inaccurate or erroneous techniques, but it points up several very important factors already active in our individual and collective consciousness. More importantly, it points the way toward the next plateau of human awareness, the consciousness of the new human.

Philosophizing as a Function of Consensual Consciousness

Historically, we always use the highest, most evolved type of consciousness that we are consensually comfortable with as the criterion of truth and to judge the other types of consciousness we exhibit. We never use, consensually, the highest kind of consciousness we have attained as criterion – although we use it consistently as an exploratory mode. That is the cause of most of our philosophical conflicts over time. That does not mean that humans, from our very beginnings, have not exhibited the ability to think logically, to project into the future, to imagine, even to experience transcendental states. It simply means that we tend to use the most comfortable and *consensually recognized*, "respectable" part of the whole spectrum of our consciousness to judge the veracity, adequacy, and "respectability" of all the other parts. As a single example: the Greeks were not at the stage of just becoming aware of reason operating in their consciousness; they were already very aware of it, used it, took it for granted, considered it safe. Their preoccupation and function was to codify it analogously to the way new systems of mathematics are developed.

Even though reason had reached the status of "consensual" as a fundamental tool to attaining truth enough to be codified with the Greeks, there were some within the population that had evolved past that point. Or, from another perspective, other types of consciousness were available to humans in general but were not integrated into the common consciousness sufficiently and with sufficient understanding of them to be considered "safe," "meaningful" and "usable." And metaphors, concepts and a vocabulary had not been developed sufficiently to communicate comfortably to allow them to be recognized. And the understanding of the universe in precise terms with regard to that kind or kinds of consciousness had not been arrived at to make it clear how that kind of consciousness fit into the universe.

By contrast, as individuals, we tend to use the most evolved form of consciousness we can experience, at any given time, as our primary criterion of truth and the primary tool of our philosophizing. In the case of the individual whose consciousness has expanded beyond the ordinary, this can often bring the individual into conflict with consensual thinking and criteria. Reason has been our primary philosophical tool. But reason is clearly a narrow band of the spectrum of human consciousness. The essence of our epistemological dilemma for centuries has been that we use that one "frequency" of our consciousness as the criterion of others. Reason judges ecstasy suspect, intuition naive, transcendent experience incomprehensible, the metasyllogistic meaningless or, sometimes, worse, insane. Many realize that the high consciousness of which the *Book Of the Tao* is a textbook and which is attainable through the use of powerful tools such as LSD or a lifetime of disciplined mediation and practice is a far more expanded and evolved kind than our ordinary consciousness. But it is not, therefore, considered the consensual modality of our primary philosophical criterion.

A classic example: in the nineteen sixties, the powerful psychedelic experience brought millions of people suddenly in direct contact with their normally screened physiological and mental processes – including the brain experiencing itself in operation. The realization was that it was a dimension of human experience which could allow one to "watch," so to speak, one's logical and other processes operating. This advantage of having a type of consciousness available that was obviously more than just logical and actually subsumed logic tended to prompt many to use this experiential mode as a criterion of the validity, adequacy and veracity of the other modes of their consciousness including reason and logic. This was taken as a direct threat by the majority of establishment scientists and philosophers although it was not the primary reason for the fear and loathing that eventually made the psychedelics illegal. Besides being seen as a threat to authority and power, it was perceived, consciously or unconsciously, as a threat because it was seen as an attempt to replace logic and the scientific method as a criterion of truth. Putting it all down as some nebulous hallucination and making it seem like bubble-headed anarchism was the way to suppress it most effectively. But it is far more adequate to the fullness of philosophical comprehension and understanding as well as scientific investigation than our ordinary consciousness. Yet, because it is unusual and a challenge to many who have come to be addicted to the cautious security of lineal, sequential thinking, it is not consensually accepted or even understood.

We can actually witness, in historical perspective, our kind of consciousness going through stages of development. The modalities for determining truth and true relationships of ideas and information gradually arise. Our consciousness has been lineal and three-dimensional to this point and therefore we have the kinds of logic and the geometry that we possess. But the new consciousness is different, more expanded, and the modalities that we have found adequate to this point will either have to expand to match or be replaced by new and adequate ones appropriate to the new consciousness. As with our previous paradigms of science, the old will be shown to be special cases of the new like Newtonian physics is a subset of relativity. Beyond recognizing the nature of concepts and metaphor as a relative function of our type of consciousness, and reasoning in syllogistic form as the means of manipulating concepts and metaphors relative to our consciousness, we must step back and consider our type of consciousness itself as a function of our state in the universe. This opens up the possibility that a new type of consciousness, sensory, physiological, mental, and dimensional, is what we should be deliberately seeking as the essential characteristic of our next stage of evolution.

The indicators of the limitations of our consciousness are manifest in different areas. Godel has pointed out that no human system of analysis has the capability of explaining itself. We do not seem to be able to escape the boundaries imposed by subjectivity. Our very consciousness and intelligence tells us there are dimensions of the universe which exist but which we are incapable of directly perceiving. We can comprehend how our unique synthetic genetic genesis as a slave race has formed our consciousness and its evolution to this point. We can observe ourselves sufficiently to realize how cramped and limiting our present state of habitual consciousness really is and be driven by that frustration to envision a new and expanded awareness of which we are not yet capable on an habitual basis. We can realize that the disciplines we have developed in philosophy, science, mathematics, psychology, to name only a few are really direct functions of our habitual consciousness and begin to envision how those disciplines will develop, change and transmute to be adequate to a more evolved human consciousness. To prevent ourselves "reinventing the wheel" in this regard it is well to examine the current crop of these philosophical gambits that propose to give us a more adequate view of reality.

Three Modern Philosophical Opening Gambits

It is useful and very appropriate to view any coherent philosophical explanation as a "gambit," as defined in chess: a kind of calculated "opening move" meant to gain maximum advantage. **The metaphor is quite precise because starting, making the basic assumption, accepting a basic principle, finding an unquestionable postulate, a level of reality beyond which we cannot penetrate, is the most fundamental problem.** Although oversimplifying the matter considerably, it is sufficiently accurate, as is done in many textbooks, to classify philosophies into two classes: those which hold for an "objective" reality, as discussed previously, and those which hold that reality is "subjective." This last category has developed due to the gradually deepening sense that it is at least difficult, if not impossible to tell whether even our concept of "objective" reality is, indeed, subjective or not.

The Linguistic Gambit

The linguistic argument says that these are only apparent problems stemming from a basic inadequacy of our languages. Science gets ahead of our means of expressing a more complex physical reality and we have to invent better words and relationships between them to adequately communicate. Linear, syllogistic logic falls short of expressing the philosophical concepts that strain the envelope of our increasingly relativistic experience of the universe. But the problems of preemptive assumptions and infinite regressions of criteria are not solved by upgrading our language as such. The problems are simply exponentiated along with the language and logic. I believe that there is an important indication here of the direction toward which we should move in that we must upgrade our language, concepts and logic as well as our mathematics, but this will be an effect of a more fundamental solution, which I will advance subsequently, rather than the solution itself.

The Scientific Gambit

It may be asked, Is science not an old phenomenon? Why should it be viewed as a modern philosophical approach? It is the "scientific method" of which I speak here. Even though most scientists find it distasteful to call it philosophical, the scientific method is based on a classic philosophical view of nature and cannot escape being philosophical in its pronouncements. In large part inspired as a reaction against the theological approach and the detached reliance on logic as if it were efficacious in itself by the philosopher, the scientific method was conceived and developed as a more neutral and precise way to approach reality. By controlling conditions, repeating rigorously strict experimental procedures, relying on careful statistical analysis applied to verified data, the scientific method was looked to as a means of attaining truth superior to the other approaches and which could eventually be codified into "laws" of nature. By it's very nature the scientific method has to *presume*, however, that there is an "objective" reality that operates according to laws that are there in the first place to be discovered. It also tends, indeed is forced to *assume* the entire spectrum of reality as "physical," as nothing more or less than measurable, quantifiable energy and matter. The scientific method, on its surface, seems to be immune from the conflicts between philosophy and theology because of its structure and orientation. But the presumptions and assumptions on which is is based, mentioned above, are philosophical: an unproven assumption of any kind, in order to even begin the process of scientific investigation, is a philosophical item. The unprovable and unavoidable assumption that the scientific method is a valid method for arriving at truth is a philsophical assumption. The only other alternative is to use the scientific method to prove that the scientific method is a valid method. This is clearly circular reasoning (the grass is green because the grass is green) and invalid in the eyes of theology, philosophy and science.

It is very ironic that, analogous to the way theological belief emboldened the religious to wield dogma like a philosophical ax, the modern version of the scientific method (meant for dealing with rigorously controlled experimental situations, when applied to argumentative discourse) is wielded in just the same fashion against those who would question the authority, adequacy, scope or competence of consensual scientific positions. This present day analog of Scholastic heretic hunting dialectic wielded, sincerely but inadequately, – and paradoxically – in naive Darwinian righteousness limits us to think in terms of narrow choice between either religious belief or scientific materialism. As E.O. Wilson has pointed out, when we "look inward, to dissect the machinery of the mind and to retrace its evolutionary history," we are preclusively pinned, syllogistically, to an interpretation of our past as little more than a series of "partly obsolete, Ice-Age adaptations" (Wilson, *On Human Nature*).

Proof, according to the scientific method, comes from rigorously controlled, duplicable experiments, the results of which are amenable to statistical analysis and independent verification. Less rigorous forms, adapted to investigations in which experimentation is usually not appropriate or possible – archaeology, as an example – have been developed over time. Because the goal of the scientific method is to determine basic truths about the subject under investigation and, hopefully, to discover actual laws that may be written, a cardinal goal and characteristic of effective science is predictability. If you can write a law about how a phenomenon such as gravity effects falling objects and it is correct, then you can predict, all things being equal, how it will effect any falling object in the future. Certainly this is a powerful and advantageous methodology and its efficacy in the limited ranges of physical phenomenon where it normally operates is not in dispute here. The practitioner of the scientific method, however, as we have seen above, in attempting to approach the universe without preconceptions and to only accept conclusions drawn from rigorously derived and analyzed data, must, paradoxically, impose a relatively elaborate a priori model on the universe in order even to begin. The scientific method simply sidesteps the more fundamental critical problems of epistemology. But, by its assumptions and structure, the scientific method cannot address the fundamental question of subjectivity. It can only take the relativity of perception and concepts into consideration in a very limited way, if at all. It is not structured to deal with the fundamental questions such as what is truth. It is useful and productive within the scope for which it was developed. But it should not attempt to reduce, a priori, the entire spectrum of reality to its limited pervue. As we explore the topic of subjectivity and self-reference the reader may agree that a "subjective" universe does not mean an unlawful one. At the very extreme the most fundamental law could conceivably be that there are no permanent laws or that the only law is constant change. That begins to sound like the dictum of Heraclitus that the only thing that doesn't change is change itself. . . Science has no methodology for dealing with those questions and, as we shall see, consciousness itself is the major problem for science in its quest for a Law Of Everything. The scientific method cannot look to itself as its justifying criterion of superiority without getting caught in the circular reflectivity of self-reference. To attempt to avoid the trap by frankly stating a primary assumption that there is an objective order of reality that is knowable is a rehash of the Scholastic principle of identity (a thing is what it is) – and most scientists who unholster that semi-theological muzzleloader, in an ill concealed contempt to waste some wretched dog of an objecting philosopher, don't even know it.

The New Age Gambit

Besides the semi-religious and intuitive orientation of many of the New Age there is a view taken by another segment of the alternative culture that is primarily philosophical. It takes a broad overview, and is articulated by the more erudite and iconoclastic thinkers at the leading edge of what is usually considered alternative culture thinking. Partly from a deep-rooted reaction to absolutistic religious dogma and partly from a deep sense of the epistemic barrier, these thinkers focus on the arbitrariness of our personal "reality tunnels." They criticize the scientific and academic and religious establishments for narrowness, hypocrisy, the persecution of those holding alternative views, power plays and a patronizing attitude. They point out the inadequacy of the scientific method and insulated academic theory and excorciate the repression and disassociation caused by godspell religious dogma. Most telling is their competent and comprehensive expose of the subjectivity and relativity of our epistemic thinking. Some offer practical systems and exercises to "break trance" and expand consciousness. In the process of breaking the godspell and its secondary but extended influence even on our science and philosophy, all these are necessary points to be made. But it is not enough to systematically and emphatically point out the subjectivity of all our thought and the damage done by rigid dogmatism. Unless there is an alternative world view available, a new definition of human nature consensually agreed upon, a positive vision of a way to overcome both the barriers between us and the epistemic barrier to our thinking, we will continue to work in a partial vacuum.

But a curious thing often happens at this point: those who would break down the robot programming in human nature, expand awareness, promote freedom of thought and action often stop here as if there could not be any explanatory paradigm that encompassed all previous partial views, no explaining of our previous explaining. And so they tend to hold in a permanent state of suspended judgment – except for the very definite judgment that permanent suspension is the best way to go. Most unfortunately, some even give signals that they have fallen into the authoritarian posture, in defense of their position, that they have condemned so vigorously in establishment circles. What is fascinating about this phenomenon is that, in our recent bout of collective epistemological enlightenment, many reached a point where they would not be caught dead making dogmatic statements – except, perhaps, dogmatic statements that we should not make dogmatic statements – saying dirty words like "must," where we gave a Pulitzer prize to Douglas Hofstadter for the dreary epistemic hypochondria of *Godel, Escher and Bach*, where we made an absolute out of relativism, etc. Is it just possible to get so entrenched on that side of the coin that we might miss the possibility of actually reaching a grand unified field law?

A caveat is required here: everything I am saying (to avoid the absolutes being questioned here) should be understood as also subjective, an exercise of my self-referential consciousness and my subjectively adopted criteria of what is truth and reality. Furthermore, in the name of relative consistency, all that is said should probably be understood as metaphors, temporarily adequate, perhaps, but metaphors, images, nevertheless, modeled according to our peculiar type of self-reflexive consciousness, and we should not take, as so often well emphasized by my good friend and favorite iconoclast, Robert Anton Wilson, the map for the territory. But that does not mean or imply that the ultimate truth is that we cannot know truth.

Anticipatory Inescapability

What all these opening moves at the very base and beginning of these systems of thought have shown us over time is that, regardless of how we begin, no beginning can be made without some previous assumption, some basic postulate having been accepted as true without proof. We need not look any further than that statement itself as an example. Just the fact that I am convinced that saying anything at all and attempting to communicate requires a vast number of assumptions about reality in the first place. The assumption is made that we exist in one way and not in another; that there is some common universe or existence that we can experience; that we are capable of communicating, understanding each other, agreeing or disagreeing; that there is some validity to using reason and logic; that whatever it is we are experiencing remains stable enough so that it will probably be substantially the same tomorrow as we agreed or disagreed on today. Those are only a few of the assumptions, usually unspoken, of the many that underlie our usual self-perception and interpersonal communications. If we were able to question and verify all those assumptions before beginning to communicate it seems impossible to do so without having made a vast number of *other* assumptions about reality – and we would have already begun the process we were preparing to begin. I have called this phenomenon *anticipatory inescapability* and, having put HIr mind through these hoops for a bit, perhaps the reader will come to agree that the terminology is apt.

This underlying, apparently inescapable, preemptive phenomenon is another facet of the epistemic barrier that we need to break. It seems we must be aware in order to be aware; must assume some reality in order to determine it; hold that it is true that we can know truth before we can determine if it is true that we can know truth; we must assume that we can assume before we can assume; we don't seem to be able to "position" ourselves and our thought "before" any discrimination of the most primal mode, before any criterion of truth, without an awareness or an assumption which preempts the attempt. Because of the different basic assumptions, those principles or postulates that had to be accepted as true, either "self-evident" (how does one know anything is self-evident?) taken for granted, without proof, in order to "start" in the first place with each of these three major "opening moves" to reality, the three classic approaches have always been at odds.

I am probably far more aware of the convoluted mess already created here in just a few paragraphs, as one who is genetically afflicted with a compulsion to think inordinately about such things, than the average reader. (For the record, these are called epistemological topics. The area of Philosophy known as epistemology has developed over the centuries through the focus on the questions about what truth is, whether it is subjective or objective and how we determine whether it is one or the other; how we formulate the questions about the nature of truth and reality in the first place. An important characteristic of epistemology, because of its focus, is its reflexivity, its self-reflexive inspection of even its own questions.) When we reach a certain point where we don't seem to be able to get outside of our own consciousness' kind of perception of the universe, even though we can imagine all sorts of other types of consciousnesses at least in a science fiction sort of way, we sense a sort of elastic envelope out of which we cannot break – and the reader who reacts "instinctively" and strongly in a negative way to the involutedness is already well into the experience of such limitations. I like to call this the "epistemic barrier" analogous to the "sound barrier."

The experience becomes most acute when one encounters one form or another of the nasty questions I pointed out above: How absolutely certain can one be that there are no absolutes? How do we know that we know that we know that we know that we know. . . By what criteria do we judge the criteria by which we judge the criteria by which we judge the criteria. . . This statement is false. (If it's true, it's false; if it's false, it's true. The statement is meaningful, if only for its significance as a type of statement, yet it oscillates between true and false.) How do we know that anything we know corresponds to objective reality except through our own subjective judgment? In fact, how do we know that it might be only through our subjective judgment that we know that we know anything except through our subjective judgment? How do we know that we know anything but the neurological impressions in our brain? Does the ultimate essence of the objective order of reality determine that we can only know it subjectively? To put it even more paradoxically, Is the ultimate objective order purely subjective? How do we know that the acceptance of any "authority," whether putatively divine or expert or trusted, as to what is true or false with regard to any matter, is the right thing to do, except by our own subjective judgment? How do we know that the criteria of comparison of subjective judgments about objective reality is an accurate one to use except by our subjective judgment?

Now many or all of these problems or seeming contradictions may be familiar to the reader, not tremendously surprising at this point in time, and viewed, often, as bothersome but not earthshaking. But, most often, the subject is often just simply dropped as frustrating, not amenable to a resolution – and probably just as well left to the philosophers who don't have anything better to do anyway. After all, a person has to get on with HIr life. . . But this barrier is critical to our progress and will not go away by the fact that some of us chose to ignore it. **The new human, free to explore the vast freedom of the universe and determine HIr own evolutionary trajectory must become an adept at the relativity and recursive nature of this kind of thought and reality in order to understand HIrself and to comfortably create HIr own reality.**

After having been preoccupied with this problem for many years I have become convinced that not only is there a way to circumvent this barrier but that the barrier itself is a patent clue as to how to overcome it. I will be the first to acknowledge that such a goal seems, in light of the way we tend to view our current planetary situation, highly utopian and unrealistic, even unattainable, to say the least. But the changes are taking place below the level of public focus on the news media and the extrapolations of the professional futurist. Because the establishment academician and scientist and theologian will not ordinarily even acknowledge a new modality because of the direct threat to their status, there is not consensual support for an individual or group to adopt it. But it is the individual, many individuals, who are making the shift quietly and with self-confidence because they have come to trust their own intelligence and judgment sufficiently to feel comfortable evaluating the evidence. And, with that comfortable assumption of responsibility and self-confidence in one's own subjective judgment comes a gradual discovery of others experiencing the same thing. Although, by the nature of the independence gained, the support of a group is no longer essential, it still is helpful even if only for stimulus to thought. With that self-confidence gradually develops a courage to encounter the daunting challenge of our current epistemic barrier. By epistemic barrier I mean the focal problem of philosophy, of all reasoned thought, that we face today so unavoidably: the specifically human form of the phenomenon of self-referentiality, subjectivity.

We are painfully aware that even our most objective judgment that there is an objective order of reality is, ultimately, a subjective judgment; that the judgment that duplication of rigorous experimental evidence by others, the "objective" writing of data in a notebook, the predictive potential of a "law," or whatever as a criteria of truth or reality or objectivity is a subjective evaluation and judgment – including the judgment that it is subjective. We cannot get "outside" our own perceptions and concepts enough to tell whether they correspond to some objective "outside" reality because we must use those same kinds of perceptions and concepts to judge whether our perceptions and judgments and concepts are correct or not.

The Criteria Crunch

From a purely philosophical perspective, the criterion by which we judge the criterion by which we judge the criterion of truth is probably an infinite regression. At least I think, subjectively, it might be. . . I would like to be able to prove that there is a level of reality at which things are known as true without possibility of doubt but the assumptions I have had to make to even make that statement cancel out the effort. In our times this aspect of it has become dominant because of the closed loops that certain of those questions seem to create. Even now those problematic closed loops, these self-reflexive questions and statements that turn back on themselves, may be understood as either dead ends or clues to the direction of further developments and expansions. Great minds have dealt with these puzzles, over time, from many perspectives and with great energy and have succeeded, for the most part, only in clarifying the problem rather than solving it. But I believe that there is a way to turn the problem back on itself to allow us to actually use it to develop an expanded, higher power form of epistemology that will allow us the flexibility and degree of certainty that we seek.

Breaking the Epistemic Barrier

If the barrier to our philosophical progress is the inescapable self-referential nature of our determination of truth and reality, how shall we develop a new set of criteria and a new modality of thought adequate to allow us to operate comfortably and confidently in such a relativistic environment? **The key to comfortable existence and full reality creation in that relatively radical freedom lies in a new understanding of self-reference.**

The cynical street wisdom that says "If you can't beat 'em, join 'em," has direct application in breaking the perceived epistemological barrier we face due to not being able to get beyond our apparent self-referencing subjectivity of perception and judgment of what is reality. I propose and advocate that the way to get beyond the apparent impasse is not to attempt to break the barrier, analogous to the way a supersonic aircraft breaks the sound barrier, but to subvert it, turn it to a positive tool against itself, ultimately to simply bypass it.

The approach I am proposing here works from assumptions diametrically opposed to those of our current scientific methodology. We use our human subjectivity as a starting point rather than as an obstacle to truth which must be overcome. How can we do that? We know, or at least think, that we cannot start without an assumption so we tentatively and pragmatically make a very simple and straightforward assumption: since we are an integral part of the universe and we are self-referencing, subjectively operating beings we assume that self-reference is the way the universe works. And then see what that gains us. As it turns out, it gains us a great deal: we gain a powerful philosophic and scientific methodology of broad spectrum practical application that allows us to gracefully exist and operate in a relativistic universe – and we gain a key to a unified field theory, a "theory of everything." Ultimately, we gain a commonality of means of freeing our thought and our action and dimensionally expanding our consciousness. Actually rather satisfactory.

The bonus I see, the adoption of self-reflexivity as the new metaphor of an adequate unified field concept is simple: **we recognize that we cannot get beyond subjectivity. Since we seem to be an integral part of the universe, generated by the universe, and are self-referential then we simply take the universe as self-referential all the way – since it is, apparently, the only way we can perceive it anyway.**

We do not have to reject the possibility of there being an objective order of reality (things being the way they are, independent of how they are perceived or understood by any consciousness). Neither do we have to make the assumption that there is no objective order of reality. For both those who would insist, a priori, on either an objective order of reality or a subjective reality, we simply adopt the pragmatic, tentative postulate that the objective order of reality is that it is subjective in so far as the universe we know exists only in our conceptualization of it. We should be very precise here. Does that mean that there is nothing, no thing, outside of our minds? No, or, at least, not necessarily or de facto. Does it mean that our minds actually create the universe? No, or, at least not necessarily or de facto. Does it mean that there is some sort of in-between ground where there is something out there yet which is constantly being modified to conform to our conceptualization of it? No, or, at least, not de facto. Does it mean that there is something outside of our minds which is actively modifying our minds to think we are perceiving it, albeit subjectively? No, or, at least, not de facto. Am I suggesting that we begin without any assumptions at all? No. That in itself would require the assumption that we could or should or both.

All arguments that would object to this approach may be countered by questioning how, other than by subjective judgment, does one know they are true, including this statement. All arguments that start with the opposite assumption that there is an objective reality knowable by us is subject to the same question. The most obvious clues are, at the philosophical level, the inability of our current 3-state logic (see Doug Hofstadter for an encyclopedic cataloging and G.S. Brown for indicated solution) to handle oscillating self-referential statements and infinitely regressive loops; in mathematics the necessity for imaginary numbers; perhaps the not too subtle deviations from the chance anticipated in raw natural selection at the biological level. The descriptive metaphors of the most advanced quantum mechanics are becoming more highly self-referential. Degree of self-referentiality may be said to be the implicit metaphor of measurement of success of complexification as one proceeds upward from the border (extension at the "size" expressed by Plank's constant) of pre-geometric potential to us.

Self-reference also seems to be the engine of adaptive change. Godel has shown, with regard, specifically, to mathematics and as has been generalized by others, any system of explanation cannot completely explain itself. But it seems to me that, positively, we may eventually be able to generalize that, by the very nature of self-reference of whatever degree, reaching the limits set by the parameters of any system reveals to that system what is "lacking" to attain to the next kind or degree of potential revealed by the nature of the system to itself. It would be oversimplified, certainly, but perhaps helpful, to say that this is source of the "deep grammar," the "genetic epistemology," the "prepared learning," the "manu-factured thresholds" – and the development of more adequate metaphors to ex-plain our explaining. All clearly manifestations of a fundamental genetic adapta-tion, as is genetic adaptation itself, to extend the self-referential metaphor to that area.

Self-reference as a Meta-Metaphor

Setting myself the task of writing a concise statement with regard to how I see the principle of self-reference being applied practically to the development of a new dimension of philosophy – which in itself provides a key to the formation of an unified field theory and law – has forced a good deal of background pro-cessing and clarification. The intensity of the effort is due to the novelty of the subject itself – I don't know of anyone else who is suggesting we develop a uni-fied field law specifically in terms of self-reference: the closest work would seem to be that of the physicist, John Archibald Wheeler – and also the inherently com-plex nature of the subject. When one deals, self-referentially, with the topic of self-referentiality, the convoluted feedback loops can become daunting. It is the significance of the seeming impossibility of disentangling those loops that is our key. In the following, therefore, I first give the background of how I came to understand the importance and central nature of the topic so the reader can gain some appreciation of my accuracy or inaccuracy in doing so. Then I develop, step by step, the way I see clues to how the principle was used on this planet and, quite probably, on Mars by the Nefilim as a most general unified field metaphor. Thirdly, I describe how I believe we can use the same approach to arrive at a unified field law ourselves. If I am correct, this is the second most important profound piece of information we have been able recover from the Nefilim legacy. (The most im-portant piece of information we have recovered from that nearly lost body of data is the nature of the gold compound which, I believe from the evidence, is the drink the Nefilim used as a maintenance tonic for immortality. I will discuss this startling find in a later chapter.)

Subjectivity as the Human Type of Self-Reference

To properly explain the concept of self-reference we must first clarify a cen-tral, key concept critical to this discussion, linkage of the two terms, self-reflex-ive and subjective. Subjectivity is simply the human form of self-reference: we define ourselves by our self-awareness and, ultimately, look to ourselves for the truth. In the contemplation of the reciprocal linkage I suggest, lies, available to the reader, a convenient beginning of the experience I have undergone that has convinced me of what I am about to propose here and which the reader will thereby be able to judge subjectively for HIrself. From one point of view – which phrase is, in itself, an automatic pun in this context – "subjective" may be taken as a sub-category of "self-referential"; subjective perception is a form of self-reference: we refer to our perception of ourselves perceiving to judge if our perceptions are accurate. If we were not able to recognize ourselves as self-referentially aware we would not be able to distinguish between what we commonly define as "subjec-tive" and "objective." From the opposite point of view, "self-reflexive" might not be a concept available to us unless we first could recognize ourselves as having "subjective" thoughts, ideas that were uniquely or, at least, privately ours.

The value of the private, subjective view is that it leads us to the potentially perplexing encounter with the doubt that arises when one allows oneself to won-der if it is really possible to verify subjective truth as correspondent with "objective" reality "outside" the mind. In this case "self-reflexive" might be understood as a sub-category of subjective, at least as in so far as "self-reflexive" or "self-referential" may be taken as derived as concepts modeled on our own perception of our own subjective perception of ourselves. The reciprocity between self-referential and subjective, paradoxically, may become clear as they become confused at this point.

How Do We Avoid Analysis Paralysis?

It may by asked, quite reasonably, How are we to function in a totally recursive mental environment? Won't we tend to mental paralysis? A sort of epistemological catatonia? A sort of neurological shorting out from overload? Not if we see it as event-oriented and self-propelling. And not if we work from a platform of species maturity that knows how to use the open-mindedness of the endlessly recursive as a way to "hang loose," against itself so to speak. I suggest that it is the temporarily adequate, appropriate modality as we move out of racial adolescence – but yet a function of our metamorphic development.

Life on the Other Side of the Epistemological Barrier

Since our consciousness is the "cause" of the epistemic barrier and also the critical barrier for the physicists in arriving at a Law Of Everything, I suggest that self-reference as a meta-metaphor is the key to the resolution of both problems. We can tolerate the apparent contradictions because we allow, temporarily, that our language is simply inadequate to express the complexity involved. We, in effect, therefore, have assumed, provisionally, a universe that is self-referential in its fundamental nature. So there are two correlative major factors to consider with regard to self-reference, two sides of the same coin: the psychological nature of our own self-referential consciousness and the relativistic nature of our epistemology, i.e., our conceptions of truth and reality and how we know them.

Let us consider the primary psychological character of our self-referential consciousness. It has two cardinal features, the relativity of perception and the relativity of epistemic criteria. It is at least difficult, if not impossible, to avoid the conclusion that the reality we perceive is conditioned by the type of senses and brain we are and use. The ordinary educated human probably knows of the ultraviolet part of the electromagnetic spectrum. But humans perceive physical reality without a direct visual perception of that wavelength although it is integral in all we "see."

The Second "Law" of Relativity

We humans perceive in three dimensions and our "truth tests" tend to be three-dimensional. Although we know something of a fourth dimension indirectly we do not integrate the physics of that additional dimension into our ordinary perception and conceptualizations or logic. For the sake of the discussion at hand we might generalize the principle governing our understanding of the universe; we might call it, *the second law of relativity: the "reality" any perceiving entity perceives is a direct function of its perceptive mechanisms*. That is almost a trivial fact in light of our common daily experience of the differences between our personal opinions.

It bears emphasis here that the garden varieties of subjectivity that are at basis of our everyday disagreements through misunderstanding, self-interest, anxiety, and ignorance flourish at least in part because of the long-term effect of the godspell conflict between absolutistic theologies and the resultant desperate notions that there is no way to tell who is right – or that one opinion is as good as another, etc. These attitudes are, ultimately, psychologically crippling.

For centuries the godspell syndrome has fostered and supported particularly extreme institutionalized forms of felonious arrogance. The Inquisition is a familiar example and the hysterical burning of a million women as witches is another. With the advent of the new paradigm and the dissipation of the godspell, the support for dogmatism will give way. But that, initially, will only clear the philosophical arena for free discussion. Free discussion will focus on determining new, more adequate criteria of truth.

The Third Law of Relativity

Just as we have generalized the Second Law Of Relativity to state the reality we perceive is a function of our sensory and neurological apparatus, so may we formulate *a third law of relativity: the criterion of truth any perceiving entity adopts, is a direct function of its type of perception and consciousness.* Even now our consensual concepts and communication are evolving to be able to express comfortably the expanded awareness of the Einsteinian four-dimensional paradigm. (The relativity of the four-dimensional Einsteinian concept is preeminently psychological: events "happen" in linear "sequence" in a four-dimensional space-time universe only in our mode of perception of them, not "really.") If one person thinks, and therefore lives, in a strictly Cartesian/Newtonian experiential universe there will be a constant conceptual conflict or, at least, misunderstanding with another person who literally lives in an experiential Einsteinian space-time universe. A quaint, primitive Flatlander scenario in 3-D. Everyone is aware that, not only do we disagree because of the differences between our individual perceptions of the world around us, but we tend to take our subjective perceptions and convictions as the only reality. As a matter of practical effect, we equate "reality," as we see it with "truth." We then use that "truth" as a means of judging the "reality" and "truth" of others. The problems arise when we attempt to impress that criteria on others and our "truth" as the only "truth."

Moving Toward Habitual Four-dimensional Consciousness

Alan Dressler, in *Voyage To The Great Attractor* (pp.134-5; Alfred A. Knopf, 1994) speculates that, if our consciousness could discriminate time to the billionth of a second we could, indeed, perceive, think and act in space-time with ease. He feels that, under sufficient demand and motivation to do so, we could develop through experience and practice at least a sense of space-time, and that the future may well see us using highly advanced electronic chip implants to enhance our ability to operate in space-time. I would suggest that evolutionary pressures are driving human development in that direction already and chip implants may only be a temporary aid toward that end for a short time. I also think that we may find very specific chemical enhancers which will be more subtle and sophisticated than implanted chips to bring us gracefully up to those levels. Psychedelics like LSD are certainly some of the first we have discovered of this type. It is not beyond our imagination to conceive of a time when we will have the ability to genetically engineer for space-time, four-dimensional consciousness. We may be seeing the beginnings of some of these characteristics of four-dimensional consciousness in some of our children who exhibit remarkable abilities.

Expanding Our Philosophizing
by Expanding Our Consciousness

We need to redefine philosophy, traditionally understood as the seeking of ultimate causes by the use of reason alone, as a dynamic process that uses as its primary tool whatever is our currently most dimensionally evolved conscious modality.

Whatever is the "highest" consciousness we can detect "rules," laws for, like the Greeks did for reason, we will install as our criteria base. Our philosophical modality, the toolkit we use for philosophizing, will be eventually understood as a temporarily adequate mirror of our most advanced consciousness and therefore of the leading edge of our evolutionary development. We need to recognize that each of the types of awareness available in the spectrum of human consciousness has its own inherent epistemology, logic, truth, indeed eschatology. It's just a simple corollary of the first step.

We clearly need to matriculate from an answer-based philosophical mode to an event-oriented one. We have crude, limited, precursory models of this in the "that is not it" technique of the East and in Zen. The "answer" to the Zen monk's koan is not a syllogistic resolution but an event: the seeker who succeeds, who "gets it," "pops" a neurological "relay" into an expanded non-linear meta-syllogistic recursive awareness that subsumes rationality (like Einsteinian relativity subsumes Newtonian mechanics). Simply put, it is an expanded awareness that includes reasoning but is more than "reasonable" and which sustains itself by a sort of oscillating suspension beyond over-simple opposites and a view that transcends linear time. Does that mean that we all need to become experts in Relativity theory or Zen monks? Certainly not; I use these as familiar analogies to illustrate the point. The point is that we are headed for a plateau of our species' general consciousness which will make Zen satori seem antique and outmoded and communication with the strangest alien we can imagine normal and natural.

The First Steps to an Expanded New Philosophy

In this context we can see the significance of Descartes turning thought back on itself, Godel turning mathematics back on itself, G. Spencer Brown turning logic back on itself, Douglas Hofstadter turning solipsism back on itself, Timothy Leary turning the human brain/neurological system back on itself in what Leary calls I^2, (intelligence squared), intelligence increase, the only unlimited game we know. As the most practical first step toward expanding our philosophizing capabilities we should recognize and adopt an expanded, four-state logic as developed by G. Spencer Brown in his *The Laws of Form*. This will immediately afford us a much more flexible and inclusive logical modality mirroring our shift to four-dimensional consciousness.

G. Spencer Brown: The Logical
One to Revolutionize Our Logic

The thought and work of this philosopher-mathematician may be singled out as a prime example of what is required for the expansion and upgrading of our philosophical modality.

G. Spencer Brown is an English mathematician and logician. In a modest book called *The Laws of Form*, he begins with the problem of beginning itself, reducing to the greatest simplicity the least number of assumptions to begin with, to develop a fundamental set of rules of thinking and an elementary arithmetic.

One of the important conclusions he arrives at is that our three-state (true, false, meaningless) logic is inadequate. We should have a four-state (true, false, meaningless and imaginary) logic to accommodate the dimensionality of our thought. Logic diagrams in four-state logic must be drawn in three dimensions and, when executed thus, allow for the imaginary components to be incorporated in such a way that they become the nexus points for integration with other logic statements. They represent a dynamic grid of potential relationships. At that point of development it is not too difficult to see at least a beginning analogy with the neural networks of the brain – or neural simulations in computer programming and hardware.

I suggest that his approach is the most adequate, so far, for expanding and evolving our formal logicizing so that it becomes a better mirror/metaphor for the way our brains operate. It affords us a way to comprehend the significance of oscillating statements, infinite regressions and progressions, and especially self-referential statements, and to integrate them into our thinking and our logicizing. It is well to remind ourselves once in a while that our formal and mathematical logic, our computer simulations in software and hardware of neural networks, our artificial intelligence programming attempts to develop thinking devices that pass the Turing test or better, are quite obviously primitive models of the best of what we do with our biological wetware to this point. Even when we aim for specialized functions in hardware or software which can outperform the human brain, the brain is still the model for the basic approach. Another way of saying it is that, no matter how far we try to move even beyond the brain's ordinary capacities we, nevertheless, are attempting to model our own, however expanded, type of consciousness. Because, literally, that is the only way we can conceive of doing it. If we could conceive of it another way, that way would be a product of our consciousness. How does it appear to your subjective evaluation?

Self-Reference as a Key to The Law of Everything

In the process of expanding our logic and our scope of perception, and developing a new dimension of philosophy on the basis of the fundamental principle of self-reference, we are drawn inescapably into dealing with our consciousness. Since we must consider and define reality philosophically in terms of that consciousness, we find ourselves dealing with our consciousness in the same terms in which it is perceived as the "problem" which prevents scientists from arriving at a Unified Field Theory understood as a Law Of Everything. If I am correct, then this resolution not only will give us a new dimension and modality of philosophy, but the key to unifying the laws of our consciousness with the fundamental laws of physics and therefore the key to the Law Of Everything. Philosophy and physics – and mathematics, logic, all the sciences, information handling – would then become unified, a single discipline. This is one of the first characteristics of a Law Of Everything, among others, which begin to manifest naturally out of these considerations. In the next chapter we will consider what a Law Of Everything might be like and what effect its discovery will bring. Suffice it to say here that, having made the tentative assumption that self-reference is the key concept and meta-metaphor for our understanding of the universe and ourselves as part of it, we should proceed by developing a philsophy and science which approaches the universe as a closed, self-referential system and all parts of the universe as a system of self-referential systems. The Law Of Everything then will be a precise description, statement, equation perhaps, "law" of how self-referential systems come about and interact. That is oversimplified for now but substantially correct at least as I see it currently.

Let us imagine some of the practical changes philosophy will go through as a result of us breaking the godspell and moving to new human status. The new meta-metaphor of self-reference allows us to cleanly and adequately map the additional dimensions of our current consciousness and pass through the (evolutionary) epistemological barrier perceived as our inability to get past or outside of subjectivity as an ultimate criterion of reality. Without process theology; without the retrograde humanistic pragmatic appeal to some secularized theology of an objective order knowable by reason to avoid the anarchic hell imputed to radical subjectivity; without appeal to the obtuse haiku physics of Wheeler's genius; without appeal to the criterion of that most recent bio-adaption of scientific methodology for genetic advantage and evolutionary change, sociobiology.

Everyone Their Own Philosopher

In the new society, everyone will be HIr own philosopher, HIr own criterion, HIr own scientist, HIr own validation. By that I do not mean that the new human will be the epitome of self-centeredness or narcissism; just the opposite. It simply means that the individual will be free to deal with the universe directly, one-on-one with no preemptive theological strictures, no mandated "reality." A correct view of self leads to a correct view of the universe and of others.

CHAPTER 10

EVALUATING THE NEFILIM LEGACY

[on the basis of research]. . . . we are assuming that the ancients had knowledge perhaps different in external description but essentially equivalent to ours in accuracy, elegance and sophistication. We believe this is possible even with regard to the frontiers of current knowledge in fields such as topology, transfinite numbers, higher-dimensional geometries/symmetries, genetics and DNA structure, nuclear, quantum and relativistic physics, optical and acoustic interferometry and holography and solid state physics. This is not to say that this knowledge was available and/or confirmable during most historical periods. In fact, it is likely that the last possible period for these ideas to have been experimentally verified would have been in pre-Christian Alexandria or Syracuse. Clearly during most other periods we have only remembered an increasingly distorted, amended, abridged, extrapolated and interpolated local metaphor. . .

Stan Tenen
Torus

Stan Tenen's evaluation of the knowledge base of the "ancients" is intriguing, particularly with regard to the questions it prompts. In view of what he has discovered and uncovered concerning the knowledge base possessed by whoever it was that invented the Hebrew alphabet, he has concluded that the areas of knowledge he lists above were known to some in the ancient past. Certainly, in the light of greater and greater knowledge which we acquire about the capabilities of those in the past who used measurement systems base on the precise polar circumference of the earth, cast the Giza pyramid blocks in geopolymeric concrete, worked comfortably in elegant geometries, used alphabets developed from the topology of self-referential consciousness, etc. we feel compelled to concur. But the question is, precisely, who those ancients were that not only knew those things and used them, but discovered and developed them in the first place. Having recognized that the Nefilim were the "ancients" of "an advanced civilization" who developed and possessed all this science and taught some of it to some humans, the past becomes bright and we can begin to evaluate and take advantage of that legacy as we expand our vision and become the new humans.

Science in the Ancient Past: Who Was
Capable of It; Who Was Responsible for It?

With regard to the discovery and development of that advanced science in our ancient past, there are two possible scenarios: humans were capable of doing so and alone discovered and developed that knowledge, or humans could not have done so in those times and must have been helped or taught it. The question, then, being by whom? It is very difficult to find a consensus among establishment scientists as to whether humans in those times could have discovered and developed such detailed science and disciplines. There are two factors that bear on the question: Were humans evolved sufficiently to accomplish that relatively stupendous feat in those times? And, if, at least, some were capable, neurologically and mentally of arriving at such insights and conceiving such concepts, would they have been capable of organizing and spreading such elaborate science so that it could be used across the world and form the basis for a number of languages in a number of different cultures?

The traditional position of archaeologists and historians has been influenced by the version of evolution understood as a continuum from the time the lightning struck the mud, through us as the current, most advanced product of that process. As a result, the predisposition has been to look back on humans of those early times as less evolved. Even though recent discoveries have brought to light thirty-five thousand year old flutes, three-dimensional carvings in ivory of human heads complete with precisely drilled irises, the startling art of the Lasceau caves, a veritable encyclopedia of ooparts (tools, toys, scientific devices of advanced technology ostensibly out of place in history) and the fact that Homo sapiens has been around for at least 150,000 years, the scientific community is extremely reluctant to acknowledge that such high scientific organization and achievement was even possible by humans only thousands of years ago.

At the time of this writing, the geophysical evidence, as presented by West and Schoch, that the Sphinx was constructed at least 9000 years ago is being resisted by the scientific community not as invalid geological evidence, but on the basis that it could not be, that the Egyptians simply did not have the capability to do so at that time. The general consensus is clearly that, in view of all the evidence, humans in those remote times, although even possibly possessing the potential, neurologically and mentally, to have been taught it, did not have the experience of the knowledge base to develop the mature systems of advanced science that we know are extant. The conclusion, therefore, since it existed and was used, is that humans did not create and develop it. And that, of course, is precisely what humans in those times declared and recorded. All the recorded history of those ancient civilizations say plainly that humans were taught and given technology, science, culture and learning when "wisdom was lowered from heaven," when the Nefilim decided to teach us and furnish us with it.

So the current academic position and reasoning is very awkward to say the least. It borders on a flat contradiction to say that the humans of the first civilizations were one step above being naive primitives incapable of such advanced thought and, at the same time, claim that they were evolved, intelligent, developed and had sufficient leisure to attain to the advanced level of spherical geometric and topological knowledge as well as somehow knowing the precise polar circumference of the earth and the existence of all the known planets in our solar system – plus one other, Nibiru, the home planet of the Nefilim, which we are only now beginning to locate.

Isolated Geniuses?

We cannot rule out the isolated genius who was capable of at least envisioning such theory, analogous to Leonardo Da Vinci envisioning the helicopter, but putting it into concrete form and widespread practical technology seems very unlikely. In the final analysis it appears quite clear from both the evidence from all of the investigative sciences and the ubiquitous testimony and recorded history of the ancient civilizations, that the simple fact is that the body of strictly human collective experience and thought could not have produced such theory and practice. This appears to have been the case even up to the advent of the sudden advanced cultures such as the Sumerians'. The fact that the first civilization was "sudden" and complete, even by our standards, is itself an indication that the knowledge and science and technology were donated by the Nefilim rather than the culmination of the collective experience and thought of humans in those remote times. And the simple, unavoidable fact is that all those ancient civilizations told us exactly the same thing: The Anunnaki/Nefilim did those things and what humans knew, they were taught.

When confronted with strong evidence for the knowledge and use of electricity (demonstrated by the recovery of clay pots with electrodes still in them, probably used for plating, from the Iraqi desert dating from 2500 B.C.) on the part of the ancient civilizations the same experts who will vehemently deny that the ancients could have known or developed such an advanced technology because they were so relatively primitive and naive will still not admit that the Nefilim were real and taught humans such information and technology. But that leaves them in a position of not being able to give any good reason at all for the evidence. The use of electricity is a single example of the attitude toward the entire topic of ooparts. Ooparts are real, the products of high technology, but out of place in time – and used by humans. If humans were not capable of creating them then, then who *was* capable and gave them to humans to use?

There is one other explanation advanced currently that, at first consideration, seems to circumvent both the human source and the Nefilim source explanations we have been considering. It says that there must have been an advanced human civilization which existed many thousands of years ago and which was wiped out in a cataclysmic planetary event, the remnants and technology from which was somehow inherited by the early civilizations known to us. It would seem that, if this were actually the case, the early known civilizations would have had some clue to that fact, would have known, at least vaguely, of that ancient civilization of humans and acknowledged it in some way as the source of their civilization. But they never did that: they said that the civilization that invented humans, gave humans culture and technology, was that of a different race, the alien Nefilim, who came from another planet in our solar system. Proponents of a theory of a previous human advanced civilization tend to point to the reference to Atlantis by Plato – which was later denied by Aristotle, Plato's pupil – as a clue to who and what that ancient advanced civilization was. I can see no conflict – and neither does Sitchin – between the existence of Atlantis and Sitchin's thesis; the time frame of 15,000 years ago for its flourishing and eventual destruction could, if it existed, identify it as a Nefilim center of high culture and technology. But this explanation, in my perspective on history, does not put it back far enough in time to be a solely human civilization that was very advanced and yet could be forgotten completely – and still leave a legacy of high technology. All these elements do not compute together. If there was an Atlantis, it was most probably simply a

Nefilim base, given the time frame, and, perhaps, one that was devoted to their highest technology placed purposely on an energy grid point that was most appropriate. To say that a civilization, advanced or not, could rise and fall, be completely forgotten and all traces lost, yet had somehow left a legacy of advanced science is, ultimately, a contradiction. A legacy of advanced science, whole and distinct enough to be recognized as such is a significant remnant that has not been forgotten. And the fact of the matter, in this case, is that the civilization that developed and possessed that science (that is so advanced that only now are we able to begin to recognize it because we are beginning to rediscover the same principles) was never forgotten.

The Nefilim civilization, on this planet for some 150,000 years before they invented us, and continuing until their phase-off from Earth probably around 1250 B.C. at the latest, is known to us through verbal tradition, written record, inscribed reliefs, architecture, monumental construction, residual science and the myriad forms of religion abstracted from our earlier service to them. It is the focus of archaeology, linguists, cultural anthropology, archaeoastronomy, and, directly or indirectly, involves many other sciences. It fills our museums and libraries and is the focus of attention of millions of tourists every year. To deny, ignore, misinterpret or warp that vast knowledge or to assign it to the category of unreal myth is not to forget it.

We can no longer look on our ancestors as ignorant and naive primitives who could not possibly have possessed advanced technology. The scientist or skeptic who has objected that humans of those early times could not have know Pluto's existence, the use of electricity through storage batteries, the technology of airplanes (vimanas), etc. is partially validated. It is quite clear that, *on their own*, with only the human resources of their times, they could not have accomplished those things. But they could have been given the knowledge and the technology. And the evolutionarily advanced nature of the Nefilim half of our gene code gave us the potential to understand it and utilize it.

Ooparts no longer have to be dealt with as puzzling items of relatively high technology mysteriously out of place in history. They are simply remnants of technologies that were known and available to at least some of the race in the far distant past as a result of copying or learning from the Nefilim, and were gradually lost when the Nefilim left the planet and their support and reinforcement was withdrawn. The scientists who have postulated some precocious, random technology based on the evidence of ooparts have part of it correct: the technology was precocious but not random, it was advanced but taught to us, not developed by precocious humans. The occultists who have claimed the sources of their traditions to be a mysterious higher intelligence and culture in the past are, within this new context, both partially correct and vindicated. The mysterious intelligence who gave or taught us this high science is now revealed as the Nefilim. The historical scientific facts show that the real situation was paradoxically both more pedestrian yet more amazing than our partial interpretations. Conflicting groups can find a common ground on which to come together.

The new paradigm furnishes an expanded context in which the entire spectrum of genuine science, regardless of its present characterizations of whatever form, can be integrated. The new paradigm opens up a profoundly important possibility to us: the door to our past stands open; we can now confidently reexamine the past freely for information that may be available and of benefit to us. We can allow ourselves the opportunity to go back and do a systematic scientific inspection of the information recovered from those ancient civilizations, the source of which was the advanced science and training given our ancestors by the Nefilim, with the intent of gaining the value of both practical and theoretical knowledge. Easily identifiable as advanced Nefilim science taught to humans is the legacy of astronomical knowledge and calculation, the spherically enclosed tetrahedral geometry of planetary geography, linguistics, and the knowledge and use of monoatomic forms of elements.

The Value of Nefilim Science

What may we gain by recovering and relearning the concepts and system of science used by the Nefilim? In some cases in which we clearly have advanced at least as far as the Nefilim had in the past, the only value is in the historical recognition of the nature and direction of the information. But, with regard to items which we are only now discovering or may even have not yet discovered, the information may be of very great value. We may save a great deal of time in some cases, in others we may even find leads to future discoveries. We will certainly want to know what is the nature of that information, what is its factual content and what practical value it may have for us right now.

But an equally important question is: What were the form or forms in which that information was organized and handled; what were the nature of the paradigms, metaphors and codes in which and through which that information was developed, formulated and applied, communicated, transmitted and preserved? We tend to think of the question in terms of how we would expect humans to have organized, handled and applied the scientific metaphors and technology in those ancient times, i.e. how we would have done it. But the questions should be asked in a more fundamental form with an expectation of possibly discovering very different modalities. Since the Nefilim were the possessors and source of that science, how did they view the universe and do science, i.e., what was their philosophy of science?

We already have at hand information, from a number of disciplines, which gives strong indication of the approach the Nefilim took to the universe in general and what forms the application of that philosophy of science took at the practical level. But, before beginning to identify and consider the nature of Nefilim science, it is well to reiterate here how, generally, we have been able to determine what that science was.

Identifying Nefilim Science

As I have pointed out in *Breaking The Godspell* and in this book, one of the primary keys to decipherment and interpretation of Nefilim science and technology has been our own achievements in those same areas which have allowed us to identify our newly discovered technologies with theirs. A single cardinal example: it was not until we had discovered the principles of genetic engineering and begun to understand and develop its potential that we could even begin to understand the clear descriptions of how the Nefilim literally invented us by crossing their genes with those of Homo erectus. Previous to our genetic discoveries, we could only understand the technical information about the genetic engineering carried out by Ninhursag to create us, as a new species, as a story of naive magic and fantasy. Once we understood what it was that really was being described in the ancient documents, the entire picture became clear. This is a single, albeit pivotal, example of how, on reaching a certain level of development and discovery, our own science affords us insight and revelation of the same science as practiced by the Nefilim.

It is also important to note that there is another resultant benefit to us in this type of discovery: not only does it enable us to identify Nefilim science but it also makes it easier for us to accept the literal reality of the Nefilim in those times when we can see clearly that their deeds were, indeed, possible and real although, previously, we could only have seen them as fantastic. The areas of their science which we have recognized for these reasons and which seem quite obvious at this point in time are the Nefilim's astronomy, physics of space, geology, geography, metallurgy, and architecture as well as genetic engineering. The information concerning these obvious manifestations of the Nefilim's science are described in the works of Zecharia Sitchin and I highly recommend the accuracy and depth of his descriptions.

Nefilim Astronomy

The breath and sophistication of the legacy of astronomical knowledge taught and left to us by the Nefilim is nothing short of astonishing even today. They gave us a complete history of the formation of our solar system, including the intrusion of their planet from the outside and its effect on the final form of the system. They taught us all the essentials of astronomical observation and set up ziggurats and other monumental buildings all over the world precisely designed and constructed to be astronomical observation stations, with astronomical information encoded inherently in their structures. Clearly the major purpose was two-fold: it gave them and us the essential tools for determining the time of return of their planet and where we should look for it when it approached the inner solar system.

The other major focus of that knowledge, as rediscovered by Cotterell, encoded in the Mayan system of symbolism, is the effect of the powerful magnetic field effects in our sun which effect the Earth periodically. These forces, manifesting in the sun as sunspots, apparently can cause polar shifts and large magnitude, geological changes in the Earth which can and have been detrimental to whole populations. In teaching us this astronomy they were also teaching us some very important astrophysics both with regard to planetary physics and gravitation, as well as providing us a means of protecting ourselves against harmful physical effects.

Nefilim Geography and Geology

When one considers the overall picture of Nefilim activity and purpose for colonizing the Earth in the first place, one becomes aware of a broad spectrum of geological knowledge. The Nefilim were expert miners, gold being their main objective in coming here, in mining extensively in Africa as well as in the Americas. Unlike the astronomers who tend to dismiss any advanced knowledge of their science as not being possible in ancient times and, thereby, missing valuable opportunities in ancient data, the mining engineers in Africa have taken advantage of this knowledge by locating ancient gold mines rather than laboriously trying to locate gold through difficult geological search, finding gold mines close to 100,000 years old.

The fact that the Sumerian vocabulary had no less than sixty terms for petroleum products, no doubt learned from the Nefilim, shows even through this second hand source, the extent of their knowledge of such a fuel and power source.

Ancient maps, often drawn from even more ancient source maps, seem mysterious to many in their amazing detail of the geography of the world, their depiction of the continent of Antarctica without its covering of ice and snow, their sophistication of method of projection techniques, etc. But the Nefilim were capable of all these things and the mystery clears up immediately when we recognize that they were the source who transmitted this knowledge to humans. All evidence indicates that they were obviously capable of mapping the entire surface of the planet in detail and actually did so both for navigational and geological purposes. Satellite scans would have accomplished the mapping well and probably provided them with the information they needed to locate gold, tin, copper and other mineral deposits. It is only recently that we have come to recognize that the precise cutting of huge stones in Nefilim constructions all over the world, and particularly in disturbance prone areas such as Machu Picchu, reveal an advanced engineering knowledge that made the construction earthquake resistant by the intricate, multifaceted interlock of the stones.

Nefilim Metallurgy

The Nefilim preoccupation with gold as well as their extensive mining and use of tin, copper and other metals is written large over the entire planet. Even after they left the deeply ingrained tradition that humans mined gold but it belonged to the gods, the idea lingers yet in the legends of peoples all over the Earth. Yahweh, not one to leave any ambiguity about anything, made it clear that "The gold and the silver are mine." But their techniques of metallurgical processing were advanced and we might learn something from them yet.

Although it is held to be mysterious by those who do not know of the Anunnaki\Nefilim heritage yet, an item like the pillar of "iron" in India which never rusts or shows any deterioration in a very humid and warm climate is most likely their handiwork, and it would be well worth a rigorous scientific analysis to determine if, indeed, there is something we might learn from such technology.

Ultimately, it may not be the scientific-industrial applications of Nefilim metallurgical techniques that will prove to be as invaluable to us as their use of the monoatomic form of gold for a very different purpose, immortality. Immortality is already on the horizon through genetic engineering and most probably through the rediscovery of the monoatomic form of gold by David Hudson, quite certainly the focus of the alchemical and some occult traditions, lost over time but remembered as the goal. The monoatomic form of gold (and other "elements") is the single-atom form of the element. The periodic chart of the chemical elements lists them in the most stable, fundamental form in which they normally occur in nature. But gold, in its yellow metal form, is always formed of two atoms.

Through a long process of investigation to solve a mining problem Hudson gradually came to realize that he had rediscovered the monoatomic form of elements and applied for a patent. Gold in its monoatomic form is a fluffy white powder, still gold, but not metallic. This white powder, monoatomic form, according to Hudson, is a superconductor at body or room temperature and has very unusual properties. In the process of discovery he became aware that this is the form of gold sought by the alchemists as the means to immortality and the extraordinary powers of the human mind. The traditions say that it was known in the past but the technology had been lost over time. Research has shown Hudson that previous to the alchemical tradition the white powder of gold was known and used by the Hebrew priests who ingested it behind the temple veil and that the

Hebrews had been taught the technique so that they could function as the producers for the Egyptian rulers. The Essenes had probably taken the technique of manufacturing it into the desert at the time of the destruction of the Temple at Jerusalem and used it as a central facet of their lifestyle. All evidence points to the Nefilim as the original source for the production of the white powder which they used themselves for longevity or relative immortality. This points to the fact that the primary Nefilim mission here on earth of mining huge quantities of gold was both to seed their planet's atmosphere with a molecular form of gold as Sitchin has postulated and also to provide themselves a source of the monoatomic form as Hudson has suggested.

Again we have an excellent example of how we may profit from advanced technology developed by the Nefilim and need only to allow ourselves to acknowledge its possibility to rediscover it and take advantage of its potential. Immortality will be the primary, dominant characteristic of the dawning new phase of our racial maturation, a matter of simple human dignity. The legacy of the technology of the white powder of gold may well be the key left by the Nefilim to that end.

Nefilim Architecture

When we come to the stupendous constructions the Nefilim accomplished, often in very difficult terrain with what seemed to be a scarcity of the raw materials needed and requiring transport over great distances, their expertise is obvious. But it is not sufficient to simply recognize the awesome size, beauty, and general impressiveness of these buildings. The topology and geometry with they were designed and built holds much for us to learn and profit from. The obvious requirements for an advanced technique for transport and lifting in the construction should prompt us to find an answer as to how they could do such things as precisely cut, transport and put into place a solid block of stone which weighed well over 800 tons in a wall such as the foundation of the temple at Baalbek. We still possess no technology with which to accomplish such feats and the speculation is forced that they had some mastery of gravity. This should be a significant clue for our engineers and scientists if they can finally open up to the idea of the physical reality and presence of the Anunnaki in the first place. Just as there are depictions of high voltage electronic tube devices on the walls of tombs in Egypt which are only now beginning to be recognized because of our technology having reached that point of development, so we should begin to look for indications in pictures and texts of specifically anti-gravitational technology. Since some of the indications, through what seems to be a remnant of it still used by Tibetan monks, are that at least small applications were primarily acoustic in nature. The monks are reported to use instruments placed precisely to create an acoustic wave that can levitate large boulders for construction purposes. Acoustic technology of this sort may have brought down the walls of Jericho.

More Advanced Rudiments and Suggestions for Our Science

It is the notion of anti-gravity technology that moves us into an even more advanced area of evidence of Nefilim science and technology that may be even more valuable to us. These areas of Nefilim science, which we are only now beginning to penetrate, are of great potential significance but are not as obvious, as yet, due to the novel concepts upon which they are built.

These novel concepts, it is becoming clear, give us even deeper clues to the philosophy of science underpinning their scientific applications and engineering.

These new concepts are to be found in the areas we know as higher dimensional physics, linguistics, communication, and cellular biology.

Specifically, lest we think only in the terms in which we normally conceive of the function and limitations of these human sciences, the novel aspects lie in an added dimensionality and an unified conceptualization of the laws of the universe underlying them. I suggest that this unified conceptualization equates to what we would consider a unified field law. This unified field law, literally a "law of everything" as it is conceived by some scientists currently, would be the basis for all their other sciences. By now, through the popularization of the notion of an unified field theory by the media and the writings of experts in the field such as Stephen Hawking, even the non-scientist is at least vaguely familiar with the idea of an unified field law.

For the sake of this discussion, therefore, we must first know what our scientists understand by a unified field theory and a Law Of Everything. It is critical to understand that there is not a total consensus about the definition of these terms by scientists. Some scientists see a complete unification of all the laws of the known forces of nature into one single unified field law as a complete explanation of the universe, automatically a law of everything because they are convinced that the universe is entirely physical, matter and energy is all that there is. Some scientists are not all that convinced that we will have arrived at a law of everything by the unification of all the known forces because they recognize that phenomenon like consciousness cannot be derived from the known forces and a complete unified field law would still not be a law of everything. We shall discuss these important and fascinating arguments in a later chapter.

The Concepts of a Unified Field Law and "A Law Of Everything"

Our science formulated the concept of an unified field theory within the recent past and is working, as a primary goal, toward the realization of this "law of everything." That is a rather awesome concept. Currently, unified field theory is conceived of basically in terms of physics, as a scientific concept: a law, a set of equations or a single equation, or principle which would precisely describe how, literally, everything in the universe "works." By its very comprehensive nature it would allow us to also predict, therefore, how things were going to happen.

This profound concept of a unified field law is most easily understood through the meaning of the word unification in this context. Some time ago, magnetism, which has been known as a curious phenomenon of nature to us for a very long time, was finally understood sufficiently to allow for its description and prediction in precise mathematical form. The laws of magnetism, i.e., how magnetic fields interacted with bodies and other magnetic fields, were written. Separately, the laws of electricity were also determined and expressed in precise mathematical form. Eventually, it was discovered that there were correlations between the two sets of laws: electrical currents generated magnetic fields and magnetic fields induced electrical flow. The two separately perceived phenomena were then taken as manifestations of a single, more fundamental, phenomenon, and a unified law in mathematical form was written for the combined forces as a single force, electromagnetism. Currently, without going into unnecessary detail, it is understood that, although they have been narrowed down considerably, there are still fields and forces in nature that can be separately perceived and measured and for which precise laws have been written which the physicists are attempting to unify in the same way that magnetism and electricity were "unified" in our comprehension of them, into the single concept of electromagnetism.

It has been recognized, over the last two centuries, that there are four and probably five elementary forces operating in nature (the strong force, the weak force, the electromagnetic force, gravity and possibly one more) at various levels of size. The strong and weak force operate within the confines of the atom; the electromagnetic force operates in macro space; gravity is pervasive throughout the known universe. The goal of unified field theory is quite simply to discover and derive a single equation or set of equations that expresses all of these forces integrated together as a single law. So far three of the four forces have been integrated; gravity remains un-integrated, at least to the complete satisfaction of all scientists involved. There are theories extant that purport to have integrated gravity and to approach an unified field theory but they are still controversial at this time of writing. If there is indeed a complete integration when gravity is incorporated then, it is anticipated, it should be possible to describe and predict any phenomenon of nature.

It may well be that, in our day, a grand unified field law may be achieved. All the separate laws of physics for the various phenomena we have known would then be subsumed into a single law which would precisely describe the universe and the way it "works" as a holistic entity. Rather impressive. It seems quite possible, even mechanical, perhaps inevitable. Having established a basic understanding of the idea our science currently holds of an unified field theory, let us return to the Nefilim version as I believe they may have developed it, up until they phased off the planet some three thousand years ago.

If they, indeed, had reached a comprehension of that profound an understanding of the universe, all their individual sciences would be special applications of that principle. Again, the same mechanism is at work here: we are just at the threshold of a unified field theory and that knowledge allows us to recognize what the Nefilim may have developed thousands of years ago. We may also profit from that knowledge by short-cutting our development time in this matter uniquely since an unified field law would be such a revolutionary achievement. A little reflection on what it would mean to live in a known universe, where we could predict and explain all phenomena, including ourselves explaining ourselves can be fascinating and startling. I believe that there are sufficient clues in the Nefilim legacy to indicate to us that the Anunnaki/Nefilim had attained the rudimentary notions of a Law Of Everything and that they had realized that the key to it was the meta-metaphor of self-reference. I take this as the most valuable contribution to our development embedded in the Nefilim legacy and it has prompted me to explore the concept of self-reference and self-referential systems fully as a means of resolving the "problem of consciousness" now thwarting science in its quest for a Law Of Everything and as a key to the Law Of Everything itself.

The Nature of a Self-Referential System

The notion of a self-referential system is not new. It simply means a closed system each part of which refers to all other parts, and all the parts to the whole. In a full self-referential system the key to how it operates and the principle upon which it is fundamentally based and developed is embedded in the system itself.

Consideration of such a system in evolutionary terms may be found in the impressive work of Arthur Young in his book *The Reflexive Universe*. Arthur's thinking is so novel and his articulation of it so elegantly clear that it is a very important and remarkable work. Stan Tenen has done a tremendous amount of work with a special application of the self-referential approach in his research and rediscovery of the Hebrew alphabet being constructed on just such a principle.

As I have pointed out above, Maurice Chatelain, formerly of NASA and the Apollo project, rediscovered a special application in the measuring systems of the ancient world which all were referenced to the circumpolar circumference of the earth. Munck, Hoagland and Torun have brought to light further special applications in the topological and geometric characteristics of the world grid. The same system appears to have been applied on Mars. So we have a body of evidence which shows that such a conception of the universe was extant in the ancient civilizations and was applied practically in various disciplines.

A Theory of the Nefilim General Approach to Science

It is my subjective opinion, based on the evidence I see, that the Nefilim had arrived at an unified field law or, at least, developed a conceptualization of the theory from the character of the individual applications of their science and engineering of which we have become aware. It may easily be argued that I am inferring too much, that they may have had an advanced science but to see an unified field theory behind it is not called for. That may well be true but, regardless, the clues that are available to us from the scientific and engineering evidence we have at hand are no less valuable. An examination of the applications known to us, therefore, will throw light on their approach to reality. The following will outline three major areas from which we have evidence: planetary physics, architecture, and linguistics. All three areas share common principles.

Planetary Physics

The Nefilim approach when arriving at a planet seems to have been thus: begin with the precise measurement of polar circumference of the planet. Using a sexagesimal system, determine a minute of arc by dividing up the imaginary circle of the circumpolar circumference into 360 degrees which, in turn, were divided by 60 to determine minutes of a degree. One minute of arc at the level of the polar circumference was selected as the standard length of what we called a "nautical mile." Smaller units of measurement comparable to our foot, yard, meter, etc. were then determined by using convenient lengths which were precise fractions of the standard "nautical mile." (We define a nautical mile as a minute of arc of the circumference of the earth divided into 360 degrees. The Nefilim apparently used, specifically, the circumpolar circumference rather than the equatorial. It does not take too much reflection to understand where we inherited the 360 degree division of a circle and the basis for our current definition of a nautical mile – although we have gradually lost, over time, the fact that it is best to begin with the circumpolar circumference rather than the equatorial or some other one.) Having determined a planet's relative nautical mile, convenient units of measurement for various purposes of geographic and architectural application were derived as precise fractions of that nautical mile. Further, locally used fractional standard sub-units of measurement were probably an expression of altitude relationship with reference to what they had determined to be the standard datum plane altitude of the polar circumference and may even be expressions of time dating relative to determinable polar shifts and accompanying variations in polar circumference.

The evidence for this approach came to my attention first through the work of Maurice Chatelain (*Our Cosmic Ancestors*, Temple Golden Publications). Maurice, recently deceased, was formerly associated with the communications division of the NASA Apollo missions as an employee of Rockwell. Having become aware of clues in the Old Testament and ancient literature to advanced technology, Maurice began to investigate and analyze and discovered that the units of measurement used by the ancient civilizations in their architectural constructions were all precise fractions of the circumpolar circumference of the Earth. The immediate question which forced itself on him was Who could have done that in the ancient past since it was not until the Apollo missions that we were able to get into space and measure that circumference precisely? He eventually had to conclude that humans were not able to measure it in those times and so must have been taught it by someone who could have. It was this fact that led him to conclude that the gods of the ancient civilizations were real and humanoid aliens who could indeed travel in space – just as the ancient peoples all claimed. The three types of cubits, the Megalithic foot, the ancient yards, etc. all revealed themselves as directly related to the Earth's nautical mile, indeed, a precise fractional components of it. That was the beginning of real insight for Maurice.

Subsequently, when the exploration of Mars by camera revealed startlingly artificial objects such as the now famous Face and generated intense interest in the possibility of a presence in the past of intelligent life living on or visiting Mars, careful measurements and analysis was begun on that data. The results of the photo analysis work of DiPietro and Molinar, NASA employees who combed the photos and discovered the Face pictures, indicated that it was artificial. Later investigation by Richard Hoagland, Mark Carlotto, Erol Torun and others turned up additional, apparently artificial constructions in the Cydonia area adjacent to the Face. Hoagland and Torun published the measurements and positions between these alleged constructions demonstrating that there was a consistent and common geometric and trigonometric relationship between them that was not found in nature. It was also demonstrated that the apparent units of measurement used in situating and constructing these monuments were fractions of the circumpolar circumference of Mars. The system of derivation of these units of measurement, based on a Martian "nautical mile" (relatively smaller than that of Earth because of the smaller size of Mars) clearly was identical to that previously discovered and explicated by Chatelain here on planet Earth. I take this correlation, in agreement with Sitchin, as indication of the operation of the Nefilim on Earth and on Mars.

Hyperdimensional Planetary Physics

There are other discoveries that have occurred since those of Chatelain's concerning the measurement systems of ancient times – those of DiPietro and Molinar, Hoagland, and Torun concerning the monumental constructions on Mars which throw new light on the science practiced by the Nefilim. If one searches the literature of conventional science for discoveries about the physics of planet Earth the information gathered, as a rule, is in terms of geologic makeup, the nature of the core, the mantle, the dynamics of plate tectonics, pole shifts and the various motions of the Earth relative to the sun and the other planets. The magnetic and gravitational field of the Earth are explained in conventional scientific terms. Seldom does one find an article on effects of fourth-dimensional physics that might be indirectly manifest on Earth in such a way that we might perceive those effects. As with most new and, therefore, almost inevitably, controversial discoveries, the hypothesis that there is evidence of the effect of four-dimensional physics observable in our three-dimensional field of perception operating on and in this planet may seem startling to some and difficult to accept. But, over time, physics has come to recognize that there has to be more dimensions inherent in the universe than just the three in which we have evolved to perceive, to this point. There is argument over how many dimensions actually exist but that there is at least a fourth dimension seems to be consensual at this point. The recent recognition of evidence for the indirect evidence of the operation of fourth-dimensional physics is a fascinating story in itself and summarized well by Richard Hoagland in a videotaped address in the U.N. auditorium (February 22, 1992, Mars Mission). For almost one hundred years, Hoagland points out, topologists, those mathematicians who are concerned with the comparative study of surfaces, have been interested in the basic question, If there is, indeed, a fourth-dimensional physics and it is effecting the three-dimensional physics we can observe and predict, how might some effect of those four-dimensional physics manifest in three dimensions in such a way that we could actually see it? Over time some topologists came to the conclusion that, hypothetically, on a very large spinning body such as a planet, the effect of four-dimensional physics would result in some sort of hexagonal pattern or phenomenon at the pole of the planet. When our space probe, Voyager II, flew by Saturn and took pictures over the north pole of that huge planet, a hexagonal pattern was discovered in the atmosphere. Hoagland, remembering an article by Godfrey in Icarus magazine in 1988 which focused on this startling pattern, recognized its potential significance as the phenomenon that the topologists had predicted. The pattern, a quite precise six sided figure was a part of the atmosphere, rotated with the planet, yet the winds inside of it blew counter to the rotation of the planet at some 300 miles per hour. The pattern was clearly linked to the bulk angular momentum of the planet indicating its probable generation by the effect of four-dimensional, hyperdimensional physics manifest in three dimensions. This is one of two major pieces of evidence for the operation of such physics.

Spherically Enclosed, Tetrahedral Planetary Geometry

The second piece of evidence pointed out by Hoagland had also been predicted theoretically by the topologists. As with the hexagonal pattern, the topologists predicted that, if, indeed, a planet existed in more dimensions than we can see and, therefore, was embedded in and effected by the physics of those higher dimensions, the major energy upwellings on the planet would occur at 19.5 degrees north and south latitudes, above and below the equator. The reason for this would be, they theorized, the effect of a very specific type of physics, best modeled by a spherically enclosed, tetrahedral geometry. It is common knowledge that the planets are not perfect spheres (ball shapes) and, obviously, they do not contain tetrahedrons inside them. But the geometric and mathematical models of the physics postulated by the topologists is convenient and quite accurate. In order to understand the startling implications of such physics it is necessary to be familiar with some basic concepts such as the nature of a tetrahedron and how it can be enclosed by a sphere. A three-dimensional tetrahedron is a basic geometric solid, a polyhedron with four faces. It looks generally like a three-sided pyramid. It's three sides and its base are all identical in shape, area, length of side and all corner angles of all sides are equal. The next step is to merge the two figures mentally. We assume that, looking at the sphere even with its equator, the north pole of the sphere is at the top and south pole is at the bottom. We then place the tetrahedron inside of the sphere of such a size as to just fit it, so all its apexes, points of all the sides, just touch the surface of the sphere, neither sticking through or falling inside the surface. We adjust the tetrahedron so that one of its apexes is precisely at the south pole. As soon as we do this, providing the tetrahedron just fits inside the sphere, the flat base/bottom of the tetrahedron will be exactly 19.5 degrees above the equator of the sphere. Obviously, if we put one of the apexes of another tetrahedron at the north pole, its base will fall 19.5 degrees below the equator. It is difficult to "see" how such simple, rigid, and abstract geometric figures, alone or nested together in this fashion, could be any kind of accurate model of dynamic, moving forces in a planet. That is a reasonable reaction if we visualize it all and only in three dimensions. But the basic assumption of the topologists, based on their theorizing, is that the three-dimensional aspects of these figures – or of planets when the model is transferred to them – are only part of the picture. They understand these solids and planets to exist in more dimensions than just the three we can perceive. If they are correct, in these higher dimensions dynamic processes can take place, giving rise to the effects which they have predicted and which manifest in three dimensions at these points. According to the predictions of topologists, the effect of the fundamental natural forces, modeled by the spherically enclosed tetrahedral geometry, working in a large spinning spherical body such as a planet will be upwellings of energy at the 19.5 degree latitude above and below the equator where the base of the tetrahedral form resides. Hoagland has pointed out that the ancient huge volcanic upwelling that has formed the Hawaiian islands on Earth, the twenty mile high volcano, Olympus Mons on Mars, the Great Red Spot on Jupiter, all fall out precisely on the 19.5 latitude. It is certainly startling and impressive to see, in these examples of colossal natural energy upwellings, verification of the physics predicted by topologists. But it does not end there: when one reexamines the location of the huge monumental constructions and pyramids of ancient times with respect to this fundamental geometry of planet Earth, one finds that the major points generally fall out on the natural energy grid that would be generated by such hyperdimensional physics. Subsequent to these early developments, the work of several investigators has contributed to a fuller understanding of the Nefilim planetary science.

Carl Munck, retired Air Force, developed the geometry of a geodetic world grid through a laborious manual process before computers were available to do so. Munck's work was valuable to Richard Hoagland, who found indication that there was a tetrahedral geometry inherent in the nature of a large spinning body, such as a planet, and which was known to and utilized by those who established the grid system. The key, major feature of this method of establishing a measurement and positional system based on the circumpolar circumference of any given planet is that it is self-referential: the system is relative to, referenced to, a fundamental characteristic of the planet.

Note that the common denominator of all these Nefilim techniques for dealing with physical reality, from linguistics to planetary, is self-reference. They approached reality as a self-referential system made up of self-referential sub-systems. What are the advantages of using such a self-referential system? Many. We have briefly touched on some already and we will consider those clues and their advantages in detail as we acknowledge their legacy without slavishly copying it, and apply the general principle to the solution of the profound philosophical problems we have confronted ourselves with as we explore our new found freedom as individuals and as a race.

Life In a Known Universe

Why is this so important to those who are ready to play their own god games? Because a "law of everything," particularly when it explains our consciousness, means life in a known universe. And that is a type of existence we haven't even thought much about at all – and which will be of a dimension of awareness and complexity that will be fitting of those playing their own god games for some time to come.

Relative physical immortality or extreme longevity, possessed by the Nefilim probably through genetic engineering and withheld from humans, is seen as a collective preoccupation. It has also been sublimated to the status of reward in the afterlife with the god or gods. If you can't get it here you will get it there if you do as the god(s) says. We need to explore the fullness of the significance and ramifications of immortality because it is inevitable that we shall acquire it. Let us do so on a truly human basis and not simply try to mimic the Nefilim version of it.

Ultimately, we are the most astounding technological feat that we know of performed by the Anunnaki. Not the stupendous architecture, not their space flight, not their advanced science, not even the androids and robots they were able to create but we, at least relative to ourselves, are the most significant legacy they left. We need to study ourselves genetically in order to identify all the clues to our heritage, both Nefilim and Homo erectus, and take advantage of that knowledge as we evolve consciously into the future, in charge of our own evolutionary direction. It may well have been that the Nefilim were advanced enough to be able to predict the stages we would move through into the future. Consider the concept of the kalpas, vast ages succeeding one another as spoken of as almost inevitable and as a matter of fact in the "sacred" texts of the East. Assuming that those texts are fairly faithful transmissions of important teachings originally imparted to humans by the Nefilim, it seems as if they could predict the course of our evolutionary development. This may simply be a matter of their having experienced similar stages in their past which would make it rather easy for them or it may have been a result of their having actually arrived at, at least, a rudimentary form of a law of everything. If we ascertain that indeed this was the case, we may find a valuable clue not only to how to explain the stages the human race has gone through and probably will go through in the future but how to, at least as individuals, step out of the determination of the process and move more rapidly ahead. The game will not be the same when we become aware of it and knowledgeable about it. In Part III we begin to explore the fullness of the potential of the new human and the new human society. Let us always keep in mind that, whatever clue or information or example we use from our Nefilim heritage, let the games we play, be uniquely and fully human.

PART III

THE NEW HUMAN

CHAPTER 11

THE NEW HUMAN

Rather, in our time, we shall learn
The sound of our own freedom,
At first disconcerting in the gentleness
Of its echo off the back wall of infinity,
Learn the intricate steps of the quaint
Dance of our oscillatory and peculiar
Kind of consciousness; re-discover
The threads of our common humanity
Woven in the tapestries of our cultures,
Struggle into the lightness
Of an unaccustomed, unassailable integrity
And prepare to take the children
For a visit to the patient grandparents.

Neil Freer
Neuroglyphs

The blurb on the jacket of Hans Moravec's *Mind Children* (Harvard University Press, 1988) invites us to put ourselves in the audience at the turn of the century as Orville Wright describes the future of flight, or Alexander Graham Bell describes satellite communications. The implication is that this perspective will help us appreciate the efforts of Moravec as he projects that, in forty years, humanity will be well into a phase in which the distinctions between machine and biological intelligence will blur. We have gained enough perspective to know that we should immediately put Moravec's projections into the same perspective, realizing that, in forty years, his vision will be seen in the same perspective as we now see the Wright Brothers' or A. G. Bell's projections. The new human will surpass the limited, extrapolated projections of the professional futurists by magnitudes.

However free I feel I am at this point in time, and ready to extend vision indefinitely into the future, I doubt that I will be immune from the same temporal effect. What I say here will probably seem antique in forty years because of the rapidity and novelty of the discoveries we are making constantly. I can only intend that what I offer here will expedite individual freedom and independence and facilitate creative individual contributions to the collective racial evolutionary trajectory.

The Demographer's View of the Status Quo

Paul Ray, vice-president of the market research firm American Lives Inc., has identified three major subcultures in America society.

Heartlanders, almost one third of the population, are the most conservative and hold on, nostalgically, to the small-town traditionalist worldview prevalent from the mid-1800's to around 1920.

Modernists, just under half of the population, tend to be materialistic, status-conscious, conventionally religious, wide-ranging in political views and focused on the immediate.

Cultural creatives, a little less than one quarter of the population, began to show in the population around 1970. This type of person is neither left nor right, demands ecologically sound conditions, authenticity in every facet of life, including relationships, politics, economics, philosophy and science. Their worldview sees reality as more than just the material world. Ray says that his statistics show that this sub-culture's orientation and demands can cause a new kind of "integral" culture, one that merges the best of modernism and traditionalism, embraces both East and West, and ushers in a new form of Renaissance.

Ray's analysis and categorizations are penetrating and concise. I take issue with his work on only one point: he does not recognize and include the one to two percent of the population that do not fit any of his three categories, the futants, those at the leading edge of our racial evolution who are investigating the future and reporting to the rest of the population.

Notice that what Ray projects cultural creatives could cause, a new kind of "integral" culture, one that merges the best of modernism and traditionalism, embraces both East and West, and ushers in a new form of Renaissance, are all recycled old things or recycled syntheses of old things. Futants, although they respect old things and keep them in perspective, do not deal in recycled old things including Renaissances, "rebirthed" old things. Furthermore, all of the items he mentions are products of our godspell cultural timidity, sourced in the ancient sublimated religions both East and West. Futants create our future demands, setting all those cultural antiques aside and acting like new humans. Only futants do that. Cultural creatives can hang creative reproductions on their walls and sew creative new versions of old costumes for the next cultural Halloween party but futants will be busy creating the future for them to recreate in years to come.

True Human Freedom and Responsibility

Having restored the knowledge of our racial beginnings in fully scientific, genetic terms at a point in our specie's metamorphosis where we are ready for that revelation, we provide ourselves with a generic, common definition of human nature. We are Homo erectus Nefilimus. **The act of breaking the godspell is the turning point: We shall restore our true history to ourselves and that genetic enlightenment will lead to liberation from separatism and repression and lead to a planetary unity that will release a social energy that the world has never seen. The question then becomes: What do you do after you have become genetically enlightened?** Genetic enlightenment, when one encounters the term as I have invented and used it herein, typically strikes the reader as a more superficial term than enlightenment as used in the classic manner with regard to Zen satori, Buddhist nirvana or alignment with the Tao in the Taoist tradition. It seems that I am using the term in a much lighter way (pun intended) than the hallowed term is usually treated. Consider, however, the novel perspective into which this term is cast when one realizes that the concept of enlightenment in the classic sense is a pointer toward self-realization, often in terms of the recognition of the divinity within, and that those traditions of consciousness expansion techniques and trajectory were most probably taught to some humans by the Nefilim – or assimilated from them by example. I suggest that the lights really go on when enlightenment is understood as a realization of the true nature of the human as being demi-god (the recognition of our half-Nefilim "divinity"). Perhaps these teachings were the most important element in the preparation of humans by the Nefilim for the time after they had withdrawn from the planet: remember, realize what and who you are so you can take on responsibility for yourselves. We haven't done a very good job of it so far, having kept ourselves in the dark so to speak in a good case of racial amnesia.

But What Do You Do After
Becoming Genetically Enlightened?

The classic Zen answer to the question What do you do after you have become enlightened? everyone remembers as "Chop wood, carry water." The Zen sense of things is that you really don't do anything very differently than you did before, it's the state of being that counts. That's fine as far as it goes: you pop the neurological relays exercised by the un-logical Zen koans, realize that you still exist, but your consciousness is capable of experiencing and comprehending in a modality in which logic is subsumed into a more expanded kind of meta-logical awareness. The dimensionality of your consciousness has been expanded. But genetic enlightenment is not only a realization but a release into a freedom that invites exploration of the universe. Continuing to chop wood and carry water, if that's what you happened to be doing previously, is certainly an option but I'll bet it won't be very satisfying for long – especially if you know you can live indefinitely. One soon realizes, besides the fact that chopping wood or carrying water or whatevre your favorite activity is currently, may not even be good ecological practice any more, that there are some intriguing options to explore in the context of the liberation of immortality. This is not some sort of polyanna scenario. All the early signs are clearly evident in our time. The only reason why an individual does not see them is because their godspell thinking will not permit them to even look, or because their personal problems cause them to project their situation on the entire planet.

Clearing Up Some Perennial Obstacles

With the godspell will also dissipate and disappear the mind-body problem springing from the resultant erroneous concepts of human nature. The alienation and the schizoid split conceived of in terms of between "spirit and matter," "body and soul," due to the ancient owning of us by the Nefilim and to the denied tensions between our bicameral genetic components, will evaporate. And we will finally be free from the pressures of others who do not agree due to conflicting dogmatic convictions resulting in a retardation of the progress and activity of the individual. Once a consensual ground is reached concerning what we are, those restrictions will be removed even socially and politically and the freedom of the individual and groups will be exponentiated. This new freedom will be perceived as the ability to control our collective and individual evolutionary trajectories and determine our own realities in a universe understood as a plenum of potentials, limited only by our ability to conceive of them. The harmonious energy released by the common view of our origin and nature will enable us to act as a single race, to restructure our social, political, educational, scientific, psychological and philosophic structures in a profound and fundamental way through a common vision that, intrinsically, will open up a degree of individual and collective freedom greater than any yet known. There is an intriguing apparent paradox here: the new paradigm provides a basis for a world unity based on a common definition of who and what we are, yet the freedom that results from that vision promotes a diversity of individual views that would seem to cancel its potential. But the consensual unity based on a common definition of man will be focused on the past, on our origins, on our original nature and the diversity of views will be about the possible futures of individuals and the potentials they may realize. We may argue about what we can or should become and do, but we will be arguing from a common base of understanding of what and who is doing the arguing about the doing and becoming.

Individual Liberation and Freedom

Once possessing a known common planetary history right down to the bio-genetic level, the individual will know what and who SHe is with a certainty that is a focusing and liberating force more powerful than any we have before known. (This is true even compared to the situation in the time when the Nefilim were actually here. We may have known clearly then, as humans, what and who we were but there was no liberation because we also knew clearly that we were subjects of the Nefilim in a very absolute sense.) The liberating aspect of this planetarily consensual definition and understanding afforded by genetic enlightenment provides the maximum latitude for individual growth and expression to develop in a radical, but intelligent and mutually benevolent, broad freedom. This release will be so profound that it will create a state of consciousness that will give rise, is already giving rise, to a new type of human being. The "unassailable integrity" of the individual involves the intrinsic element of a radical freedom and that also brings with it a deepened responsibility. The new human assumes, demands that exponentiated freedom and responsibility and operates in that expanded dimensionality as a natural environment.

Paradoxes of Real Freedom

In an unhampered direct relationship to the universe the individual has a full appreciation of HIr being an integral part of that universe. Being part of that universe means being subject to and effected by the laws of the universe from gravity on the largest scale to particle physics, the physical laws of the physical universe. But comprehension of being determined by those laws is balanced by the perception that that same universe gives the ability to and causes HIr to be aware of the possibility of controlling and altering HIr own state. The part of the universe which is SHe, through self-awareness, is able to self-modify. To the fullest degree that we seem to be able to understand our situation at this stage of what we consider to be an evolutionary process, the best we can say is that we are determined to determine our own determination. There does not seem to me to be any better or less confusing way of stating the situation. There are two main ways in which we can view the activation of our ability to determine ourselves: one, internal, is the way we self-modify our own actions, thoughts, physical body and, eventually, our very genetic makeup; second, external, the way we can influence the world around us as far as events, collective decisions, even physical reality are concerned. We have come to call this being "reality creators." As we begin to explore the fullness of the radical freedom of a universe experienced as a benign, neutral plenum of unlimited potential, the forms of the god games we will find it worthy to play take shape in front of us. To choose the ones we want to play we have to make a judgment according to adequate criteria. The criteria of the ancient godspell context are no longer either valid or adequate and the most reasonable of those that are valid are being transcended. In order to determine criteria adequate to our dignity and freedom we need to explore what lies beyond any limits we have known before. The characteristics that mark the new human are: physical immortality; an unassailable personal integrity; relativistic epistemology; profound compassion; robust depth of informational data; understanding of the universe in terms of a full unified field; broad-spectrum competence; transcendental competition; a facility for graceful dimensional shifting; preference for dyadic operation; a profound ability to enjoy, to play the "games" most enjoyable and satisfying to generic "gods," an expanded capacity to literally have great fun creating new realities; a space-time orientation evolving to habitual four-dimensional perception and consciousness. These characteristics all find their expression and are integrated in the two "god games" the new human will play and we can already begin to play – in reality simply major facets of the focused life of the new human – which we explore in this final section: immortality and conscious evolution as reality creation.

Immortality and an Evolving Life of Reality Creation

The most important thing we can do is begin to think about life under those almost unthinkable conditions now, rather than when they are already upon us. How will we live then and why? All the potential exists within each of us. Living in a known universe as immortals as members of stellar society may seem like astounding and astonishing activities right now, but I suggest that we keep these games of stellar adolescence in perspective. Our science fiction is simply our imagination working out there ahead of us, letting us explore the future possibles on a philosophical basis in a safe virtual form. It won't be that long before we will want to explore the higher-dimensional realms directly, perhaps take a spin as an intelligent pure energy field, maybe play on vacation in some cosmic Disneyworld as a shape-shifter, visit the past or the future or even another sub-universe. And even these projections are going to be pitifully antique in a very short time.

Becoming Conscious of Our Ability to Evolve Consciously

It is an inherent effect of becoming self-aware enough to be aware that one is self-aware that, gradually, one realizes that one has at least some ability to control one's fate. At first it is by realizing that we can anticipate and control to some degree what is going to happen to us by our choices and actions. But, with increasing knowledge and insight one can also become aware of the potential to determine by intention, practice, and expansion of one's consciousness and reflection that one can influence one's own personal development, even modify one's basic genetic tendencies. We reach a point where we see clearly that we can consciously control our own evolutionary trajectory. At first our efforts are amateurist, halting, intermittent and groping. But the process is one of learning through feedback and becomes more efficient, intelligent and faster. The effect eventually feeds into the collective consciousness and, clearly, the human race is attempting to take greater and greater charge of its own evolutionary direction and to orchestrate our own future in the solar system and beyond. Even before the recent dawning of the new paradigm and the awareness of our true genetic history, our technology had reached a point where we were beginning to become aware of the power it afforded us to control our genetic makeup, eliminate disease, repair injury and work toward the elimination of death. As marvelous as those potentials were to control our fates, there were still major problems to deal with in that the Evolutionist vs. Creationist conflicts obstructed and drained energy from these developments. We had no common, generic definition of human nature and quarreled about it besides so, literally, we did not know what we were talking about. The new paradigm affords the generic, consensual platform from which to greatly accelerate and realize the fullness of conscious evolution, the intelligent, deliberate determination of our evolutionary direction and choices.

Darwinian Defectors Define Their Own Evolution

We no longer see our development as simple-minded Darwinian survival of the fittest, nor the Lamarkian transmittal of learned skills, or the manifestation of some mechanism of self-organizing criticality, but a very deliberate, controlled and calculated choosing of the trajectory of our expansion into the universe. Self-conscious evolution is characterized by a radical shift from a natural selection leading to the survival of genes that are most successful in adapting to local conditions, to genes that are most successful in choosing the best evolutionary trajectory. That choice may be influenced very little by local conditions, may be predominantly cultural and may include expansion into higher dimensions in the sense that we already see ourselves attempting to understand and, eventually, directly perceive, the fourth dimension. With the dawning of the new paradigm and its definitive clarification of the real nature of our special case evolution, we are enabled to evaluate the possible variables present in our genetic makeup because of the self-interest of our Nefilim creators. This will greatly facilitate our understanding of our nature and, therefore, how we can determine, select, direct and modify the path we wish to take.

Are We Pre-Determined to Determine Our Own Evolution?

A very basic question must be asked at this point. Is there really only one set pattern of evolution for any given species or race, an optimal path of adaptation, adjustment, growth, expansion and type of consciousness? This would imply an objective and predetermined order of the universe that we simply had to discover and it would seem to negate any real creativity or freedom on the part of the individual. But if we understand the universe as a plenum of potential rather than a clockworks in which we can be ground up if we are not careful, we will see that our evolutionary choices are simply the form the exercise of our radical freedom will take. Having reached a generic understanding of who and what we are, evolving self-consciously and comprehending that one may choose the trajectory of one's personal evolution and contribute to the choice of the evolutionary trajectory of the species, is a profound realization. The actual experience of it adds another dimension to the human condition. The uniqueness of human consciousness displayed in the type and scope of our self awareness is the field in which we will direct our own evolution. Let us look at our actual experience with our own personal and racial evolution. We know we are aware of being aware, we witness the race and ourselves changing and evolving over history and even in our lifetimes. We can identify our collective planetary development at this time as a stage in that process. We do not choose to devolve backward into dumb brutes, we choose to evolve into more and more intelligent beings capable of entering stellar society. We witness and experience ourselves becoming more and more intensely focused on determining the most beneficial and satisfactory trajectory into the future. And the experience is of a more and more enlightened comprehension of the process itself and our being able to exercise free decision and choice in these matters with the conviction that we have the capabilities to accomplish those projected goals. At the same time that we recognize that there are laws of physics we seem to discover and work by, we also see that we have the ability to innovate within the context of those laws to increase our capabilities and to use those laws to further our own evolution. So, in one sense, we can say yes, the physical laws of the universe restrict us to certain possible scenarios of evolutionary development. But, as stated above, we can also say that apparently one of the fundamental characteristics of the universe that determines both those laws and our type of consciousness is that we are free to determine our own determination. We, as self-referential consciousnesses, recognize that as our primary, quintessential characteristic.

Transcendental Experience as Conscious Evolution at the Leading Edge

We may redefine personal transcendental experience in evolutionary terms. The term transcendental experience has long been defined in religious, philosophical, metaphysical parlance. It usually is thought of in the context of an individual experiencing some kind of divinity, higher state of awareness or consciousness, a metaphysical truth, or generally in the vague popular sense of "religious" experience. Previously such experience was generally thought of in passive terms: in the extreme such experience was considered, particularly in the Christian context, as something given by God, or a divinity, something that was not under one's control but gratuitous. In less passive contexts where practices such as yoga and chi kung meditation are pursued, the descriptive metaphors are still in contemplation of higher states, oneness with Being, etc. and lack the precision and scientific understanding of higher dimensionality and conscious evolution that we now possess. **We redefine personal transcendental experience as the individual's direct experience of her or his own personal, unique evolutionary trajectory at the leading edge of the collective racial expansion. We define that leading edge currently as expansion into habitual four-dimensional consciousness.** The leading edge in the far future most certainly will be radically different. We redefine collective, racial, transcendental experience as the dynamic expression of human nature expanding at the leading edge of its consciously determined evolutionary development. Having redefined transcendental experience in these terms we must immediately consider the question Does this redefinition reduce any and all human experience to simply physics? The answer is certainly No. This is not a reductionist concept: what I have said might well, and should be, understood as more advanced, informed, free, intelligent and dimensionally expanded than anything we have ever known. Consciousness, at this level, is constantly self-reinventing in a very literal sense. This is self-reflexive consciousness working at its highest level. By the intrinsic nature of the process there is no possible way to reduce the complex process of self-referential transcendence of one's transcendence to a derivation from the laws of physics as we know them now. Where is the leading edge of our self-conscious evolution right now? It is manifest in those individuals whose lives are focused on exploring the potential of conscious evolution in a constant expansion into new dimensions of awareness and reality. It is from that platform that the new human launches into the future. In the far future the generic nature of the process will remain although the focus of the expansion and exploration may, indeed probably shall, change.

A typical reaction to these considerations is, more often than not, a disconcerted, sometimes cynical and pessimistic, sometimes despairing and fearful, refusal to engage. Often they are branded as "head games," useless intellectualizing. But this is not a sophomoric display or a trivial philosophical exercise; it addresses the very essence of the process by which a self-aware species, once it has reached that point of recognition, goes about controlling its own evolution. Is there anything that can be more important to our species' future survival and prosperity at this point in time?

It is fascinating that, once one realizes that one is able to take charge of one's own destiny and evolutionary direction, one realizes that both the ability and the responsibility are inescapable. Whether one embraces the intellectual challenge or ignores it, takes it as critically important or as trivial, accepts the definition of human evolution as self-aware and modifiable or mechanistically determined by the genes, one has already contributed to and modified the process to some degree. This is one experiment where the experimenter cannot avoid modification of the experiment's conditions by participation no matter how much one attempts to remain outside the experiment. One is the experiment. As we determine what criteria we shall use to judge one potential direction of our evolution as preferable to another, we are determining an evolutionary direction, that of choosing certain criteria as opposed to others as a means of judging one path from another.

This unavoidable characteristic is another facet of the inescapability of the reduction to self-reference we have experienced in our philosophy and consciousness in general. It is no accident that our becoming intensely aware of the epistemic barrier of self-referential subjectivity has reached its critical point at the same time in our evolutionary development as a species when we have become aware that our awareness of the evolutionary process puts us in a position to modify it. The realization of the inescapability of participation in the evolutionary process in its broadest context – we could, theoretically, deliberately opt to genetically engineer ourselves into a species that would be resistant to evolving for reasons we saw as the best for survival although that would still be choosing an evolutionary path – also brings us around full circle to the limitations which are involved. We are a part of a process which we do not even fully comprehend, reacting in ways, including self-aware ways, which are dictated by that process. We are that process and every attempt to step completely outside of it may be understood as part of the process itself. It is at that fundamental level which we face, most profoundly, the question of how completely we are determined. And it is at that point, when we confront the question most self-consciously, that we realize that, if indeed we are completely determined by the process, we clearly are determined to be self-determining. It is argued that this is simply a delusion caused by our inability to accept the notion that our free will is an illusion. And our current degree of development of the processes of reasoning does not help; it traps us between too-simple opposites and does not mirror the relativity of our consciousness. It does not permit us to deal easily with the apparent paradox: determined to be self-determining. The problem is in the reciprocal relationship between our consciousness and the logic that is – or should be – a mirror of it. Our consciousness evolves, perceives and attempts to deal with more and more complexity and express it and the relationships between its parts. The primary tool, to this point, reason and logic, invented to express reality precisely, always tends to lag behind the experience and the expression. Once we recognize logic as a metaphor and a tool that must be constantly expanded and refined, we will overcome the problem.

A Law of Evolutionary Development?

If the essential characteristic of our conscious evolution is as unlimited a freedom as possible for the individual and it is inescapable also, is it possible to discover a law of evolution that determines and governs that peculiar dynamic process?

Evolution by Extrapolation as Inadequate

Certainly, we do reflect on the way we are now and how we seem to have evolved to this point and project ourselves and our desires and potential into the future. That may be called extrapolating our future from what we know now. I am convinced, however, that we will soon move from this disorganized and controversial way of trying to determine our future to a far more intelligent, consensual and intense modality. Eliminating the godspell syndrome will free us from the imagined absolutes and definitions that separate us, and that freedom will afford us the opportunity to get used to a common, consensual definition of what and who we are simply as humans. Having established that base for discussion we will move to debate what we, simply as humans, want to evolve toward. But that will be only a beginning because self-conscious, self-determined evolution will inevitably give rise to a systematic, scientific study of our peculiar evolutionary development with the purpose of attempting to predict and define not only how we will, but how we should evolve. At that point we will have arrived at a new plateau of thought. Evolution by extrapolation from our current state of affairs is patently inadequate. Can our evolutionary explorations be facilitated by disciplined scientific investigation, experiment, philosophical considerations? At first this novel process may seem to be a wild debate that may never end because each person will insist on an evolutionary path unique to HIrself. We witness some beginnings of such a debate even in our time but it is often neither focused nor even in an evolutionary context. But this insistence on maximum evolutionary freedom for the individual will, paradoxically and inexorably, force a consensual recognition and definition of the sociopolitical phase of the evolution of the race: we will have little choice but to recognize that the evolutionary path is to develop highly refined and sophisticated means, perhaps even institutions and agencies, laws and constitutions to encourage, foster, preserve, protect and abet that core, inalienable human right and freedom of the individual. In the next chapter we will examine the situation regarding when we will have to develop a second Constitution for this country. That will be only the first small step toward the new human society.

A Criterion for "Right" Evolution?

To speak of criteria in this regard is to enter deeper into the subjective web. It is our peculiar type of consciousness and experience that prompts us to adopt the criterion of maximal freedom for the individual and the race to explore the maximum number of possibles in the greatest number of dimensions. It is our consciousness that prompts us to set the only cautionary limitation being that the individual and the race avoid adopting a path that would destroy the social context which supports such freedom, injures or prevents the individual or the race – or another race – from such freedom of play in the universe. It is our consciousness that has gained sufficient apparent insight to recognize that it is self-referential, could possibly be peculiar to this planet, even unique in the universe, that it can know only subjectively although there seems to be an objective order which it is perceiving in that manner, that it seems to itself that it is evolving, that the criteria it sets for its own maximal evolutionary benefit and expansion is subjective – and may well change as it evolves and learns more about the universe. So let us say that our criteria are relative and subjective and pragmatic and that our consciousness is its own criterion. It is far easier to frankly set pragmatic working criteria for our evolutionary development than to discover some "objective" scientific law for it. How could it be possible to develop criteria for and to write a law of human evolution if it had to account for our desire for complete freedom of self-determination? That would mean that it would have to encompass every possible thing which we could attempt to explore or experience, every possibility of which we could conceive now and in the future. It would seem that the only law comprehensive enough would be that of the unified field, the "law of everything." Clearly, that stretches our minds about as far as we can manage right now. But that is the way our consciousness operates. Our type of consciousness is open-ended, expanding, capable of dealing with the fact that it cannot "get outside" itself, of evaluating and changing itself, of conceiving of itself in a stage of evolution that we are not at yet, and working toward. As we have discussed previously, we will have to expand our conception of the universe to allow for a kind of "law" that is itself capable of self-modification and of mirroring and, eventually, predicting, whatever our consciousness is capable of doing or evolving to. It is not difficult to see that the terms of our consciousness in which we are speaking of here are the same terms in which we considered the nature of the unified field Law Of Everything in a previous chapter. This is simply because the LOE can only be known to us in terms which our consciousness can grasp, and therefore must be in terms our consciousness invents. But since our consciousness is at once the product of the LOE, and the source of its formulation in terms it can understand, it is inevitable that the "laws" that govern our consciousness will seem to be reciprocally related to our notion of the LOE. Inside the web of this subjective field, all we seem to be able to do is to state a general working rule of right conscious action: we are free to do as we will, to explore any possibilities but, in doing so, we should not infringe on that same freedom in others – because that seems to be the way things work out maximally best for ourselves and others. Sounds strangely like the Golden Rule, Do unto others as you would have them do unto you. It sounds almost trite but, just as we are rediscovering the deep significance of the Golden Mean, so perhaps is it time to reevaluate the significance of the Golden Rule. . . Are we close to that enlightened situation at present? Not very. There are some individuals who are there already, but we have a long way to go to dispel the godspell obstacles before the race reaches even a consensual generic definition of what a human being really is. The individual must simply trust HIr own judgment, learn and understand HIr own unique genetic potentials and dominant tendencies, decide how SHe wants to interpret the seemingly inescapable process we are involved in, determine HIr unique evolutionary path and follow it. It is difficult to do this now because the godspell mentality militates against it in many ways, but one does not have to wait for the rest of the race or anyone. The groundswell of a myriad of self-realization schools and techniques that have sprung up under the initial impetus of the sixties' liberation through the use of the Western yoga of LSD may easily be recognized as an early thrust in this generic evolutionary direction. It has been hampered and confused by the use of outmoded and inadequate Eastern metaphors and attempts to modify it to keep it within the context of godspell religions and philosophies. But, regardless of the obstacles, the blind leads, or by the fear of going alone into uncharted territory one can and may become their own conscious evolutionary artist. I did not say "must" or "should," I simply respectfully recommend it.

Some Practical Problems to Come

Even after we reach a consensus as to the reality and nature of conscious evolution we will never be able to be democratic about it in the sense of a majority rule, because that would be directly contradictory to the uniqueness of individual personal evolution. This will be a problem in the early stages of the assimilation of the notion of the individual contributing (by the exploration and direction of her or his own unique evolutionary path) to the direction that our collective racial evolution should take. At that point also, at least for our race, we will probably have to contend with a very significant conflict: there will be a tendency for some to move toward a planetary collective hive mentality downplaying the creativity and importance of the individual. Opposing, of course, will be those who will want to maintain the individual as primary. Both should be given every opportunity to explore their preferred contextual option as long as they do not try to impose that option on or obstruct anyone who does not wish it. We have projected and explored the extreme, aggressive form of the hive mentality in the fictional Star Trek in the Borg whom we portray to ourselves as devolved in their robot-like, predatory "assimilation" of all other cultures. Even the most enlightened hive-type society is not the gene pool of which I am, but it may well be for others.

Meantime, let's talk about what's coming next. I can only respectfully present the limited vision that I see and, hopefully, the reader will find a familiar face in this mirror. Because the freedom of the new human is radical and primary, to impose boundaries, conditions, considerations or even a great deal of description would be contradictory and counterproductive. But let us consider the general nature of the games we demi-gods shall play in our first halting expansion into the universe truly on our own, as our own validation.

CHAPTER 12

THE POLITICS
OF COMPASSION

I am not an advocate for frequent changes in laws and constitutions. But laws and institutions must go hand in hand with the progress of the human mind. As that becomes more developed, more enlightened, as new discoveries are made, new truths are discovered and manners and opinions change. With the change of circumstances, institutions must advance also to keep pace with the times. We might as well require a man to wear still the coat which fitted him when a boy as civilized society to remain ever under the regimen of their barbarous ancestors.

Thomas Jefferson

Are the modern civilizations to remain spiritually locked from each other in their local notions of the sense of the general tradition; or can we not now break through to some more profoundly based point and counterpoint of human understanding?"

Joseph Campbell
The Masks of God: Primitive Mythology

Homo sapiens is on the threshold of discovering that expanding contelligence is the goal of the trip. That pleasure resides not in external materials but inside the time envelope of the body; that power resides not in muscles and muscle-surrogate machines, but in the brain; that the evolutionary blueprint is to be found in the genetic scriptures; that Higher Intelligence is to be found in the galaxy.

Timothy Leary, Ph.D.
Info-Psychology

No one is more aware than I of how preposterous the claims and statements I am making in this book will seem to some. When it comes to describing the new human and the coming new post-godspell social context beyond war, want, alienation and boundaries, the reader may experience the most acute sensation of my seeming indulgence in pure fantasy. Doesn't everyone just know without any doubt that human nature left to itself will act in a depraved and selfish manner? That without religion man will revert to the primitive predator state? That humanity on this planet is going downhill faster and faster, fouling its nest to the point of it being almost uninhabitable? That greed is the driving force behind all human social interactions and compassion and love are only the rare exceptions?

I am not suggesting some Pollyanna kind of paradise about to happen to everyone's surprise. I fully acknowledge the potential for and presence of injustice and barbarity: it happens every day and one does not even have to watch the TV to learn of it. But that same TV news brings us the early signs of a gradually more humanistic and compassionate society, the coming of which will be highly accelerated by the recognition and assimilation of the new paradigm.

I am going to say here that it is undeniable that we have seen all those elements operating in human civilizations over time, sometimes to a hideously grotesque extreme, but that the underlying causes are not an inherent depravity and brutally rapacious greed, but the ancient programming to subservience, dependence and fear, causing a kind of racial psychosis, the most prevalent symptom of which is a passive stupidity. The simple-minded, red in tooth and claw, naked ape Darwinian human is a fiction. If depravity, and an inherent tendency to unbridled evil and rapacious greed are the natural driving characteristics of the human being, how many humans around you do you see operating with that engine under their hood? Do you operate that way, human? How many consider someone who does operate that way normal or do we separate them from the rest of us, declare them criminal or insane?

War No More

Wars and rumors of war. "Drum and trumpet history." Almost as if a threat was essential for progress or change. Humanity seems to harbor a secret conviction that somehow that is the way things are supposed to be, indeed, must be. In fact, the Nefilim did want it that way. They instructed us in ways of war when using us as disposable units in their wars against each other; they played "GI Joe" with us. . . There seemed to be a bit of ritualism in their wars, as if there was an accepted protocol of combat for political succession. We learned war in the form that we know it from them. Sitchin has identified the first recorded event of that training: the formation, at Edfu in the (Egyptian calendar) year 363, or shortly thereafter, of a corps of human *mesniu*, male human "metal people," forgers and wielders of iron weapons as foot soldiers. The broader view of our real history shows that, previous to that arbitrary conscription, humans had generally not acted in a warlike manner – but that is no certain indicator of our inherent racial nature because we were never our own people until the Nefilim phased off the planet and, since then, we have not truly found our own unique racial character.

Identifying A Truly Human Politic

Since we were never our own people from the time of our creation and over the hundreds of thousands of years that the Nefilim dominated the planet, it is difficult to identify a purely human politic. Being subordinated, we were peaceful, unwillingly or not. When conscripted and ordered to make war we were warlike, willingly or not. It is, likewise, difficult to distinguish the inherent propensities for violence and warfare even over the brief three thousand years since the Nefilim vacated the planet. When left, finally, on our own, we continued to act as we had been taught, still fighting each other in the name of our now absent local "god." The Biblical wars and enslavements, escapes and subjugations, the religious factional conflicts, persecutions of Jews, Christians, pagans, Moslems, the

Christian crusades, the jihads, are all repetitions of the political conflicts of the Nefilim in which they used us as pawns. Even the ideological, "for God and country" slogans of our horrendous modern wars still echo hollow and robot-like as the war cries of the *mesniu*.

The greatest paradox is that it is religious fanaticism which has given rise to the most horrendous persecutions of humans by humans. The problems and the great war conflicts can almost in all cases be traced to acute ideological differences, giving rise to seeing the other as evil, wrong, in error, and therefore a threat religiously, politically, economically or genetically. The wars of conquest and imperial expansion were not just motivated by economic advantage, but justified by a view that those conquered were inferior, not as evolved or less than human by some religious or theological belief. We need look no farther in the past than the actions of the Muslims in the Mediterranean region or the Catholic Spanish conquistadors in the Americas for examples of such behavior.

Against the context of the new paradigm, we may have inherited a tendency to violence more from the Nefilim than from Homo erectus. But the Nefilim had dealt with it in their nature and culture, by, at least to some degree, ritualizing it. The gross nature of warfare on this planet may be due to our as yet unrecognized Nefilim training rather than a biological drive running deep and wild which we have inherited from either our Nefilim or our indigenous roots. This perspective affords us the opportunity to understand and appreciate a major fact: war is not necessarily the end product of some deep, uncontrollable, biological imperative in the human.

My point of deep conviction, based on the patterns of history, is that all critical indications are that the general direction of human development is toward the more compassionate rather than towards the brutal and the competitive, and that the vast majority of individuals in the human population the planet over are oriented to peace rather than war, to compassion rather than competition. Once free of the absolutistic ideological Babel factors of the godspell mentality which have artificially divided us to this point, we may disagree quite dramatically over issues, but we will no longer kill each other in the name of some God, seeing each other as a threatening, almost alien species. An immediate reinforcement of the truth of this view is in the contradiction of the view that, in proportion to the degree that mankind separates itself from the subjection to religious dogma and doctrines and authority, the individual and society will become anarchistic and depraved. Even now the grip of religious authority is slipping more and more, regardless of the reactive backlash of radical fundamentalism both East and West, and yet the worth and integrity and moral stability of the individual shines through to a greater and greater degree.

The overall pattern and direction of history, even taking into account the day to day aberrations, if graphed, would look very much like a rising stock bull market. It has been pointed out repeatedly that instant access to worldwide news, open and freer access to information, a gradual increase in educational standards, the expansion of our physical presence and mental focus into space, the sophistication of our technology, the greater difficulty of keeping secrets whether military, industrial, social or religious all contribute to an eventual better world. It is obvious that all these means of control and progress can be and sometimes are abused. Even the greater awareness of healthy diets, avoidance of toxins and pollution, and the need for intelligent exercise can be subverted and people exploited by dishonest individuals in the field. But the positive direction toward a more evolved society is clearly indicated by these individual developments in the sciences, medicine and social reform and especially by the attitudes of the general population which desires, supports and reinforces these particular developments in the first place.

Perhaps the most accurate evidence of our inherent bellicosity or altruism we have so far, may be found in human civilizations most removed from the Nefilim city-centers and, therefore, we assume tentatively, most removed from Nefilim domination. The American Indian societies such as the Hopi have traditionally been peaceful and altruistic. Nomadic tribes such as the Blackfoot were more warlike. But the character of each is clearly a product of their basic lifestyle and survival pressures. This would make it appear that there is no inherent tendency to violence, much less warfare, in humans in general. Until we break the ancient godspell discipline we shall not truly know how inherently warlike or altruistic we really are.

I am confident in saying that, in view of the shift enabled by the potential for planetary unity springing from the new paradigm, the ascendance of the importance of the individual, and the fading of the militant dogmatism of religions from which so much war has come, war will be no more. We are reaching that point already in a halting way due to the destructiveness of the weapons and the small size of the planet, and that is an encouraging sign. But, at a later stage, war and violence will be seen simply as a pathological social aberration and highly trained, highly professional police, eventually international and limited to preserving individual and societal rights, will replace the military. This will be a process of transformation that will see its share of mistakes, overcorrection and under-correction, certainly even corruption, but it will take place and sooner than we think. The process, although it sometimes may be difficult to see, has already begun through the pressures of simple pragmatic expediency, but the fullness of the consciousness that will live beyond war will be a long time coming.

It is tragically amusing that it was only a few years ago that a president of the United States was threatening to "kick some butt" if the dictator of Iraq did not withdraw from Kuwait.

At a certain critical point of evolutionary development violence, killing, war, or threat of violence will no longer be a permissible option to settle disputes or to realize one's individual or political or national goals. Personal self-defense against an aberrant violent aggressor will be recognized as even more justified than it is now because the worth of the individual will be the most esteemed value, and the option of physical immortality the premier right of the individual. Violent aggression will unambiguously be considered pathological. Nationalism, exposed as springing from the ancient roots of Nefilim politics when we were used against each other as pawns, and the even more primitive programming to primate grouping, and the warfare it spawns, will have faded into decadent history. Just as we have reached a point, on the level of individual and group social interaction, where aggressive acts of physical violence and even non-physical violence, toward another is interpreted as sickness by a consensus if not by all, so shall we achieve on the level of international relations. And war will be no more. Even as we see the irruption of "brush wars," smaller local regional conflicts occasionally, the great world wars are ended.

It will be argued that international law already criminalizes aggressive war and that already the only justifiable war is one waged by a people or a nation in self-defense. Note carefully, however, that what is not allowed as moral, ethical, legal or justifiable behavior on the part of a person or corporation in the business world, killing, physical violence, intimidation, blackmail, sabotage is recognized generally as necessary if not legitimate between nations. Certainly we recognize the legitimate need for corporate security patrols, police intervention, even the use of force in appropriate response to criminal intrusion, theft or personal attack. But what we already condemn is the Mafia way of "doing business": the use of threat, harm, force or violence to intimidate and coerce as a means of winning the competitive economic game. But what we condemn at the personal or corporate level we still condone at the national level – although corporations are acting more and more like sovereign states and governments more and more like international corporations. Covert operations, black-ops, political assassinations, instigation of violent overthrows, mind-control, lying and false propaganda, lethal experimentation and massively destructive weaponry, do not even make up an entire list of our inhuman acts against each other.

We always try to justify what we do in the name of nationalism as self-protection. So far warfare is not considered a sickness in the same way as individual aggression and physical violence. We do not look on warring nations – at least in general – as we look on individuals engaging in a barroom brawl or a domestic quarrel or a gang slaying. It is easier to ignore the psychopathology at the national level than deal with it. Certainly, we recognize the legitimacy of self-defense against aggression but that is the only common element on all levels. The United Nations exists but is used when it is expedient or convenient, and ignored when it is not. We are headed in the right direction but there is a great distance to go.

We no longer accept without question the concept of a "holy" war. A jihad, on the part of a Moslem nation or people is looked on as aberrant. If the Pope called for a crusade in this day and age the battle-call would be met with bewilderment, skepticism, probably some docile response from those given to fanaticism, protest and even theologian criticism, and a good measure of simple amusement.

Slavery as a Learned Practice

Slavery is a degenerate practice that can be attributed solely to human innovation only if we ignore our real history. We were invented as slaves, treated as slaves for millennia on millennia. Even when human intermediaries (kings) were installed by the Nefilim to manage the subordinate human population we were still in bondage to the Nefilim. Human armies were used by Nefilim rulers to capture human slaves from other Nefilim rulers in other areas. In short, we have a long history as slaves, as slave takers, and the example of the Nefilim in this matter to look to. Would we not have invented the practice, at least in isolated instances, if we had evolved as an independent species? It is difficult to say since the judgment is made on how we act now, influenced by the Nefilim.

In addition, in the United States for example, our practice of slavery often has been based on rationalizations that looked directly to a Nefilim example. If we understand the Old Testament of the Bible as a history of a particular tribe, the Hebrews, under the despotic rule of the Nefilim ruler, Yahweh, it is easy to see where the justification of using a fellow human as a slave because he or she was a follower of another Nefilim ruler (a "heathen" worshipping other gods in the language of the Bible), or was considered to have no "soul," came from. Our practices of war and slavery are clearly according to the Nefilim example and tradition, and were continued by humans after the Nefilim had left the planet as an effect of the godspell mentality. Our Nefilim genetic bicameral heritage coupled with whatever aggressive tendencies we have from our Homo erectus bicameral genetic heritage has done the rest.

Race as Rationalization

The most sensitive topic, currently, is racial prejudice. A friend whose opinion I very much respect has stated flatly that there is no way to deal with it. But I disagree; we have at hand the historical information to see the roots of slavery and the subjection of other humans in the internecine politics of the Nefilim and the concomitant use of humans as disposable units in that context.

The element of skin color is extremely superficial. It is generally recognized that it only takes about 10,000 years for such adaptations to climatic conditions to evolve. As a single example: the gradations of protective skin pigmentation from the blazing equatorial sun regions to the relatively weak light of the midnight sun regions of the far north is well defined and its purpose understood. Those who still cling to racial prejudice on the basis of skin color or physical racial characteristics choose to remain oblivious to the simple fact that racial characteristics are all survival adaptations to local climatic conditions.

Cultural set should be distinguished from the mentality that allows subjection, but there is a close connection. If we see it in the context of various cultures springing up under the auspices of the various Nefilim personalities, it is easy to understand the differences in cultural set. But the prejudices against color among whites in the U.S. has its roots in the psychological element of having to see the black person as something less than human in order to live with the fact that blacks were slaves or that in the past one's family or oneself had held slaves. Humans in the United States by the time of the Civil War had reached a point of sensitivity that they may have denied or suppressed but their reactions and words, at least those of the more evolved, belied their philosophic convictions concerning bondage.

We learned of slavery by experiencing it: we have tended to perpetuate it by political tradition and because of its advantage to the slave holder. But we may no longer simply mimic our ancient masters. Slavery's gradual demise already may be understood as part of that gradual process toward fully human, dignified and just treatment of humans by humans. Over the two hundred plus years that the United States has been ruled by the Constitution we have seen slavery not only abolished in this country but come to be considered a barbarous evil the world over. It can be found in isolated places and pockets in cultures and disguised as the practice of "indenture" in others. But these instances mark the progression of slavery from an open, accepted practice through the stages of unacceptable, illegal, immoral, detestable, barbarous, to unthinkable, as the intrinsic worth of the individual becomes more and more recognized and respected.

An understanding of our true nature clarifies what and who we are and facilitates a universal comprehension of human nature and its intrinsic dignity. An understanding of the origins of slavery, conquest, warfare and political dominance in the context of our use as subservient inferiors and pawns by the Nefilim makes it clear that we have inherited and continued, as unquestioned institutions and practices, ways taught us by them for their benefit and not necessarily in our best interest. It is past time we eliminated from our society the old slave mentality, the ancient training to conquer, kill, plunder, take and hold slaves, to insult, injure, or destroy our own kind on behest and behalf of the master or surrogate – formerly the Nefilim themselves and, subsequently, whatever human king, ruler or president we have recognized as either directly representing (divine right of kings or emperor-gods) the ancient master's order or some sublimation of that concept. The godspell of violence and war already is broken. But so far it has been overcome through the sheer force of our, mostly unconsciousness, genetic developmental momentum. We are already beginning to regain our rightful racial self-respect – and, therefore, more respect for each other.

Because we are half Nefilim and half Homo erectus will some of the social and psychological elements inherited from both be, ultimately, only half good for us? Are there ways that are specifically and uniquely human that we have not even explored yet or may be suppressing? We need to do a great deal of analysis of these factors which may effect us very deeply. Such questions will occupy reflective thinkers and scientists for decades into the future and the constructive debates will be a great contribution to our racial self-consciousness. The identification and detailed tracing of the genetic and learned elements of our social and political interactions and institutions to their Nefilim and Homo erectus roots will be an ongoing major occupation for scholars and scientists into the next century.

Power Plays and Power Politics

Breaking the godspell has a far deeper meaning than a simple iconoclastic severing of adolescent ties to parental customs or the integration of parental traits with our modern racial identity. It may, ultimately, mean the literal salvation of humankind from an adolescent racial suicide. In working toward a complete and final recognition and integration of the elements of our bicameral heritage, it is well that we recognize that the sociopolitical manifestation of the godspell has three main characteristics: worship, servitude, imitation – and, therefore, prejudice, power seeking, violence and war. It is also well to recognize that, in the cycles of our racial development we have reached a kind of political menopause. We are embarrassed about our history, our past, less than our human politics and wars, persecutions, fratricidal insanities and general treatment of each other. Although some are still mystified about why we have acted so and some have never considered that it could have been different, the inevitable ascendance of the new paradigm will take us out of the dreary, repetitive drum and trumpet history of wars and rumors of wars and the racial amnesia that clouds our racial memories. We are due to transform our sociopolitical format of prejudice, violence, war and power struggles just as we are due to transform our personal lives.

The Tradition of the "Divine Right" of Kings Revisited

Certainly a form of governance or control through the leadership of a king in a male dominant society is common throughout the world and does not always look to some higher power for its office and authority. Kingship seems to be, in at least some instances, as is often asserted by sociologists, a natural development from simpler forms of human social organization in earlier times. But an obvious Nefilim-based tradition still with us is that of the institution of royalty based on the "divine right" of kings. The institution of kingship has a long history traceable back to the explicit decreeing of the initiation of it when "kingship was lowered from heaven at Eridu." The institution of human foremen, kings, as go-betweens between the Nefilim ruler and his or her human subjects, was brought about by a decree from the off-planet Nefilim "high command." "Lowered" is both an awkward translation for "handed down from above," and is yet a very precise description of the event. Kings, under the Nefilim, never acted in their own name, always in the name of their Nefilim ruler. As historians and mythologists have pointed out repeatedly and emphasized, "theological offense" on the part of kings was the cause of many royal downfalls and even deaths. Since historians and mythologists persist in considering the Nefilim fictions, then the mistake or wrong action "offense" committed by a king is considered by these scholars to have been against a spiritual or mythic god. Since of this nature, they call the offense "theological," meaning against a "supernatural" god of some sort. But the simple fact of the matter is that the king, being just another human and subject absolutely to his Nefilim ruler, was held to strict rules of conduct. If he transgressed seriously he could be banished, punished or executed. And examples of all those forms of censure are abundant in the recovered records.

Regardless of the original subservient position of the king as go-between foreman, with the retreat of the Nefilim, the human rulers assumed power, sometimes reluctantly, sometimes ruthlessly, and the tradition of their being appointed by the gods remained. Over time the notion of divine appointment transmuted as the idea of flesh and blood humanoid gods was transmuted into the concept of a single cosmic and omnipotent Being. The divine right of a king coming from this Divine Being – or the position of the king or queen to be the head of a Church as with the current Queen of England – is still with us today although its unquestioned luster is fast fading and, eventually, will be relegated to the history books. Its origin, nature and influence in our societies need to be made clear so we can relegate this ancient godspell custom to its proper place in the museum.

Because it was instituted to solve a problem caused by our godspell heritage which will no longer exist, our current system of a republic under a Constitution is also due for reconsideration and rewriting.

A Second Constitution

The most immediate example of how we will move to the next sociopolitical plateau in the United States is found in how we will transform our Constitution. If we knew what and who we are when America was formed, the Constitution would not have been written in the way it is and our society would not be like it is. It is necessary to write a new constitution and develop a new society. The constitution was built to resolve the problem of conflict in a society of religious pluralism.

The general acceptance of the new paradigm will mean the natural obsolescence of the United States Constitution as we know it. It will not simply be because we will no longer be able to use the "under God" theme and base the concept of fundamental human rights on some objective order of reality held in existence by the thought of a supreme being. The current Constitution will be outmoded and fall into obsolescence because we possess a new worldview within which the consensual reality it supports will be superseded by a far more comprehensive, robust and planetary understanding of our common human nature. We should be grateful that that time has come.

The familiar and successful Constitution we know has been so because of the fundamental assumption that its principles are the best way, so far, to maintain a situation where differing religious views – differing views of fundamental reality – can exist side by side with minimum strife and in the maximum productive peace. That essential characteristic is what has made, and continues to make the Constitution viable. When we market "democratic capitalism" (whatever that really means if it is not just a manipulative buzzword) to other countries, as if there was some intrinsic relationship between democracy and capitalism, the emphasis is on economics. It is supposed to be good because of the opportunities it affords the individual. But everyone with some perceptivity knows that, in reality, it is simply the preparation and development of new markets for the American economy which is based on constant expansion and growth. But the theme that is promulgated to make this vulgar intent palatable is the appeal to the supposed support this touted system provides for social equality and human rights, opportunity, and the fundamental thrust of the vitality of the Constitution. Even though the record makes it clear, that has not necessarily been the case, even in the United States.

The writers of the Constitution perceived the problem clearly: a variety of cultures and, particularly, religions were represented, flourished, sometimes co-existed and, more often than not, came into conflict in the original colonies and a context was needed which would keep them from each others' throats. Some groups had already begun to aggressively persecute others with which they disagreed. The challenge was to forge a set of principles which would be as consensually acceptable to as broad a spectrum of worldviews as possible. That they were able to distance themselves from the consensual reality sufficiently to accomplish as much as they did is still impressive. That they succeeded as well as they obviously did in creating a situation where religion and the state were separated, religious freedom and the individual's worldview, secular or religious, were respected, is very impressive indeed. And those men were not unsophisticated about the philosophical premises upon which they relied to gain broad consensual agreement; they spoke of the fundamental premises of the Constitution as being self-evident and those self-evident rights being those of the individual. Not based on religious revelation, not scientifically dictated, not arbitrarily or pragmatically chosen, but self-evident. That was just about as philosophically neutral and fundamental as possible without bringing into question the notion of an objective order of reality. They mentioned God because of the religious mind-set of probably the majority of the people at that time and, although the atheist citizen got shorted, that was something they could get away with. It fit the situation best in their view. That was shrewd – and effective.

The Constitution will never be completely successful because of the ongoing rapid evolution of human nature, and the fact that the godspell differences still hold sway, even if dissipating. Jefferson, remarkably foresighted, knew that would happen and recommended constructive and peaceful periodic revolution.

But there is no indication of any anticipation that there would ever be a resolution of those differences, no anticipation of a common definition and understanding of human nature. It is very clear that the writers of the Constitution never envisioned a time when the fundamental problem of religious differences, of radically conflicting world-views would be solved; they never anticipated or expected the situation to change radically to a positive one in which a fundamental, obvious, common, planetary consensual understanding of the nature and beginnings of the human species would be available. We will soon be compelled to either rewrite the existing Constitution or to write a more evolved document. But that will be only a transition phase toward the time of the post-American human.

Violent Sport as Psychopathology

If one surveys the panorama of our brief history on this planet one can also see the same gradual process happening with other primitive phenomena. We are a mix of two separate gene codes and our development over time as a synthesized species has seen the emergence of traits that have their source in both genetic backgrounds. We also see practices that are easily traced to either an indigenous Homo erectus, Terran genetic source or to a Nefilim source.

In the slow-motion process of evolving from a combative and competitive social structure to a more benevolent and compassionate one, we have found it necessary to evolve ritualized and controlled forms of the more primitive tendencies inherited from both our Nefilim and Homo erectus ancestors as transitional easements. The history of violent sports shows a gradual lessening of brutality from the gladiatorial games down through boxing of today.

The Roman circus where humans, criminal or otherwise, were slaughtered were the first transparent forms of vicarious participation and release. Dueling has been outlawed as well as gunfighting but they still continue outside the law. But, within the law, we have sublimated brute conflict into the transitional body contact sports like football, boxing and the "martial" arts.

The next stage has begun already: virtual reality games find their precursor in the video games involving violent contests that are bloodless and injury free but still satisfy the primitive drive in the young for violent contestation. Those brutal, injurious, violent sports that have been the ritualized safety valves for so long will eventually also be rejected as below the dignity and worth of the individual. Some may be even made illegal as we reach the end of the transition to the new human. It will not be soon, since the beer swilling crowd, cheering for the football players, boxers, choreographed wrestlers and the brawling hockey players is still in the majority. It will take a long time and a great deal of evolution. But it will come. And it is here for some already.

We have gone from ritualized combat to death in the arena and the settling of quarrels of "honor" by duels with sword or pistol to boxing matches and televised "live" courtroom drama. When social conditions become so deteriorated as to create great desperation in the individual there is a regression to gang warfare, but these aberrations are now beginning to be seen for what they are. Muhammad Ali, admired as the greatest in the sport of boxing, now is a palsied icon of the wages of such inhuman contests and there is a growing opposition to legally allowing them.

It is the slowness of such processes, conditioned and determined by the inertia of the mass consciousness, that causes us to think that "nothing is happening." But the leavening of the mass consciousness, though painfully slow and measured, in some cases in thousands of years, is perceptible and its direction certain. Made conscious and explicit, our racial self-image and self-worth restored, we will consider it depraved to mistreat each other. Only then will we be recognized as mature and safe enough to be allowed into stellar society.

The Post American Human

Just as a second Constitution will be a very elementary step toward the new human society, so will the post-American phase of planetary society be a very early and elementary foreshadowing of the new human. We may take these easily comprehensible developments as indicators of the trajectory toward the fully new human and the new human social context.

The politics of the post American society will be the least evolved element in the new human social structure. This is almost inevitable at this stage of human development because public policy and decision and action involve the myriad opinions extant in the entire spectrum of the population and, therefore, will be the slowest moving and developing segment of the general evolutionary process. Nevertheless, the mechanics of the politics of the post-American society will not only be direct (each individual having a direct voice in the local and planetary forums), but the consensus-forming modalities will not be simply by majority rule. We have already learned that the majority does not always represent the best interests of the whole and the minority or minorities often represent a more advanced and enlightened view. The electronic medium of instantaneous input of opinion combined with a policy of complete openness will mark the new politics as radically different from the politics of our times. But the vote will not be a simple yes or no, the options allowed will be carefully tailored to include complex and conditioned opinions and degrees of emphasis and conviction across the spectrum. The compilation of inputs on any given issue will be analyzed in depth for both a clear expression of the wisest, most compassionate and enlightened views, and the discovery of the most constructive, new insights.

Once the new paradigm has been recognized and accepted, the dominant characteristic and power of the new politics will be its being based, not on some divine right, sanction, or religious principles, not on some abstract philosophical or sociobiological base, but on the nature of the generically human and the primacy of the individual. I have postulated the full nature of the new politics on the acceptance of the new paradigm as the critical factor because without a planetwide understanding and agreement about who and what we are, there will not be the possibility for the planetary unity prerequisite for a truly human politics.

Notice, however, that there are already indicators, directional vectors, recognized by some futurists pointing toward those dawning conditions. But these are, in some cases, predicted for our society in the United States and only much farther into the future for other parts of the globe, or they are identified only as trends recognized as long-term to be achieved with difficulty. Two examples will illustrate.

John Naisbitt and Patricia Auburdene in their books, *Megatrends* (for the decade of the 80's) and *Megatrends 2000* (for the 90's) (Morrow, 1990) identify, through extrapolation from established statistics and testimony of "experts," major trends toward the ascendancy and independence of the individual, the turning away from rigidly dogmatic institutional churches, a movement from representative democracy to participatory democracy, global life-styles, the emergence of free-market socialism, among others. Even though they are working on the basis of ten-year periods, very short-term indeed, it is interesting to see that the overall political trend they identify, with which we are concerned here, is toward the recognition of the ascendancy of the individual, direct participatory democracy, and a greater and greater planetary unity. To reiterate my point negatively, although they identify these strong trends for what they are, they feel that one would have to be almost "a bit mad" to even predict such trends for even ten years, and they apparently do not see either the burgeoning new paradigm or its power as a factor. Logically, limiting themselves to a myopic ten year perspective by statistical extrapolation, it may be another ten or twenty – or more – years before they would or could recognize it.

From almost the other pole of prediction, Timothy Leary, in his book *Neuropolitique*, working directly from his own direct experience as a political prisoner and his insight as a professional psychologist, has clearly indicated the crucially needed shifts from the Judaeo-Christian guilt-and-sin, authority-based ethos to the age of personal responsibility, from representative government to participatory democracy, from secrecy to complete openness, from a static politic to an evolutionarily oriented one, from the ascendancy of the state to the sovereignty of the individual. Again, only the planetary unity afforded by the new paradigm's acceptance will provide the platform for such change if it is to be planetary.

Inalienable Human Rights Revisited

The concept of inalienable human rights is in itself under critical scrutiny at the present time as a part of the larger philosophical discussion of direction of political change. As we have seen, previously, in the reexamination of notions of objective and subjective reality, it is argued that if there is no objective reality, as assumed by the theological traditions that are founded on that concept, then there is no basis, no fundamental substrate, in which anything inalienable can be said to be embedded.

It is feared by those who rely on an objective order as determined and sustained by some cosmic divinity that anarchy and evil are the inevitable results of rejecting – indeed even questioning – that view. If no "higher" or real order other than the whim of the individual exists, then no authority or common agreement can be invoked to regulate human society. It does not seem as if these thinkers, who more often than not are of the conviction that human nature, if left to itself, will tend to disorder if not evil, are looking at the existential facts. Even though successive generations in western culture tend to be more and more removed from authoritarian principles and the concept of an objective order of reality, the general evolutionary trend is toward a more humanitarian society. We need only appreciate the greater and greater abhorrence of war and violence; the recognition of the inalienable dignity of the individual human; the sociopolitical movements of peristroika and glasnost, etc. None of these phenomenon are founded on the old absolutistic objective-order religions or philosophies.

The New Politics

The restoration of our true history resulting in the achieving of an unassailable personal integrity based on a consensual knowledge of who and what one is sweeps away the Babel factors, the points of division between humans based on the myriad religious and concomitant philosophical definitions of human nature. The elimination of those deep rifts will heal many of the wounds of the world and allow an empathy and compassion far beyond the crippled semblance of such we know now. There will certainly always be differences of opinion as long as we are human, but the shift coming may be best understood through a very pale shadow of something like we are witness to today. In the relatively short time of writing this book the world has seen the almost unthinkable shift in relations between the United States and Russia. The key characteristic of the change is the elimination of the radical concepts that had caused, or had been used to sustain, the confrontation for so long: the ostensible philosophical differences that could generate moralistic slogans like "capitalist imperialism" and "evil empire" that cast the other in terms of degenerate oppressor or atheistic monster. The respective propaganda implied or stated that the other was to be suspected, feared and/or opposed because of an erroneous and probably evil definition of what a human being was, a potentially exploitable economic loser in the capitalistic monopoly game, or a disposable pawn in the game of the godless welfare state. There are still very real and deep differences between the two superpowers but, once the relationship was moved from the level of pseudo-philosophical absolutes to the level of pragmatic politics that inferred or implied no real radical moral differences, the pressure was immediately relieved and disagreement could be processed without resort to primitive survival mechanisms. Although, certainly, an anemic indicator, this international phenomenon gives some idea of how the elimination of deep religious and concomitant philosophical Babel factors, the result of the ancient godspell, will bring about the consensual planetary concept of generic humanity, a common definition and understanding of human nature, its characteristics and mechanisms, its evolutionary direction, options, potential dimensions and place in the universe, a lens through which differences of opinion will be seen as superficial, transient and constructively conducive to change and growth. That condition having been established, a level of respect, cooperation, concern, mutual support, empathy and compassion between all humans will be facilitated and advanced that will become the general rule, the prevailing psychological climate – far beyond the plucky "we are all in it together" banners waving at the vanguard of our best efforts today. The cynical will no doubt continue to argue that human nature is intrinsically degenerate or so myopically selfish or genetically deficient that such a seemingly Utopian condition is unattainable. I submit that, even now, the direction of our metamorphic evolution is clearly in that direction and it will only take the full acceptance of the new paradigm and the elation and sheer relief it will bring, to establish the beginning of that stage. It is obvious that humans do not want to be separate, alienated from each other, at odds, and that it is only the unresolved stresses of age old imposed discords that divide.

Inevitably, the influence of a planetarily consensual understanding and definition of what a human being is will be tremendously powerful and compelling. It will be about the expandability of human nature, about whether and how this species can transcend any given state of existence and being it has already reached. In a real sense, we are already well into the phase of redefinition. *Breaking the Godspell* and this book are intended to stimulate and contribute not only to that redefinition, but to open up the vision of how we will be and live once that redefinition has been worked through. The realization that the nature of the universe, of which we are a product and intrinsic part, has given rise to a manifestation of itself which is capable of and primarily focused on constantly redetermining itself, will be the other side of this spinning coin.

Society, any social context, is a construct that springs from the interrelatedness and interdependence of individuals. That is why human societies vary a great deal over the planet: they are forms that spring from long adaptation to local survival conditions. The most artificial ones are those that are based on godspell "religious" tenets – the reason why they must be absolutistic and repressively authoritarian: they must be held together by force since they are contrary to human nature and the individual must be coerced into subservience to the state or to the religion. Social structure is simply a means to an end and will exist in the future clearly to support the individual and the collective, it will not be elevated to some sort of independent status and canonized as sacred with a life of its own.

The race and the individual will strive to create a social structure, a social context which is maximally conducive to the exploration of the potential of the universe by the individual. The race will recognize that it must maximally support the individual in this, disallowing only those actions on the part of the individual that would immediately and directly destroy the race or destroy or obstruct the exercise of that same freedom by other individuals. I call this the Callen Principle. Horace Callen was a professor of philosophy well known for his ability to cut across the intellectual arguments to get at the essential mechanisms of human government. I was fortunate to have done graduate studies under him at the New School For Social Research where, at the age of ninety, he was vigorously holding forth on these fundamental principles of individual freedom. At first this principle of maximal freedom will be observed in the perspective of our current usual three-dimensional view of reality. It may be seen to be a socio-political application of the simple Golden Rule. The Golden Rule, though paradoxically propounded by the traditionally recognized great teachers through the ages, has always been limited and obstructed in its application by the very religions supposedly springing from those teachings. The proprietary godspell claims to absolute authority and custody of the truth contravene the Golden Rule. But, inevitably, as we come staggering out of racial amnesia into the clarity of genetic enlightenment, the simple truth of it, for self-referential entities as we are at this stage of things, will finally become clear as we realize our generic humanity. We will come to regain an appreciation of its almost naive simplicity as a facet of the Law Of Everything. Gradually it will expand to mirror the habitual four-dimensional view and experience we will attain.

In effect, then, what will happen is quite simple: as the individual explores more and more possibles in the universe the data base of the individual expands and possibles get classified as desirable, not desirable, interesting, not interesting, essential and non-essential, consciousness expanding and non-consciousness expanding. The individual will feed back HIr personal experiences, discoveries, preferences and rejections into the common knowledge pool. The individual will learn to carefully submit them as HIr personal reactions and preferences on a self-referential, subjective basis. The racial knowledge base will accept these data on a non-judgmental basis, discriminating only on the basis of immediate danger to the race, the individual, or the freedom of others. These data will be open to the inspection and consideration of anyone and they will be free to explore the same phenomenon, ideas, evolutionary directions or to incorporate them into their own personal experience as they see fit. The spectrum of experience, concepts and phenomenon will obviously be broad and will include legitimate things which society is uneasy with, or rejects as intolerable or wrong, such as suicide, on the basis that an individual has control over HIr own existence. This may seem contrary to the basic rule that the race should not allow the individual to jeopardize the existence of the race, but a single suicide or a mass suicide of any really conceivable proportions would not seem to threaten the existence of the race. The self-extinction of half the world population would seem to be at least questionable in this regard but cases would have to be investigated and evaluated on an individual basis.

Murder, violence, rape, enslavement, torture, all the things that we already designate as forbidden and punishable and to be prevented would obviously still fall under the same category as they do now. There would still be police but the police would be of a much higher caliber and would have high level, instantaneously available data bases of information and judgment, expert systems, to call on instantaneously at the scene of the crime. Only invasive acts will be formally defined as criminal.

There are a few developments that we are rapidly moving toward that are clear, early precursors of the new human society. The first general development that we are already working our way into it, however reluctantly, is a new economics of plenty which will facilitate a true leisure society.

Social Compassion: The Economics of the New Human

Our science has already opened up economic possibilities for the present and near future that are astounding when first perceived. The ability to mine the asteroids and the Moon, to set up manufacturing and habitation facilities even on Mars as well as in orbit around the Earth and on the Moon, all of which will afford us two very essential resources: we will have virtually unlimited resources of raw materials and we will have a virtually inexhaustible source of energy with which to process and manufacture these materials into durable goods. Gerard O'Neill, the author of *The High Frontier*, put that vision in front of us already in the 1970's, along with long range scenarios as to how we might accomplish intelligent expansion off planet Earth. The realization of these possibilities will eliminate the primitive competition which plagues us now. Buckminster Fuller in *Spaceship Earth* already demonstrated clearly that the basic engine of competition, the belief that there are only limited resources and not enough to go around, was a myth: with correct and enlightened management, the Earth was capable of supporting even the huge projected multiples of population for a long time.

But that was not enough to change the mind-set of the general population and especially those who had a vested interest in there being winners and inevitable losers in the manipulated competition of Wall Street and Main Street and international trade. Even the possibilities of manufacturing from unlimited raw materials in space will not be enough to overcome the short-sighted self-interests of those who find competition profitable in itself. But the concept of generic humanity and planetary unity, the elimination of wars based on religious beliefs, and the establishment of the primary and inherent dignity, rights and importance of the individual will tip the scales. We shall attain an enlightened and compassionate system of provision for all and for the ecological well being of the planet.

Currently, futurists like Jeremy Rifkin in *The End Of Work* are already pointing out the near term elimination of ordinary tasks, and even the tasks of the professionally trained and the highly skilled, up through the ranks of the architect, the engineer, the computer programmer, the accountant and the high level manager, with the looming possibility of the elimination of even higher level management. The use of robotics, better and better computers and expert programs, with the gradual implementation of artificial intelligence and nanotechnology, will phase out all but the essential human directors in industry and business and even the engineers and scientists who have created the first generations of robots and smart computers in the first place. Robots and computers will self-replicate and self-service and self-manage under human supervision to a degree that will make the completely automated Japanese car factory seem as primitive as the first Wright Brothers' airplane.

We will know the new society has begun to emerge when the myth of competition for limited supply has been overcome, when democracy and the other primitive forms of pluralistic governments we now witness will have been superseded by direct input discussion, voting and proportional respect for minority positions, when the basic survival standards of living will be guaranteed to all without infringing on their dignity, and when the law has been revised and concerned only with invasive crime, violence, even in sports, has been recognized as psychopathological.

An Enlightened Science

As we have discussed, the problem currently facing science, specifically physics, is the problem of consciousness: even the most advanced theory of physics cannot account for or explain consciousness. Once physicists have expanded both their theory and their consciousness sufficiently to resolve this dilemma and include consciousness in their universe they will, by that very fact, have reached a point of enlightenment which will automatically transform science. The new science will be a compassionate discipline that will be inherently ecological, altruistic and holistic.

Habitual Space-Time Perspective

The focus of science of the new human society will mirror and support the focus of the mind of the new human, habitual four-dimensional consciousness.

The habitual life focus of the new society will be no less than the cosmos from the perspective of four-dimensions. The "medium" or, at first, the mental perspective in which the new human perceives and operates is space-time.

It is a very significant step to understand the nature of space-time in which all things and events "are" and are related, and in which linear time sequences are really simply a function of our mode of perception and experience. But the new human will have moved to the next step beyond the novel initial conceptualization to the habitual experiential perspective. The new human perceives, experiences and judges in the perspective of space-time.

As we have noted, it has been speculated that if we could differentiate time to a billionth of a second rather than the coarse half or quarter second we are physically capable of now, we could perceive space-time directly. We routinely employed devices in scientific experiment and procedures which can measure to very small intervals of time but I am talking here of our unaided bio-mental perceptive abilities. But it would seem, at first, that direct perception of and "in" space-time would require a macro rather than a micro view. If Einstein is correct, all things and events are simultaneously existing and in space-time rather than the illusion of past, present and future. So it would seem that it would require increasing our time perception and discrimination by magnitudes to encompass eons rather than to discriminate a billionth of a second in order to directly perceive the space-time continuum. But reflection shows that it is not a perception of time as we normally experience it expanded to encompass huge passages of historical type time that is required but the ability to develop a sort of intensity of time perception that is essential from which the cosmic perspective flows.

By the very recognition of that possibility, coupled with the scientific knowledge we have developed about space-time, we open up the potential to achieve it. Curiously, the contemplation of that kind of perception brings one closer and closer to it. The assimilation of such a concept seems to lead to subtle neurological adjustments that enable one to adjust one's perspective to at least perceive, from the point of view of what one conceives, space-time consciousness to be like. Certain experiential modes, such as facilitated by LSD and other powerful psychedelics, are quite certainly of a much higher speed of differentiation: but the psychedelic experiential mode is more sophisticated and profound than just very high speed differentiation and perception. One may perceive one's own brain perceiving and cognizing in modes that are ordinarily not open to reflexive inspection and certain of those modalities quite certainly are very robust space-time modes of perception and cognition. Metaphorically we are climbing down from the Cartesian-Newtonian "monkey bars" (the classic three-dimensional geometric axes usually represented as the x,y,z coordinate system in textbooks) and shedding the sequential, linear Newtonian type of time perception. Some have said that they have attained such insights routinely when using LSD. I believe that this is partly due to a heightened sensitivity of the neurological system to the body signals of others and subliminal signals embedded in events. But, beyond this, I believe that LSD can and does neurologically facilitate actual fourth-dimensional perception both temporarily and long term, through the habituation to the evolved consciousness.

The space-time perspective is the ideal orientation from which to most intelligently determine one's, and contribute to the determination of the collective, evolutionary trajectory. This, at first, seems to be a deep paradox: if the space-time perspective affords one the view of past, present and future all at once, does that not negate the notion of personal and racial evolution and serious effort to determine the ideal trajectory? The answer addresses an even more generic form of such a question that asks: If one attains the enlightenment of awareness beyond opposites are not then the perceptions of usual three-dimensional consciousness simply illusions as taught by Eastern philosophies? From the point of view of fourth-dimensional consciousness the constructs of three-dimensional, lineal consciousness are seen as limited and time-bound, but not illusory per se. Even though fourth-dimensional consciousness can see the sequential nature of three-dimensional consciousness and even where its logic will lead, that does not mean that the necessity for three-dimensional logic, consciousness and events to work themselves out will be superseded because that is inevitable just by there being three dimensions. It does mean that those who attain four-dimensional consciousness, perception and perspective will be able to view the sequences happening in three dimensions as past, present and future in a real sense simultaneously. This affords an advantage of major consequence. One may avoid otherwise unforeseen undesirable consequences, gain a degree of control and mastery of one's fate otherwise unattainable, recognize patterns of very long periodicity otherwise impossible and, in a real sense, predict the future. The law of everything is the ultimate resonant harmony of everything and the intensity of its beauty may completely stun if not kill the first to grasp it if they have not developed enough to be comfortable, at least with fourth-dimensional consciousness. But I can easily conceive of a time when we will as easily and comfortably operate in a known universe fully aware of the law of everything as we now do with the Newtonian laws of physics.

Those who do not yet have the four-dimensional perspective see those who do as indulging in wishful thinking, put down their predictions that come true as luck and wonder how they can see into the nature of events and persons as well as they do. Four-dimensional consciousness is such a novel and important concept that we will consider it again in depth in a following chapter on consciousness in a known universe, i.e., in a universe which we know through the discovery of a "complete" unified field law.

All these early indicators and evolutionary markers we have been speaking of are but glimpses of the mature new human society which will inevitably develop once we have reclaimed our racial birthright. I am totally confident that we, as a race, will achieve that status – which in its turn, will only be a major phase in our open-ended racial development. I already see all the prepotent seeds of such a state in some humans and in myself.

The New Technology

Besides those expanded capacities, almost incidentally, the new human will be one who is living a life which has resolved and transcended both the ordinary and the extraordinary obstacles, problems and traumas which we take for granted now including sickness, want, war, etc. Many will deny that that could ever happen; many will cynically deny our capacity to even handle it if it did come about. I am convinced that it will come about much sooner than we generally think. All the early signs are already present. If that sounds a bit rich for your current blood type, a bit incredible according to current standards, compare our contemporary fundamental living status to European standards only two hundred years ago – and then begin raising those standards by powers of ten. And those are just standards of ordinary living. We are talking here of the result of a profound planetary change of worldview and advances in technology.

Those applications take the form of a new psychology, a new logic (and, therefore, of computer programming) a new mode of science, a new mathematics, and, in the most general terms, a unified field concept of the universe that is the lowest order mode which constitutes the base of all of these sciences.

The leading edge, currently, is in the contemplation of self-reflexivity and the recognition of it as the key to a new mode of human philosophy and science. (The Nefilim use of self-reference as a meta-metaphor for a unified field law of the universe, only now becoming apparent and recoverable by us, is the most novel and least appreciated element of the Nefilim legacy.) We are pushing the envelope of our own consciousness in this area just as we were pushing the envelope of our consciousness in the area of atomic and genetic science only a few decades ago. Even if we expand our consciousness it still is in terms of self-reflexive awareness. The new human will determine HIr own reality. The new human will be HIr own philosopher to a degree that will rival the depth of sophistication and training of the greatest philosophers the world has known so far. Philosophy as a discipline will be taught systematically in the early teen years when the full rational faculties mature. We are now at a point where we can develop a new logic, philosophy and science by turning the self-referential barrier into a positive platform for a unified field approach to the universe. The new human will operate in this unified field as naturally as we now operate in the classic Newtonian and Relativistic context. We will discuss this expansion of our knowledge and consciousness in the next chapter.

We have outlined briefly in the introductory chapter to this third section what the characteristics of the new human are: the new society will be a mirror of the new human, it will exhibit the characteristics of the new human.

Philotropic Humanism

The key concept to understanding the new humanism, the new social context which new humans will create and in which they will live, is integration. I have called this humanism, this philosophy of life, philotropic, meaning love/wisdom seeking, as wisdom in the form of information, and love in the form of compassion, will be the dominant characteristics.

In contrast to the fragmented, often contradictory world view and resulting personality of the godspell human, the new human will have integrated all facets of human existence and all facets of HIr own personality. This integration may be understood as the elimination of conflict between contradictory philosophies and motivations, the making conscious and integrating of all instinctual and sub-conscious impulses suppressed through the repressive artificial slave morality of godspell religions; the adjustments to priorities, objectives, perspectives, values, and relationships required by the potential for relative immortality and brought about by the profound liberation into personal freedom. This gives a whole new dimension to an individual becoming "their own person" in a social context.

I will be the first to say that that sounds like an almost fantastic science fiction prediction. We seem to be so immersed in the shallow soap opera dramas of ordinary daily human life that this level of maturity and integration must be a dream. But we must not underestimate the radical, planetary transformation that will inevitably occur due to the dissemination of the new paradigm. It is not that there is no presentiment of this level of racial maturity or early indicators of such a new planetary plateau. On the contrary, decades ago Marilyn Ferguson identified the groundswell rolling over the planet toward a new compassionate humanism as the Aquarian Conspiracy. The myriad self-realization schools, sects, teachers, New Age and New Science philosophies and techniques, from bio-feedback to Transcendental Meditation, from self-hypnosis to past life regression, from Indian philosophies to American Indian prophecies, are all early indicators of the coming synthesis under the new paradigm's redefinition and the liberation from the godspell. Why should it require the comprehensive umbrella of the new paradigm to achieve planetary unity? Because it finally establishes a universal definition of human nature and dispels the Babel factors that have kept us separate and at odds and unable to trust and cooperate.

A New Category of Citizenship:
The Immortal In and Out of Society

Immortality involves either living indefinitely from the time of one's birth, or living, then dying, then being revived and living again, perhaps more than once. Once this is an available option to humans it will eventually create a new category in our social contracts. Even now, those who are anticipating or have actually been put into cryonic suspension are causing this new category to be discussed and explored.

This brief discussion of the legal and social ramifications of immortality, whether attained through revivification, soon to be developed genetic techniques, nanotechnology or a combination of any or all of these methodologies, serves to elucidate how such a potential or achieved status will impact our social contracts, especially our consideration of the individual's sovereignty.

The Futant as National Treasure in the New Society

Fear slows our evolution. Fear of being first to do something differently, of jeopardizing one's security, or safety, or social status, or one's job, tenure, fear of espousing or propounding new concepts, going against established order, power structures or institutions. But there is always a small percentage of the population at any given time that can and shall, indeed can hardly do otherwise. I am not speaking of political radicals who advocate violence, or "futurists" who can only predict one chess move ahead based on myopic extrapolations. I am speaking of futants, future "mutants," the bearers of change, the bearers of the vectors of evolution. To this point in our strange history they have almost always been resisted, rejected, persecuted, often killed by the godspell religions or the devolved. A race subservient to powerful masters who have created it as a slave race for their own practical purposes is not interested in precociously independent servants who think for themselves.

A race left on its own and trying to appease the gods it thinks have abandoned it for doing something wrong will tend to suppress – even violently and ruthlessly – any innovative member who rejects the traditions taught it by those yearned for masters. The futant individual or group, regardless of how creative and constructive and beneficial their contribution, has been consistently and unmercifully persecuted in human society for those reasons for as long as our history is known to us. There is a clear pattern over our past history and futants have come to be quite cautious about things like drinking hemlock, being bricked up in castle walls, burnt at the stake, consigned to the gulag, or, if lucky, only being put under "house arrest."

But our time marks a turning point with regard to the position and treatment of the futant. The degree of authority found, as examples, in the Catholic Church or the Muslim religion, that claims to speak in the name of God, is the kind of absolutism that justifies the unquestioned suppression of novel ideas, expanded freedom, untried directions, and greater independence of the individual represented by the typical futant message. But that authority is in decline and weakened, and the futants of these latter days have gained enough perspective to possess enough political and evolutionary skills to prevent, deflect or at least avoid in great measure the attack of the hive guardians.

With the slave ethos outgrown and because we are collectively ready for it, we are now at a momentous time in that we can put the futants in perspective, critically evaluate the content of their message and encourage them to participate, to contribute. This is a radical change and the importance of it should not be overlooked or minimized. The futant is an integral part of our collective planetary growth and contributes to it in a very crucial way. The true futant is a national, indeed, planetary treasure. In the time of the new human, the futant will be integrated into society and acknowledged as ordinarily as we honor a Nobel Prize winner now. Just as the Japanese recognize and honor their most gifted in the arts and crafts as national treasures so shall we come to value and honor the true futants among us, the bearers of the early signs of the direction our evolutionary trajectory should take. In the fully new human society the futants will be recognized and appreciated and their talents utilized.

A Mature Sociobiology

In the process of reading the entire human gene code and studying human psychology and neurophysiology we shall learn a great deal about the way we are programmed genetically, not only in the sense of growth and development but also about the way we are programmed to act physiologically and psychologically. The future will see us able to analyze and explicate large patterns of animal (including human) behavior through discoveries made in neurophysiology, the study of neurons and their circuitry, and in the study of the molecular basis for cellular behavior. Even now the revelations of our beginning discoveries of that determining programming sometimes makes us uneasy. Are we actually completely determined with any kind of freedom of choice or is that only an illusion? The goal of sociobiology is to be able to eventually predict human behavior in at least the social arena in detail according to amassed data. It would seem clearly not so by the very fact that we can consider the question, but the gradual expansion of this knowledge of how we really "tick" will cause us even more unease and raise awkward questions. Ultimately those questions will help us define the nature of the true freedom we potentially can enjoy, but the transition period will be uncomfortable for some, a stimulating challenge for others.

What's to Come After Four-dimensional Consciousness in a Known Universe?

What's after that? Follow your own deepest intuitions for an answer. The more proficient we become at being conscious creators of our own reality and learn to influence rather than control nature in a harmonious and respectful way, we will perhaps experiment with modifying our physical makeup into whatever form we wish, learn to communicate with any other entity we find interesting and perhaps even learn to tune directly into the experience of other entities through translational devices or directly with our own consciousness. Direct mind to mind communication would seem to be a natural development. We will most likely be living in a universe understood through a law of everything, a radically different orientation, and we may eventually learn to travel literally from sub-universe to sub-universe. Such experience and development will make immortality very unboring as well as essential. Time travel and anti-gravity technology, far beyond our crude science fiction imagining of such capabilities currently, will be commonplace. And I am sure that some of the new model intelligences coming off the assembly line now will be able at fifteen to envision developments far beyond these hints and suggestions. I salute them and recommend that we soon learn to recognize the value of listening to them from the time they are born.

There are those who will refuse to consider even the possibility of the new human society because they see nothing of its promise in themselves or those around them. There are those who will allow the possibility because they see the desire for it in themselves but feel they must look passively to others to prove it to them by bringing it about. There are those who are already reaching new human status, who see the vision, the potential, the desire and the power to bring it about already manifesting in themselves and who will work with others of their kind to make it so.

The New Human as Reality Creator

We all have unique roles to play in reality creation. We are doing it all the time whether we are conscious of it or not. But reality creation involves conceptualization of what might be, what could be, experimentation with both physical and mental theoretical constructs and the exploration of short and long term effects and ramifications of such constructs, actions, paradigms, and logic sequences and theories.

Some are prompted by their genetic makeup, environment, training and conditioning to play a part in the very immediate actions to improve, restore or at least preserve human dignity and our environment such as Greenpeace, civil rights movements, women's' causes and children's interests. All these are early signs of the growing post-godspell new human consciousness of the uniqueness, rights and dignity of the human being. This is not the propounding of a religious dogma of human unity, regardless of how high sounding any of those philosophies and teachings seem. The more robust any religions are the more they are exclusive, divisive, often castigatory, and politically or nationally enmeshed as a logical outgrowth of their source in the ancient subordination to the Nefilim.

I cannot emphasize too strongly that, until we reach a common, consensual definition and understanding of what a human being is, we will not be able to attain the planetary unity which is already potentially within our reach. Until that time, even the most compassionate and elevated teaching, attitude, and community will remain separated by doctrine, world-view, and most fundamentally, definition of the human from at least some other groups of humans on the planet. We need only contemplate the differences in world view even within the New Age movement, between groups and schools and individuals due to godspell attitudes, resulting in the varying interpretations of what a human being is or should be.

This new type of existence has been attainable by anyone breaking the godspell completely, even in the past, but the possibility has been far more difficult because the historical evidence to move us out of racial amnesia has been long buried and corrupted. When an individual must struggle through to the new human consciousness and status in isolation, progress is always much more difficult and often slower because of the lack of social reinforcement.

The transformation to the time of the new human will be gradual, if only because of the vast numbers of people on the planet, although perhaps not as slow as we might anticipate. The worldwide web of communications we possess accelerates the fact that archaeological and anthropological science is converging toward the new paradigm's explanation. The reinforcement comes from many quarters of specific research such as the geological work on the Sphinx's age, the analysis of the monuments on Mars, and the convergence of the modern astrophysical theory concerning the formation of our solar system with the same information contained in the *Enuma elish* and other ancient narratives. It becomes rapidly more and more robust.

It is one thing to disseminate the information accurately in detail concerning the new paradigm, the new interpretation of our true history, and a clear and accurate generic definition of what a human being really is, and another to have the ramifications assimilated and implemented. Even the fact that we were genetically engineered would be rather easily assimilated without much shock. After all, on the day of this writing the announcement of the cloning of a mature sheep from a single cell was the lead topic on the TV news. The speculation that the technique could be perfected so that the cloning of a mature human might be only six months away is only mildly exciting and amazing to the general public. Just getting the information out to the world would be simple and easy at this point in time. But the assimilation of the ramifications, the radical changes entailed, will be the thing that will cause the process to be very gradual. And some individuals, regardless how robust the information is and how well accepted by the rest of the planet, will never acknowledge or accept it. But that the population of the planet will eventually receive it, perhaps even with relief and enthusiasm, as the next planetary plateau, I submit, is certain. And what shall it be like then?

Living in a compassionate society with a mature sociobiology, an enlightened science and technology, an advanced and humane economy and politics, with the possibility of indefinite longevity or immortality, the occupation, indeed preoccupation, of the new human, will be personal reality creation integrated with reality creation on a collective scale. This activity bears a great deal of consideration.

CHAPTER 13

THE IMMORTALITY THRESHOLD

LIFE BEYOND THE GILGAMESH FACTOR

The drive to immortality is not the result of our suddenly acquiring the genetic engineering skills that may put it within our reach but the deep drive to achieve what was withheld – and which a few of us have already arbitrarily been given – by the Nefilim. The quest today is only the leading edge of that perennial drive. And we are getting close.

Frank Clinton
Diary Of A Bewildered Politician

Do not go gentle into that good night.

Dylan Thomas

With the passing of the macabre winter shadow
We shall see, in our astounded lifetimes,
The obliteration of the event horizon
Of the death sump; the elegant and pitiful,
Classic rage against the void;
The gruesome romanticism of the mystic
And the honest horror of the materialist
All erased to a clear glass into the future.
At this anticipated but unfamiliar threshold,
There is, no longer, an adequate archetype
For the fullness of the human but the human.

Neil Freer
Neuroglyphs

In this season of our evolution, the most profound god-game we are going to play is immortality. As we free ourselves of the inhibiting embrace of the godspell mentality we will begin to take advantage of the possibility of physical immortality through genetic engineering, nanotechnology and even more advanced technologies as they becomes available.

Immortality is clearly the major characteristic of philotropic humanism, the next plateau of human metamorphosis, the next stage of our meta-evolutionary, conscious, racial development. The relative profundity of its dawning impact demands that we consider it fully from all perspectives before it, suddenly, is available to us.

We already have experienced the confusion and ambiguities of perspectives and philosophical positions regarding genetic engineering in general. A single example is the controversy surrounding the releasing of synthesized organisms into the environment. The potential for genetically engineered immortality will far overshadow that controversy in breadth and intensity. The advent of that potential will be sudden by the nature of genetic research. And when the genetic discovery of the molecular keys takes place major sociological upheavals will occur. Although, at this writing, we have not quite yet achieved it, human physical immortality is simply a foregone fact. Be it by genetic engineering, nanotechnology, electronically enhanced yogic type skills, advanced hormonal techniques, chemical means or whatever, one way or another we shall achieve it. It's as inevitable as death and taxes used to be, to make a very bad pun.

Currently, however, physical immortality is now looked on by the general population with curiosity, even a bit of superstition, perhaps as a remote but probably impossible thing. But the drive to it is deep, powerful, attractive and inexorable. The entire planet exhibits an extraordinary variety of preoccupation with and claim of successful mental control over death or at least aging; we know that we can influence our bodies with our minds. We know that we probably could control not only our physical component far better but even our immediate physical surroundings. And yet there is still the sadness of death. The simple, harsh fact is that the lack of physical immortality and the concomitant inevitability of death colors, conditions, determines every facet of our lives – while our advanced Nefilim gene component has already driven us rapidly to a point of development where it is required to fulfill the potential of our consciousness – and we don't have it. Yet. We know we don't want to die, but we still let the fear of dying or at least its inevitability prevent us from really living. Many of us refuse to consider the possibility simply because we refuse to let our hopes be raised concerning such a painful matter because, at present, we could not bear to be disappointed. Even if we are not afraid of death's inevitability, it's still the final inconvenience. And we are ambiguous about doing something about it. But sooner or later death will be looked on as a disease. Everyone knows in their deepest thoughts that the lack of control over our lives represented by death, coupled with the fact that the universe treats us as if our situation were not important or significant for us to even know why, is the pivotal factor in our lives.

Of course, it is argued, recycled, that immortality is not the will of God ("Immortality is Immorality"(!): can you see the bumper stickers coming? Will the right-to-life people – supreme irony – be the ones to protest?); that it is unnatural; that it is our ecological duty to die; that progress will be halted if some live forever not making room for the new; that we do not have the resources to support it; we would get bored and want to die; reincarnation is taking care of that already; it's the supreme "ego trip" and a mark of the immature personality; it is the intrinsic nature of the universe that our type of being be born and die; evolution has not produced it so we should not do it ourselves; and besides it's not possible to achieve anyway; etc. The special interest groups of priests, prophets, politicians and profiteers are going to go all out against this one. Our programmed beliefs from childhood get in the way, our fear gets in the way, our dogmas get in the way

– and the universe seems unconcerned and silent. It may be the ultimate taboo. But each one of us knows in our most private thoughts that the first person who attains it will be – you guessed it – immortalized; the second and third will make the headlines and a TV documentary and then there will suddenly be large immortality industries appearing on the stock exchange.

There is an obvious reciprocity between our philosophy and science in this regard: when a profound goal like immortality was hopelessly out of reach we resorted to the rationalizations that it was not desirable, humans were not worthy of it as such, or sublimated it into a state in the afterlife which, of course, had to be won by a perfection of life cast in terms of subservience to sublimated gods. Any technology that sought to achieve it was branded as evil materialism. When the technology eventually began to give indication that it might actually be able to deliver, if only in the reasonable future, then attitudes began to change: immortality becomes a legitimate and desirable goal and the religious, spiritual and afterlife sublimations began to fade. Philosophical controversy will spontaneously combust at the same time that a hectic scramble to obtain the means to immortality occurs, and a fear and religiously motivated backlash reaction activates. That would be far too late for either measured and intelligent evaluation of its philosophic potential or a fully prepared scientific technology for its practical implementation.

As preposterous as it may seem, but judging against previous experience with fundamental rights, immortality will have to be legislated as a legal human right because there will be opposition from some who will attempt to limit and control the lives of others.

A practical consideration, when immortality first becomes a reality, is that there may be danger from those who are driven by jealousy, envy or disapproval and who might be tempted to injure and deprive, for those who are immortal can be killed – certainly an ultimate threat. But it is already clear, with the coming of the genetic ability to eliminate aging and eventually to alter the genetic code for immortality, there will also come, concomitantly or soon thereafter – perhaps even before – the abilities to repair, rebuild, restore, limbs, organs, complete cellular structures, and, indeed, life itself. We are already in the beginning stages of that sort of capability but the full capacity will see us able to literally grow a new human from a few preserved or retrieved cells. The instructions are all there in the DNA. It will take us a reasonable amount of time to learn how to do such a complete restoration but there is no doubt that we shall. The theme of molestation of those possessing or anticipating immortality takes a strange and disconcerting twist in recent times when scientists work with mummies in an attempt to sample DNA, determine pathological conditions and obtain other information about the luckless person whose mummy they are examining in the interest of archaeological and historical research.

The Egyptians are well known for their dedication to mummification. They could still have been aware of the tales of the Nefilim's technology for repair, restoration and revivification of at least their own kind or were the recipients of traditions derived from that history and they copied it. It is even conceivable that the Nefilim taught us the concept and technology of mummification and future revivification. The practice was sufficiently developed and sophisticated to achieve a success witnessed by the excellent state of preservation of many specimens even to our present day. The Egyptians obviously had a goal in mind in using the relatively time consuming and expensive process. It is generally construed by scholars and scientists to have been motivated by a "religious" belief in an after life. That seems innocent enough on the surface but the underlying assumptions that it is perfectly legitimate to exhume and haul around and invade these ancient remains because they are ancient – and it is assumed that the beliefs of their owners were quaint and naive in the first place – is certainly questionable if not very wrong. The implication is that there is a difference between the body of a human preserved or buried in olden times and that of a human just interred after yesterday's funeral. Certainly scientific progress is served by such examinations but our treatment of such remains should be reexamined and revised. As I have said in other places many times, it is a horror that anthropologists and archaeologists routinely disturb and sometimes destroy the mummified remains of those humans who, already in the ancient past, may have been aware of the possibility of and intended their own revivification, perhaps when the Nefilim came here again.

The advent of general immortality will certainly force a change in our understanding of their intentions, but in the meantime there is a peculiar problem that sharpens our focus with regard to deliberately preserved or mummified bodies of humans. Consider it in terms of the new paradigm. The ancient traditions, known to the Egyptians, portrayed the Nefilim as living tremendously attenuated life spans. The concepts of immortality they inherited were of a physical nature and the mummification they practiced was a product of those concepts. If we take the practices of the Egyptians as an imitation of those of the Nefilim then it may have been that the Egyptians had reached a stage of psychological evolution and technological understanding which made them do two things: think of themselves as important enough to deserve immortality and consider it a strong enough possibility in the future to do something about it. If that was so then using the best means of physical preservation that they could manage, mummification, was the practical solution they used. In that context it would equate to our resorting to cryogenic (deep freeze) suspension of our bodies and/or brains with the anticipation that nanotechnology and genetic science will be capable of restoring us, perhaps even furnishing a new body through cloning or some other technique, at some time in the future. The mummification process is apparently able to preserve the body well enough to conserve at least some of the DNA after thousands of years. That is probably almost as efficient as our cryogenic method is, at least at present. Perhaps the Egyptians were banking on the goodwill of the Nefilim, when they returned periodically, to provide the technology, perhaps they were banking on their own technology to develop over time to allow them to accomplish the restoration of life. In the latter case they would have been in the same mode we are but thousands of years earlier. Once the technology of genetic cloning from DNA samples is perfected or some even more advanced technique for recreating a living human is developed, the use of samples of an individual's DNA from a mummy may well be all that is needed. At that point, the skeletal and other preserved remains would serve as a convenient reference check for accuracy but would not be essential. If, indeed, those humans were following this practice in anticipation of the development of a technology of revivification or even waiting for the return of the Nefilim in anticipation of such a possibility, then their remains should not be violated or interfered with in any way. To do so would be felonious. If that was so then our tampering with, destruction of or even displacement of mummies is an atrocity. It is equivalent to doing the same to the bodies of those in cryogenic suspension. Our scientists, anyone involved, should cease and desist.

Eventually, longevity and immortality will become major industries. The research money should be carefully scrutinized as to its source and control and the industry monitored carefully. Although the United States patent office has already stated that they will draw the line and not patent a new type of human, the process of immortalization well may be a patentable item. The control afforded to inventors and corporations currently by the patent process is certainly an incentive in the first place and some guarantee that the costs of development and a reasonable profit will be realizable. The negative aspect of that proprietary control manifests, however, when a new cure for a critical disease is priced out of reach of those most in need of it. At that point the issue moves beyond simple economics into the dimension of basic human rights. We only raise a fundamental question here: Is everyone entitled to immortality regardless of their financial resources to pay for the scientific services that may be required? Does death, understood as a disease, fall into the same category as any other fatal disease? Regardless of the opinion any individual may hold, the entire subject will go through profound metamorphosis, philosophically. Ultimately, inevitably, evolutionarily, inexorably, immortality will be taken for granted as an inherent condition and right of human nature.

Processes such as cryogenic suspension (freezing at extremely low temperature) are only temporary stopgap techniques to preserve the individual until the actual techniques of restoration and immortalization can be perfected and applied. At first, these technologies will be relatively crude and will then rapidly progress through the usual stages of refinement. But it is not inconceivable that, in the not too distant future, the entire informational, experiential and sub-brain memory capabilities of an individual may be read directly from the brain, neurological system and body into some sort of adequate storage medium. Such a medium, in terms of what we know now, might be a highly refined and compact laser disk but it should be assumed that far more adequate and subtle mediums almost beyond our imagination will be developed. But the type of storage is almost irrelevant: it is conceivable and probable. Our science fiction already takes molecular or atomic transportation or transfer by total information of a human being (Star Trek transporter) as a given. Our current science is already talking about it at least as a possibility. The concept of restoration of an individual in the early phase of technology would involve a biological cloning of a body with the stored informational, experiential and memory content fed in with no more difficulty than restoring a backup in a computer. Later, more refined techniques will be far more elegant than the crude projection described here.

The Gilgamesh Factor

When I wrote, in *Breaking the Godspell*, that "the psychology of immortality remains to be written" I intended it to be a provocative statement not because of some revelatory impact but by its long range effect: to stimulate the interest, discussion and effort required to achieve just such a coherent body of information. The general subject of immortality has been with us for a very long time; by the psychology of immortality I mean the systematic study of the mentality of those who will have it actually as a possibility, an option, in their life. But before we examine the positive elements of the mind-set of immortality, it is necessary to resolve the negative historical attitudes deeply embedded in our traditions. I have called the constellation of obstacles to immortality the Gilgamesh Factor after the famous king of ancient times. It is written of him that, genetically, his mother was pure Nefilim and his father was pure human. Knowing that he did not possess it, he sought immortality at the Nefilim spaceport as a legal right due to his mother being Nefilim. Although it was not granted to him – the Nefilim authorities gave him some longevity herb as a compromise – Gilgamesh's situation embodies many of the essential elements of our current predicament as well as the arbitrariness of the Nefilim judgment about granting it and the clear indication of the potential we have to receive or achieve it. The psychology of immortality will ultimately be a subset of the tenets of the philosophy of immortality. It is the philosophical attitude and assumptions we hold that make it easy or difficult for us to adjust to immortality as a concept.

If we continue to run true to the past patterns of our performance as humans, the academically accepted psychology of immortality, as with any "scientific" discipline, will be a product of the consensual world-view, the consensual "reality" shared at any given time which can be considered "certain" (read "safe") enough to transmit to – or foist upon – the children by a sufficiently powerful majority to maintain the status quo.

But it is also clear, from past performance, that the majority-sponsored "reality" is only a subset of the "reality" constituted by the collective "realities" of all those who generate them, viz. all individual humans. So have philosophies, world-views, gone for thousands of years – or so we have been taught to believe. But there is evidence that may not have always been the case. And the evidence bears directly on the subject of immortality as a primary focus.

The dictionary treatment of "immortality" and "mortal" are fascinating and revealing. Only a cursory reading of the definition of mortal is sufficient to impress one with the ancient psychological overtones of identification of "human" and mortal, that which is subject to death, that which dies and by implication, with unrelenting hostility and shame. The definition of "immortality" as being "exempt" from death is immediately exemplified as associated with the "gods" (of the Greek and Roman pantheon since Webster's minions are still of the Cambridge/Harvard mentality of 50 years ago, allowing history to begin only with the Greeks). It is also associated with lasting fame. We well might ask why these arbitrary associations are so universally accepted without question.

The sectarian definitions of immortality are even more fascinating. In many religious contexts the spirit of man is essentially immortal and only the body is mortal; the spirit is relatively eternal (relative in the sense that you have to come into existence, get created, but from then on your spirit is around permanently, either to get rewarded by bliss for doing the right thing or punished for doing the wrong things). Some religions hold the spirit part of the human is fully eternal (no beginning, no end) in that it is simply a "part," a partial manifestation of the One, the Eternal Principle or Source, etc. The door to that new dimension of human existence may lie in our immediate future but the key to that door lies in our past. It takes little objectivity or reflection to recognize the common elements in whatever context "immortality" is examined from whatever source: "immortal" is what the "god" or "gods" are and what humans, at least in one part or another or totally, are not. And that long tradition also states that immortality, if bestowed on a human, always comes from "the gods" or a "god" no matter what context in which the action is set. The conflicts among the varieties of interpretations of those common themes and the various schools of thought that contradict them claiming that man is totally mortal, deny the existence of the "gods" or "god," seem an unresolvable and perennial situation that will hamper the attainment of extended longevity and, eventually, relative immortality (relative in that one must be born first to attain it) for an extended period. True, there is already a popular magazine devoted to Longevity, at least from a semi-clinical viewpoint, and books have been written on it extensively since the sixties. True, a number of able scientists are working on longevity from a number of experimental perspectives and talking about immortality. But we have not had, to this point, a comprehensive and intelligent context, a world-view, adequate into which we can integrate all those disparate points of view – and, therefore, immortality as a new plateau of human existence. Until now.

We have at hand extensive evidence from the last one hundred years of archaeology that those embedded patterns of our thinking have a root source that is at once very familiar and almost unthinkable. As we shake off the grip of the ancient godspell, the dawning genetic enlightenment makes it very obvious that our preoccupation with immortality arises from the fact that the immortality-mortality dichotomy was a crucial variable in our genetic creation. By the decision of the Nefilim, although it was a possibility, we were not endowed with immortality. We have been seeking it ever since. Although the Nefilim, possessing the genetic engineering knowledge required, had secured relative immortality for themselves, they chose not to include it as an attribute of human nature when they created us in the laboratory or later, when they gave us the ability to procreate. It is easy to understand that the Nefilim, when they first created us, did not include immortality as a characteristic of humans because they only needed slave-animals for very simple purposes. If we, as modern humans, decided to engineer some kind of cross between a gorilla and a human with the intention of imparting an additional quotient of intelligence to the ape form to make it smart enough to take our places in the mines or the fields we most likely would deliberately exclude any unnecessary advanced human features in the gene transfer for obvious reasons. We would want to keep things controllable, the new worker just smart enough to get the work done efficiently but not smart enough to cause problems, to question, to rebel, to innovate. From purely pragmatic considerations, why create a new creature that, with the best of intentions on our part to prevent it, might become so intelligent and self-aware that we could not ethically or morally destroy individuals for good reason (genetic defects that caused large scale violent psychosis, uncontrollable new diseases, etc.) or the entire experiment if necessary. We already are facing similar questions of lesser degree about some of our known animal species. The records we can read show it reasonable to assume that the Nefilim realized that they might not have been able to predict exactly what we would be like after a time, literally how their experiment would turn out. It seems clear that they did not completely foresee that we would advance as precociously as we have evolutionarily. Substantiating examples abound in the ancient records. It seems clear that they were only concerned with us as disposable units, as a rule, although there is also record of their having chosen to actually bestow immortality on a chosen few humans – thereby demonstrating the possibility of what we are only now beginning to consider – perhaps as long as 300,000 years ago. There is the intriguing record of the king, Gilgamesh, claiming immortality as a legal right because his mother was Nefilim and his father human.

The Nefilim are not on the planet as pure Nefilim as best we can determine at this time so recourse to their legal system is pointless – and would be a denigration of human dignity anyway. On the other hand, when we look in the mirror, we are literally looking at a continuation of Nefilim heritage. But if the potential is there even for an existing mature individual to have immortality bestowed arbitrarily, there is a clear message for our science. In fact there is a great deal of information and clues scattered throughout the ancient records concerning longevity and immortality. A reexamination of the history of the exploits of Alexander the Great should bear fruit in this respect as well as that of Gilgamesh and the information concerning Adaba (Hebrew: Adam), the "first perfect model" who is said to have been granted it. Restoring our history to ourselves means opening up a source of information that may well shorten the process of discovery for our scientists.

The first reaction on my part was to resent what appeared to be a pragmatic, even coldly selfish, decision on the part of the Nefilim to withhold immortality. But it now appears to me that there is a positive and more profound aspect to the subject. I am awed by the possible thought that went into deciding whether we should be given that potential or not by the Nefilim. It will be one we face when we encounter a less developed species somewhere else – or decide to create a synthetic species ourselves. There may be rules of advanced societies dealing with those questions already. If we were not generally ready for it psychologically it could have been disastrous. Perhaps it requires many generations of neurological and psychological development to produce specimens of sufficient quality of intelligence to be able to handle it for themselves.

Kicking The Methadone Metaphor Habit

We have passed the stage where the godspell mentality, as has been the case for thousands of years, has provided us with desperate rationalizations as to why we should accept death, submit to such an annihilation. The Eastern religious psychology of "be here now" and become reconciled to death when it comes, or the Western "God wills it" are simply the best we could muster up when no means to overcome death were available and the terrible despair that leads to suicide lurked everywhere. So deeply ingrained are these attitudes that any objection to or questioning of them is usually interpreted as indication of spiritual immaturity or imbalance. The doctrines of reincarnation, metempsychosis, immortality of the soul (only), transmigration of the soul, karma, purgatory, heaven and hell, are all offshoots of the racial psychological phase when we became self-reflexively aware enough to evaluate the absolute finality of death and were forced to explain our situation to ourselves in terms with which we could live (tragic pun). It is clear why the reward for the "good" life, i.e. docilely submitting to the will of some deity known through the rules of whatever authoritarian religion one subscribes to, is always after death. And why "eternal life," "eternal bliss," pleasant immortality is the reward. Immortality is always the key concept even when the kind supposedly due is a punishment; "hell" in the Christian sense is described as painful immortality – of the "soul" and the body as well. We need to be free of those methadone metaphors that we have clung to in order to maintain our sanity through the transition period since the Nefilim left us on our own – without immortality. It will only be within the context of the new paradigm, this new understanding of human nature as a genetically created species rapidly seeking its full potential, that we will be able to gracefully and intelligently integrate immortality. It will require at least that much of a comprehensive base to then explore the dimensions to which we shall surely aspire beyond physical immortality.

If, however, we now have a context, an adequate paradigm which frees us to intelligently pursue the immortality that was deliberately withheld from us from the beginning, how shall we view it? In the greatest perspective, perhaps we should recognize from the outset that immortality will be both a new and awesome plateau of human existence offering as yet undreamed potential – and at the same time, without denigrating that potential at all, ultimately just another "trip," just another step in our meta-evolution, the rapid metamorphosis we have been undergoing since our beginning. Within those extremes there is the greatest latitude for the inevitable expansion into a dimension which will allow us to become far wiser, individually, through greater experience, greater learning, and the ability to witness the patterns of repetitions of extended periodicity. Eliminating the pressure of a short life span that influences our choices and cramps our lives will not just give us the practical potential to travel easily between star systems and send the insurance companies into the reedit mode; it will change our perspective and our social interactions radically.

We are philotropic (fusion/wisdom seeking as, by analogy, plants are light seeking) beings intrinsically – but we have become overqualified for this planet relative to the status quo of the noosphere. We are like keen-witted, sensitive, intelligent children who drop out of a school system – or family – which is obviously atrophied or dysfunctional, but who seem apathetic because they are not yet informed enough to express a positive vision of their own. They tend to flirt with overdoses, violence and hang the edge on dangerous machines so the race flirts with pollution, war and atomic destruction because of boredom, not stupidity. I have never met a person who was "stupid" about something concerning which SHe was vitally interested. These days the intensity, luminosity and clarity of the information must be of very high quality and resolution or we just don't bother. The sophistication of the philotropic tracking sensors with which the newest human models come equipped are indicative of the overqualification and the readiness for the next plateau of expanded humanity. The new view of humanity as integral, "generically" unified, self-determining and free of the ancient master-slave godspell is the only adequate context which holds the potential to overcome the lethargy and apathy which masquerades as the apparent "stupidity" that seems to prevent us from grasping our individual and racial destiny. Once that vision is absorbed, immortality, an integral part of it, will be taken for granted – and a great deal of stupidity will vanish.

I define physical immortality as a capacity, as the ability of the individual to remain in radiantly vigorous health without aging at a point of maturity of one's choice. It should be the individual's choice to determine at what "age" one wishes to remain, as a personal freedom, rather than to set the point of constancy at full maturity or some other predetermined point. There being possible, eventually, 600 year old "children" of "age" 14 by choice, whimsical or otherwise – the laws concerning minors may have to be changed – certain adjustments of our perspective will be required. But freedom of the individual is at point here and adjustments will be made.

As for dying, I fully realize that my view of it is subjective and arbitrary – and the subjective position relative to dying of each individual must be honored. My concept of freedom in a universe that is worthy of my bothering with, however, continues to make me insist on at least the option to take as much time on my own terms to explore and learn and examine as I want to before I step off this part of the wheel onto whatever part I, at that point, have determined I think is another – or not to at all, just never die. Any universe that denies me that freedom I hold constricting, disrespectful and ecologically mismanaged. I suspect the universe is not; I suspect that it is, to use the metaphoric jargon which is a function of the limits of our current conceptual calculus, a hyperdimensional holographic plenum informing and "permitting" itself to be informed by our consciousness – and a lot of other types also – within dimensions of intrinsic freedom which are also reciprocally determined. The reason I suspect it of being so is because I see myself as wanting it to be so. If the nature of the universe is also my nature as an intrinsic part of it, it should have at least the potential for the degrees of freedom I can conceive of, at least the capacity to allow the degree of whimsical play I respectfully demand, at least the degree of respect for my wishes I extend to other beings. If I am, in a real way, the universe, and the nature of that universe is, in part, to be created by me then it should have sufficient flexibility to accommodate my creations. I say "should" here not in the sense of a logical "must" but in the sense that my reality creation requires it to meet minimum standards which I determine I want to create. Another is to be never boring; always to be able to expand one notch of potential beyond that which I have been able to move into and to always be more than my concepts of it. A major standard is the subsuming golden rule, of course, that it, I, and others should treat each other as we would like to be treated – as long as we determine it that way in the first place and agree it's good for all of us. No doubt such talk will shock and horrify those who are under the godspell, the posture of subservience to either some sublimation of the paper tiger godlets we still fear or an extension of that sublimation toward the universe in general. If I am egotistical by those standards, the universe is also, as manifest in and through me, and it would be a sniggling, cramped, boring universe that would not be able to accommodate itself with appropriate humor. The real difference between Western religion and Eastern non-dogma is perhaps tested here. Is there a radical freedom of absolute relativistic reality creation or not? Or shall we create that reality before, or by, asking that question? Have we not been attempting to live "on" or "outside" the universe rather than in it for some time now?

The most critical question centers around one single, simple concept: control. We are at a transitional phase where we all, more or less, tend to subscribe to the "healthy" norms of taking charge of one's own life, self-determination, consciousness raising for social contribution on a benevolent basis in order to support and foster self-determination by others and an enormous amount of effort to use biofeedback, yoga and innumerable other techniques to control stress, pain, disease, etc. But there is this curious thing we do when it comes to death. We do not want to recognize it as arbitrary, as an outmoded evolutionary gambit that probably comes from the level of plants dropping their leaves and finally themselves for the sake of building up a thin layer of dirt on a rocky planet for the sake of survival. We do not want to deal with time like we deal with space. There is a totemtaboo deep enough in the common psyche yet to cause the most conscious and precocious to utter glazed-eye robot platitudes about it not being in the class of a disease but the way it should be, as if there is some unspeakable inherent moral deficiency in anyone even profaning death with a question. At the time of transition we may, indeed must, in the mind of so many new age thinkers, control our minds, our bodies, our destiny, but we must die without question, without any determination, without protest or whimper. Laugh our way out of disease, fight and manipulate the body like Stephen Hawking, control any energy like electricity for harmonious purposes but not our own, not even presume we know best when we might indeed prefer to die because of a reason we consider overwhelming. At this point of transition, some admire "enlightened" gurus who, apparently in normal health, predict death to the day or hour, even choose the moment, but do not think enough of themselves to take charge of the last check point. Strange.

I assume that immortality will be an option among options; that the necessary physical vigor will be concomitant; that quite obviously we shall work out the expedient adjustments of our resources, work, ecology, economics, education, etc. as incidental facets of the new dimension once the vision has stimulated us and given us sufficient reason to break trance and outdo ourselves.

It is obvious that there are venerable traditions of life extension which are effective. There seem to be three levels of manipulation foreshadowed in classic doctrines: Taoist harmonious alignment for maximum age due to minimum stress and proper life-style; yogic control over the autonomic nervous functions leading to extraordinary powers of body and mind; conscious dimensional shifting through compassion and realization of the identity principle. We are all familiar, to one degree or another, with the blossoming longevity industry whose products range from techniques of positive thinking to the latest chemical key. But none of those techniques achieve physical immortality. The engineering of our own immortality is the best current positive answer to our problem of racial survival, our racial vote for or against survival as a species, given the tools of annihilation we have provided ourselves. Consider how quickly and well we would achieve that goal if it had the urgency of the development of the atomic bomb for purposes of annihilation . . .

How then can we go about attaining physical immortality by technological means directly? Within the range of, at least, known possibility, the approaches can be seen to fall into two broad categories.

The general approach envisioned by science is that of externally applied technique. A single example: a genetically engineered virus that will benignly invade the human system and precisely alter very specific instructions in the genetic code that control aging. The age of nanotechnology described by Eric Drexler in *Engines of Creation* (Anchor Press, Doubleday, Garden City, NY, 1976) will afford us the facility to manipulate structures at the molecular and atomic levels with ease and precision. That facility will surely be applied toward the achievement of immortality perhaps in ways we have not even thought of as yet.

I prefer an approach much more under the creative control of the individual and which would allow a far broader scope of change, accommodating a fuller implementation of the potential offered by immortality. It is one thing to alter the aging process; it is another to modify the entire organism/mind to express the uniqueness of an individual in control of one's own physical destiny (including the option of terminating that physical mode at will) and pursuing the fullness of that exponential leisure. In order to accomplish the more comprehensive and personally controllable mode I suggest we pursue the development of what might be called – I coin a word here – genetaffective chemicals. The purpose and method of use of these is best described by analogy with the psychedelic chemicals such as LSD. Well known and documented psychedelics clearly provide us the ability to suspend, amend, and reimprint the deep neurological circuits, but they do not let us penetrate the encryptions of the genetic code. Genetaffective (affective rather than effective because the action of the "mind" would be best understood as affective on the miniaturized intelligence of the DNA) chemicals (which, to the best of my knowledge, do not exist as yet) I define as those which would, on introduction into the ordinary human system, allow precise and controllable influence and manipulation of the genetic code by the exercise of our immediate and active intelligent direction. I am not suggesting that someone proficient in the use of such chemicals could turn HIrself into some other organism. I am suggesting, however, the ability to scan, evaluate, amend, delete, improve or add genetic instructions. The degree of control reached at that point would be quite awesome. It also would be commensurate with the degree of freedom and dignity which we are rapidly approaching and which it will require, indeed, demand. Although I distinguish here between the above mentioned example of an externally applied technique such as the introduction of a specially designed virus, and the second example of a genetaffective chemical tool, it seems clear that both approaches will fall within the field of nanotechnology.

I recommend funding for both types of approaches. I suggest very careful scrutiny and selection of the sources of such funding; obviously it should come from those interested directly. I suggest we immortalize the scientist who accomplishes the means for us by naming a star possessing planets of a suitable nature after HIr – and then making a gift of a round trip there which SHe would have provided the lifespan to make.

So far, however, we have only considered the extension, albeit indefinitely, of physical human existence as we know it by some sort of external intervention, however subtle, on the human organism after it has come into existence. Will it be possible to genetically engineer a human being who is physically immortal from conception, whose genetic code is such that it determines the person to be immortal? I assume, just from what we already know of genetics, that will be a definite possibility in the not too distant future. It raises some interesting questions. Should a person who is genetically determined to be physically immortal be distinguished from humans who are, for whatever reason, not, on the basis of biological determination as a different, new species? On the basis of having different inalienable rights? If we assume that parents, who are already genetically determined to be immortal, will produce children with the same genetic potential, what category will a child fall into if only one parent is of immortal genetic type? But those considerations are superficial compared to the decisions we will face when deciding to actually use whatever technology on an existing person to give them an indefinite life span or to genetically engineer the first immortal. And do it we shall. Certain of us will demand it, fight whatever authority, interest group, religious sect, or philosophy that opposes our choice, represses our freedom. But that is just "difficulty at the beginning." Over and beyond the scientific manipulation and control we are rapidly achieving over the physical body, is there a mode of control and manifestation open to us – or opening to us – that is an integral function of our own psyches, our own minds? Is there indication, anecdotal, scientific, or traditional, that human beings can modify their own physical being and its concomitant field to any extent and particularly to overcome the dying process? Perhaps. Are there dimensional arenas in which some sort of essential individual "monad" can flourish that afford independence of the temporal limitations of physical existence? Perhaps. There is at least a small percentage of the population, however, which is already ready, eager and probably overqualified for immortality, an indefinite life span. Overqualified in the sense that their consciousness is already evolved sufficiently to encompass it and ready to subsume and move beyond it. That may sound a bit strange, initially, in view of the fact that we have not yet even achieved it. But I assume that, sometime in the future, we shall discover, explore and expand into a type of human condition which subsumes even physical immortality. (And, if we are not quite careful and enlightened, the physical immortalists party will try to prevent it as evil or at least make it illegal.) I project, on the basis of the essential continuous expansion of our consciousness and the required development of information gathering facility, we will evolve to, at least, a four (or more)-dimensional physical and mental form that will subsume the three-dimensional physical and mental form we know ourselves as now. Our sensory capabilities will modify, therefore, and expand to accommodate the information gathering and processing capacities required. Physical immortality may be subsumed at this stage perhaps because we may simply evolve to a form, though still physical be definition, which is basically energy rather than matter and perhaps not subject to the organic rules. It certainly is a major element in our thinking if only, so far, in our science-fiction – which has shown itself to be a rather reliable indicator of what actually will happen.

As a sort of lowest common evolutionary denominator we need to attain immortality even if only to prove to ourselves that we can and that nothing is going to happen to us if we do. We need to know that we are not going to be confronted by some cosmic tyrant, finally provoked out of hiding, demanding what the hell we think we are doing. It seems ridiculous to even have to say that here, but it is one of the most profound effects of the godspell that must be dispelled. But much more essentially, we must attain physical immortality as a basic right, an ordinary condition, indeed, a quality of human existence which is a matter of simple human dignity.

An important perspective is gained by recognizing the historical roots of our attitudes. We are coming, now, out of the phase where physical immortality was simply not available to us and beginning the phase where it is at least a possibility. We have many adjustments to make as it becomes a reality, even conceptually, at first, and then as it is actually available as an option. In this intermediate stage we are already confronted with choices concerning the viability and feasibility of the temporary measures available to preserve our bodies or brains until the real means of effecting immortality are completely developed. Cryogenic suspension (controlled freezing and suspension at very low temperature) is already an option. In the thinking of the scientists concerned with the subject today, two essential components of the individual must be preserved, the DNA genetic coding and the brain. The anticipation is that a small sample of the DNA of the individual will provide all the information and potential necessary to eventually reconstruct a completely new body but the life experiences, memories, perhaps personality, requires that the brain be preserved also. It is assumed that those essential components, the basic life data bank, resides in the brain. (In even the most advanced thinking of today, mummification would not be adequate because it is believed that the life experience, the memories, would not be preserved.)

What is most fascinating about the transition period we are now going through, however, is the way in which individuals react to even the possibility of preservation of the body or the brain. Some find the concept of deep freeze of either the entire body or just the brain physically repulsive – as if that were a concern after you are dead. Some find it too "cold," too clinical, (let's hope for very precise measures of both) and turn away. But the most revealing aspect of the matter is that individuals very often reject it not for any physical reasons, but because they do not want to be able to come back, they do not want to attain any sort of relative immortality, that this life is difficult enough without doing it again. The inference, if not the frank admission, is that just getting through this life to an ordinary death is more than a person should have to cope with. At first this seems very strange indeed. If death is the inescapable finality that human beings find impossible, at times, to accept and against which they struggle, then why is not even the possibility of being suspended, after one has died, until science can work out a way to restore one to indefinite life, not greeted with relief and joy? There is a valuable truth to be learned here about the current state of human affairs. The disconcerting negative reaction most often turns out, in actuality, to be not to cryonic suspension's potential or aesthetics but to current conditions of human life. Not having thought it through, the person anticipates life will be no different in one hundred and fifty years (the projected time of suspension until scientific methods can achieve complete restoration) than it is now and, therefore, it will be no more tolerable to them then than it is now and they reject it out of hand. In the largest perspective perhaps that sort of reaction is to be anticipated and understood for some. But for those who have the foresight to see that conditions will inevitably be forced to change to accommodate the inherent dignity of the human being and to adjust to support large segments, at least, of any given population living indefinite life spans with unique, very long term goals and needs, there is another vision.

If a person works very hard over their lifetime at staying vibrantly healthy, eating a good diet, exercising and even taking advantage of plastic surgery and the most advanced anti-aging treatments, the obvious question is whether they just intend to go full bore until they drop dead or whether their logic will not permit them to consider the possibility of physical immortality.

Once the uncomfortable arguments are over, we have the capability, and immortality is an option among options, we may choose to be immortal or not, as easily as we choose to dine out or in, and immortality is a part of our concept of what is essentially human, though novel, what then?

Even the possibility of physical immortality, as with a person who has currently established a membership in a cryogenic suspension service, brings with it a profound change in attitude toward the universe. Goals change: shall one learn new languages, new skills, take on long term projects impossible previously? What is the real focus of life for an immortal? One tends to reevaluate present values in light of centuries of life. It changes relationship perspectives. Those who do not wish to live indefinitely will not be there when you resume. You must plan without them as much as you would like to be with them and share the new kind of existence; they will be permanently gone at some point while you live on. But what if it is gradually established even by scientific experiment and investigation that reincarnation is, in fact, literally true? As I said above, even cryonic suspension is a best-bet, stop-gap measure opted for to carry one over to the time when the geneticist and the nanotechnologist have provided the means to immortality without dying. If one does not have sufficient data by the time of near death to be personally fully convinced that there is some sort of existence after physical death that one would wish to experience, then one should be free to opt for whatever method and technology will carry one over to the time when a decision can be made with sufficiently robust data, to be comfortably convinced beyond doubt either positively or negatively. And even if one has sufficient data to convince that there is some sort of existence that one would consider satisfactory or challenging, one should still be free to choose or not choose it. The fundamental principle that I am holding for is simply the unconditional freedom of the individual to choose the future that he or she wishes to create, including the criteria and standards by which those futures will be judged.

A Possible Amazing Solution

In this chapter we have considered physical immortality as a novel possibility, a controversial topic, an opportunity which, eventually, on it becoming a reality, may change the entire fabric of human existence. We have spoken of it theoretically, of necessity, and considered the inevitable arguments which arise by its very consideration.

But, just within the time frame of the writing of this book a development has taken place which may already alter the entire perspective and context in which immortality has been considered and discussed. David Hudson, mentioned previously, is an Arizona man who has, quite possibly, rediscovered the substance which the Nefilim used to give them physical immortality and to maintain it. This substance is a special form of gold; it is, quite certainly, the very "white powder of gold" mentioned so often and sought so long by the alchemists who knew of its properties to give immortality, extrasensory powers, the ability to levitate and to disappear and to reappear – but who could not, except for possibly a very few who regained the secret, prepare it. The technology, known to a handful of specialist humans in the remote past who probably were taught the techniques so as to manufacture it for the Nefilim rulers, was available to the Old Kingdom Egyptian rulers, known probably to the Hebrew technicians who worked for them in Egypt, known to the Rabbi's before the destruction of the Temple of Jerusalem. And then, beginning with the destruction of the Temple, which may have driven off those who knew the techniques to form the Essene community where it may have been preserved and used, the knowledge of the processes and technology for making it was gradually lost to humans over time. By the time of the Middle Ages only fragments of the tradition was left, scattered in documents and tomes, the real knowledge gone dark. But, almost inadvertently, the secret has been restored to the point where the monoatomic form of gold has been the subject of a patent application. This white powder of gold, as it is described in alchemical and occult texts, was said to have extraordinary properties. The properties, as described by the person who has rediscovered it and analyzed it, are extraordinary indeed. Recall that the periodic table studied in high school chemistry class is said to be a chart of "elements." I, apparently among many others, understood that to mean that what was being described when the atomic number, atomic weight and number of electrons surrounding the nucleus were listed was a single atom of whatever "element" was represented. This is not the case. When gold is described as an element in the periodic table, as example, the metallic form is always made up of two atoms of gold. Other "elements" represented require a minimum of six or nine or some other "magic number" of atoms to make up the elemental form as represented. The white powder of gold under consideration here is, however, a single atom form, the monoatomic form of gold. It is still gold but no longer the metallic form; it is a snow white fluffy powder. After laborious years of investigation it has been determined that it is a superconductor.

Superconductors, materials which conduct electricity without resistance or almost no resistance, were theorized about some time ago and only produced in the recent past. The superconductors produced in laboratories were expected to operate as such and, indeed, did operate as such only at temperatures close to absolute zero. It was necessary to use supercooling techniques involving liquid nitrogen to bring them down to working temperatures within a few degrees of absolute zero. Obviously this made their practical application difficult.

It has been determined that the monoatomic form of gold and other "elements" such as rhodium and iridium and some nine others are efficient superconductors at ordinary room temperatures and will go into that monoatomic state very easily. They can be found in many substances in nature and are especially abundant in certain herbs and plants such as certain types of aloe vera.

The extraordinary properties of the white powder of gold as a superconductor manifest when it is placed on a highly sensitive scale in a closed, controlled atmosphere container and the heat applied to it is manipulated. Under certain conditions is loses 4/9ths of its weight even though it does not sublimate into a gaseous condition or liquefy. At a different temperature it gains three hundred times its initial weight even though there is no moisture or other compounds in the closed container for it to bond to or adsorb. Under other conditions it not only loses all its weight but the scale itself weighs less, indicating that the white powder is displaying anti-gravitic properties, according to Hudson.

It is theorized that, since the cells of the body communicate by means of some sort of superconductivity (as determined by experimental work by the Department of the Navy) although the superconducting material could not be identified, the ingestion of this white powder, monoatomic form of gold enhances cellular communication and may allow the genetic DNA in the cells to correct itself. This may be the basis of the restoration and maintenance of a person's body at a young adult age indefinitely (it is said that a person has to renew the initiating protocol only once every fifty years) and the basis for the extraordinary extrasensory powers that it bestows as well as the increase in intelligence recorded in the alchemical literature. A forty day regime is outlined in the old texts which includes a preliminary fast and ingestion of the white powder mixed in water in a relatively high dosage according to a prescribed schedule. The effects of this protocol of ingestion apparently were well known and can be expected on schedule if the protocol is followed properly.

These findings are very new at the date of this writing and will be verified over time. But only slight reflection on this phenomenon causes one to realize that we may well be on the threshold of an entirely new approach to immortality and that it is not a possible of the next future half century but may be within our reach within only a few years. With the information I have at hand right now I would judge tentatively that the discovery is real, startling and highly significant. I would judge at this time that not only will it give us physical immortality but will be the tool to the next major step in evolution. Consider the capabilities of a person who has undertaken the full protocol and literally turned their cellular structures into superconductors. They can levitate, vanish here and reappear there at will, exercise all the various extrasensory powers we know of and their minds will operate at a level we have only known briefly at high dosage levels of LSD. When a number of individuals have become accustomed to their transformation and can begin to communicate and interact and operate this way habitually, we will see a gradual transformation of the concept of human society that may surpass that which we have previously only attributed to the "gods." Since this discovery and development is so new and unproven there is a great deal of detail that must be thought out and worked out. I must let this simple recording of this profound event suffice until such time as more is known with certainty and actual experience with the substance has borne out the claims in the ancient records. I may be incorrect in my judgments made at this time about this matter but it does afford us still another perspective from which to consider at least the possibility of immortality and the ways to attain it.

When one has attained physical immortality or the certain prospect of it how will one live? This is an entirely new subject in the realm of human consciousness and we need to explore the possibilities because it will soon be upon us. We shall live indefinite life spans with all its awesome potential for expansion, growth, exploration, experience, understanding and love. The immortal has time, choice, dignity and control. There is a certain radical freedom from the pressure of aging, of a lifespan always too short for the positively oriented. It is not just a matter of the time to take the time; not just time for planting a garden with trees that will mature well past the ordinary contemporary lifespan; not just the time to really "be here now"; not just time to start over as many times as you wish. Those benefits will certainly become a reality but, beyond those relatively superficial elements, there will come the time to experience actual phases of evolutionary transition, to see and understand the cycles and patterns of the universe that are so attenuated that their recurrences and boundaries extend far beyond our current lives' brief tenure. Time to integrate into one's life the entire spectrum of human experience, to attain a fundamental wisdom and depth of knowledge almost inconceivable to our minds now conditioned to brief and uncertain lives.

"Immortality, anyone?": think about it. Are you ready for that? (Please take all the time you need to answer).

CHAPTER 14

CONSCIOUS EVOLUTION IN THE UNIFIED FIELD

It is at least conceivable that the "real" (the closer we get, the more universal and inescapable the puns become) "law of everything" should be, by nature, absolutely and inescapably self-evident – once we have reached the critical point of awareness where we can recognize it.

Frank Clinton
Retrospection in Retrospect

. . .a complete theory will have two components, physical laws, telling us about the behavior of physical systems from the infinitesimal to the cosmological, and what we might call psycho-physical laws, telling us how some of those systems are associated with conscious experience. These two components will constitute a true theory of everything.

David J. Chalmers
The Puzzle of Conscious Experience Scientific American
December 1995

If the universe is the answer, what is the question?

Leon Lederman
The God Particle

This chapter should begin: "Life in the Unified Field will be a major preoccupation, a major god game of the new human," a straightforward declaration of the point of this Part III, but a great deal of preliminary discussion is required before we begin considering this god game. Although I introduced them briefly in Part II, many are not familiar with the details of these concepts of a Unified Field Theory or a Law Of Everything. This is quite understandable since these notions, which are novel, indeed, are even now being worked on and argued about at the leading edge of our science. Because it is inescapable, however, that these concepts, eventually laws of physics, will form the scientific view of reality, the context, in which the new human will operate and conduct HIr personal evolutionary drama, it is of the utmost importance that the reader become aware of and understand the basic concepts.

There also are peculiar characteristics of this current science which, again, prevent me from beginning this chapter with simple statements. Ordinarily it would be good explanatory technique to devote the chapter just previous to this one exclusively to a thorough, clear discussion of the notions of a Unified Field Theory and the Law Of Everything to set the stage for this chapter in which we will discuss how the new human will operate in such a context and how SHe can take advantage of this new knowledge of the universe to consciously expand into it and persist in it, if desired, as an immortal. When we speak of the new human we are speaking of the evolution of the consciousness of the new human. Not only is the critical, central problem for the scientists working towards these laws "the problem (derivation) of consciousness" but these scientific concepts are so new that the new human consciousness can actually contribute feedback to their formation. So this chapter must deal simultaneously with the quest for a Unified Field Theory and a Law of Everything and how new human consciousness will contribute, ultimately, in its new dimension of freedom, to its own definition. After all, this is the way that the new human will live. One takes all the information available at any given time, assimilates it, uses it to determine two critical things: the degree of freedom one can conceive of for oneself and others, and the evolutionary direction one will take in light of the potentials in that freedom. That kind of conscious, participatory evolutionary god game will scare some, temporarily disconcert some, challenge and delight many. How will it effect you?

Don't make up your mind until you really understand the concepts and their ramifications. You do not have to know advanced mathematics or have a Ph.D. in philosophy or be an atomic physicist to handle the concepts. In fact, as one of the most brilliant physicists, Stephen Hawking, has pointed out, that when we arrive at a true Law Of Everything it should be so elegant, so simple, everyone can understand it almost intuitively. It's the groping work to discover the hyper-obvious that is the difficulty because it demands that we not only do the science but expand and sharpen our consciousness recursively as we go. Notice the constantly recurring idea of self-reference throughout these discussions: our consciousness, having to improve itself in order to define itself, changes itself and the definition of itself in a feedback loop in the process. There is nothing more vitally important to the new human who needs all the information about the universe available to make informed judgments about the direction one wishes to consciously move when determining the direction of one's personal evolution.

Just the concept of a law of everything is awesome it itself. The ramifications of attaining such a law of everything are profound. Although some scientists hold that its discovery will bring, at least at first, only modest gains in our understanding of the universe and some are more enthusiastic, clearly the gradual application of such an all-encompassing law will bring complete understanding and even prediction into every realm of science as well as into every aspect of daily life, including human psychology and sociology. What if we could predict all human action with ease? It is for this reason that I dwell at length on this startling concept since it may be achieved in a very short time; we could be very close. It will bring a profound change to human existence. I suggest that the process of discovering and refining the law of everything will stretch the human mind considerably more and will accelerate our evolution toward habitual four-dimensional consciousness. It is easy to sound very silly or even stupid when speaking about the Grand Unified Theory or the "Law Of Everything." If I say that the Law of Everything is critically important to everyone because it effects literally everything, it is the truth, but it certainly sounds inherently trivial or silly – even though most people haven't even begun to consider the ramifications of how things will be if we do, indeed, attain the Grand Unified Theory. Living in a known universe will eventually become the major god-game. We need to examine the ramifications, the potential and the evolutionary implications now, before it is on us.

In this chapter we will discuss consciousness in two ways. First we will consider the "problem of consciousness" as it impacts the search for a Unified Field Theory and Law Of Everything. We will discover that the expansion of the dimensionality of our traditional philosophizing, our understanding of our consciousness as well as our conceptions of human freedom and how we will operate in the future, is directly linked to such a search. Then we will consider how we will operate, how our consciousness will deal with life after we discover The Law Of Everything. Such a discovery will have profound ramifications for all of human existence. Then we take up the consideration of our consciously directed evolution in a known universe, i.e., once we have discovered the Law Of Everything. The reader's indulgence is requested here: I will repeat the same notions over more times than some readers will find stimulating but only because the ideas of a Grand Unified Field Theory and a Law Of Everything are so new and so mind-boggling that many readers will profit, as I have, by coming at it more than once from different perspectives. Such an approach is also valuable because it will give us opportunities to become aware of the controversies surrounding it and the startling ramifications for human existence and thought that just the contemplation of its possibility suggests, but, first, let's really understand what the awed scientists are working towards.

The Status of the Search for the Grand Unified Theory and a Law Of Everything

The leading edge of our science is the search for a Grand Unified Field Theory, a "law of everything." The goal is envisioned by the physicists engaged in the search as the unification of the four main forces they have identified as fundamental: the weak force, the strong force, electromagnetism and gravity. The assumption is that, having arrived at a single law which subsumes and unifies these four, that law will explain and predict all the known particles, all the phenomena of the subquantum level and, ultimately, why the universe is the way it is and the way it "works" from the most elemental level to the macro. Things are not simple in this case either: some physicists judge that the GUT, the Grand Unified Field Theory, once finalized, will be the Law Of Everything, others hold the opinion that a Law Of Everything must be more than "just" the grand unification of all the known forces. That is why, throughout these chapters, I have spoken of both a Grand Unified Field Theory or Law and a Law Of Everything, primarily because I think that the grand unification will most likely be just that, but will only unify the known forces, and consciousness will still need to be dealt with.

The most comprehensive and powerful theory currently being worked with by the physicists as to what is the ultimate nature of our universe is that of superstrings. Strings are described, for lack of a better image due to the fact that the images we naturally choose tend to be three-dimensional, as a sort of disruption in the fabric of one dimension. One dimension?

It is difficult for us to visualize how things would be if we were Flatlanders living in a two-dimensional space. Two-dimensional space would be like living in the plane of a sheet of paper which had no thickness at all. You could move in two directions since there was length and width but you could not move up or down.

Reduce this by one dimension and our imagination begins to balk at the effort to image the situation. But at least grant the existence of one dimension by itself for the sake of this description of superstring theory. We must take it as a fact that somehow the theoretical physicists, using very powerful mathematical tools, have arrived at the theory that superstrings are "like" tiny tears in one dimension. These tiny tears, probably self-generating, vibrate. Visualize a guitar string vibrating and imaging the wave of the vibration of, say, middle C, causing the string to undulate. The theory says that, on a vibrating superstring the configuration at certain points on the string, certain nodes (loosely like overtones on a guitar string), become (perhaps the image of densification will help) the known particles previously identified by physics. Some very esoteric work expedited by the discovery of the field equations for strings and a tool called modular equations has enabled physicists to realize that the Standard Model of physics (three fundamental forces: the strong, weak, and electromagnetic) is actually generated as an intrinsic part of superstring theory, as are the laws of Einsteinian relativity and gravity. It has also been found that the superstring theory works, at least in this universe, only if one posits ten dimensions or twenty six dimensions, a bit of a mystery that demands extra refinement.

Currently, in light of the most recent findings our universe may be defined, almost unbelievably, as a superstring driven ten-dimensional plenum of which only four dimensions are completely unfurled. The four dimensions we are familiar with are length, width, depth and time. It is believed that the remaining six dimensions are curled up so tightly that we cannot perceive them. It is also theorized that our universe is one of many making up an almost inconceivable meta-universe and that these sub-universes like ours interact something like the way soap bubbles collide and deform.

Now one of the most fascinating aspects of these discoveries in physics is how strange the universe begins to appear. Whether it be through the investigations and experiments carried out in atom smashers or on paper by theoretical mathematicians (the mathematicians are real: their work deals in theory. . .) physicists consensually agree that at extremely small limits of size (Planck's constant says it's at 10 to the minus 10 followed by 11 sets of 3 zeroes, centimeters), space and time as we know it drop out. Think of it as dimensionality, length, width, depth and time, no longer able to exist. It's somewhere "below" those limits, where space and time are only potentials, that superstrings generate the universe as we experience it. There is obvious oversimplification in that description but it is accurate for an understanding of the concepts involved without having to know either advanced particle physics or theoretical mathematics as a professional. It is accurate to say, in common parlance, that, at those limits and fundamental conditions, the universe gets very squirrelly indeed. The reality level of quantum mechanics is far stranger than the relativities of the level of the space/time of Einstein. So our perceptions of reality are tested, our consciousness is pulled and stretched, and our explanations of space, time, freedom, determination, free will, even existence are called into question.

The Physicists' Approach to Reality

Taking a bottoms up approach, the physicists assume that all other sciences can be derived from this law of everything, as chemistry has already been seen to spring directly from atomic physics. There are unresolved puzzles and problems and questions about the existence and reality of some particles, mostly due to the fact that cyclotron type machines, used to break atomic particles into their component parts by collisions at astounding velocities, have not yet been built powerful enough. They assume that literally everything in the universe from quark to consciousness is matter and energy and nothing else. But there are variations on that general theme. Yet, even with such an almost fantastic conceptualization of reality that seems to defy the imagination in its esoteric and transcendent nature, physics cannot account for consciousness.

There are some thinkers and physicists who hold that everything, including consciousness, in the universe can be explained by the laws of physics as we now understand them. There are some thinkers and physicists who believe that we will have to discover further laws of physics in order to fully explain consciousness. There are some who hold that consciousness in itself may be an irreducible entity of this universe and we need only discover the laws by which it operates and then integrate them with the laws of the other fundamental forces already known. There are some who hold that consciousness is something over and above the physical and cannot be explained by any physical laws whatsoever. They distinguish between the neurological perceptive and processing mechanisms in the brain and consciousness as such.

There are some who hold, usually with deep cynicism, that whatever consciousness is, when we discover its laws and they are accepted by the physicists, those laws will be declared to be physical laws by the physicists regardless of how bizarre or how non-physical they may seem to be. In effect the physicists alluded to here simply define the universe as completely and absolutely physical, matter and energy only, and anything that is discovered, phenomenon, entity, event, field, epiphenomenon, law, is then subsumed under the category of matter and energy.

Regardless of this reductionism, the main obstacle for the physicists in arriving at this highly desired goal is consciousness: they simply cannot account for, derive or explain consciousness and, specifically and especially, our consciousness by the known laws of physics. But the goal is lofty, and the fact that we think that we may discover this law of everything is disconcertingly amazing. The attainment of such knowledge of a law of everything will effect every facet of our existence. This process of scientific searching is already fostering a change in the scientific consciousness itself, simply because of the necessity of inspecting and analyzing our consciousness.

As elegantly – and diplomatically – articulated by David Chalmers, the widespread belief, ". . .that physics provides a complete catalogue of the universe's fundamental features and laws," is not the problem itself, but a major contributing factor because it is the fundamental reason why we insist on trying to reduce consciousness to a type of phenomenon which our current physics can manage. This widespread belief is uniquely prepotent for two major historical reasons. One of them is anti-theological and the other is anti-philosophic. Both are, seemingly paradoxically, based on philosophic premises rather than scientific.

The Bias Against Religious Dogma as a Factor

Science, because of the assumptions accepted or made in the formulation of the problem as deriving consciousness from the physical laws and, by its posture of authority, invites close reexamination. Science and the scientific method came into being not just because of more rigorous than religious top-down theodogma, but also as a direct reactive challenge to religious authoritarianism – and, therefore, by association, to that, allegedly non-physical, unquantifiable horror, the "spiritual." Although scientists are now too cool as a rule to indulge in Victorian blasphemies, there is still a deep hypersensitivity to the slightest hint of what would seem to be reversion to any concept even vaguely scented like the "spiritual." Consciousness, against this background, is sensed by the consciousness of those who have this allergy as carrying an amperage destructive to the hegemony of physics if any voltage is applied by recognition as an irreducible phenomenon which it cannot fully explicate in its own vocabulary.

The Bias Against Philosophy as a Factor

Scientists seem to have particular difficulty interpreting the neural correlates of what generalists (a term of indulgence used by the scipower elite for philosophers when forced to grudgingly call on them) have been doing while in exile since the Scientific Revolution. They tend to be in denial about the facts that the assumptions that there is a discoverable, lawful objective order of reality and that logic is a valid tool for scientific sleuthing and discussion, are philosophic postulates unprovable, without dizzying circularity, by the scientific method. Yet our usual concept of an "objective" reality is sourced in the religious belief system of Christianity. Our dialectical logicizing is both a product and a weapon of that cramping paradigm. The residue of that heretic hunting dialectic wielded, sincerely but inadequately, – and paradoxically – in naive Darwinian righteousness limits us to think in terms of narrow choice between either religious belief or scientific materialism. Some scientists tend to think of philosophers' willingness to consider questions that have no parts number in their catalogue as intellectual perversion and epistemology as something weird that philosophers insist on getting all over their hands. Even though the scientific method came into existence partly because it was conceived of as more precise than reason's ivory tower imputed virtual reality, it tends to overlook that it still thinks it reasonable to use the tool of reason borrowed from the philosopher's garage to beat the living abstraction out of him. But the scientific method cannot look to itself as its justifying criterion without getting caught in the circular reflectivity of self-reference. The initial judgment that the scientific method is a valid and accurate tool is either based on an assumption that is not scientifically demonstrable or eased in sideways on the basis of pragmatic trial of its own validation – and both are circular. To attempt to avoid the trap by frankly stating a primary assumption that there is an objective order of reality that is knowable is a rehash of the Scholastic principle of identity (a thing is what it is and is not what it is not) – and most scientists who unholster that semi-theological muzzleloader in ill concealed contempt to waste some wretched dog of an objecting philosopher don't even know it.

The Result of These Major Historical Biases

The "existence of consciousness" problem is phrased, therefore, as such by scientists working from the implicit fears of a theological renaissance and/or an imagined invasion of the lab by philosophers who, as immortalized by the great Bill Cosby, insist on wondering why there is air. More banally, the proprietary phrasing also arises explicitly from the political determination to maintain the domination of science from the sincere conviction that the universe is only matter and energy – regardless of how bizarre some of that stuff may be.

Physicists may uneasily allow that, if the existence of consciousness cannot be derived from the physical laws, there is no way to have a "law of everything" (henceforth and mercifully, LOE) in only physical terms. They will not even consider allowing the correlative opposite, that, if the existence of the physical realm cannot be derived from the laws of consciousness, there is no way to have a LOE in only consciousness terms.

It is important that the philosopher should see that neither the scientist nor the philosopher – much less the theologian who would borrow both their tools to try to repair the Popemobile and has his hands full suppressing the Dead Sea Scrolls, women dissidents and other annoyances – has a means powerful enough to write the LOE. It's just the philosopher's logical turn to contribute because the piece of the action unavoidably required right now at this wailing wall of mutual frustration is epistemological and that's the philosopher's cup of mixed metaphor tea. The philosopher sees the problem as serious as the scientist and is generally okay with being called in just to jump start the old Empirical when the data battery goes dead, so they can both ride again.

The contribution needed is epistemological (dealing with the nature and grounds of knowledge especially with regard to its limits and validity: Webster) because we are at that point of unanticipatible inescapability beyond which we cannot penetrate, the point which science, philosophy and psychology cannot reach beyond or behind, where we have to give up and say that anything before this we are not conscious of, anything before this we certainly cannot determine the existence, or nature, of by science or philosophy. We all hover at the threshold of the LOE, at the initial point of primary cosmological differentiation. Philosophers and, more recently, scientists have been being pulled kicking and screaming closer and closer to this epistemological omega point for some time now by the gravity of their cumulative discoveries. They are doubly reluctant to even admit its proximity because they think that its going to mean that both their laws break down and their lights are going out if they dare this epistemic Swartzschild radius.

Once both scientists and philosophers have acknowledged and surmounted their antique estrangement and can resume conversation, we will gain a freedom of cooperative action allowing us to reexamine, reevaluate and restate or rephrase the questions and the possibilities and possibly arrive at a LOE "comprehensive" enough to truly explain and predict everything.

The Consciousness of the Physicist as Both the Problem and Its Cause

The reason why the hard problem is hard is because the hard problem is us, our peculiar kind of consciousness. Consciousness and mind are unique phenomena to deal with for us because they are us. **The degree of evolvement of the consciousnesses of the physicists involved is the major factor causing consciousness to be a major problem for them in determining the Law Of Everything**. In the attempt to define mind by mind, of consciousness and mind by consciousness, the intrinsic self-reflexive nature of the process continually self-modifies by the very process of trying itself and by the very definitions at which we arrive. When the physicist attempts to explain consciousness through the forces his consciousness has discovered and defined, detached objectivity (which his consciousness has also defined and chosen as the unquestioned, essential basis of science by assuming an objective, lawful reality) is inherently ruled out of the question by the very conditions of investigation that the physicist's consciousness of reality has caused to be set in the first place.

John Horgan (*The End Of Science*, Addison Wesley, 1996) has recently afforded us vignettes of the personalities and consciousnesses of quite a number of scientists who are considered the leaders in their fields. These, mostly men, in personal interviews with him and in conferences, have spoken frankly about their most deeply held convictions about the ultimate nature of reality, the possibility of arriving at an unified field law and what it might be like, the future of science and the future of the human race. Because of the nature of the interview context these people, many of whom are undoubtedly possessed of quite brilliant basic intelligence, speak philosophically rather than in the rigorous language of mathematics, physics, or whatever discipline in which they are involved. As philosophers speaking about whether science would be adequate into the future to give us the answers to the most profound questions of life, whether a law of everything could be reached, what our intelligence and consciousness would evolve to in the future, the most consistent and striking feature of their statements and expressed convictions is not that they radically disagreed with one another but that they disagreed – and sometimes severely denigrated one another – primarily on the basis of their individual personality traits rather than on truly scientific grounds. Translated into the simplest terms, the degree of evolvement of their consciousness showed painfully through their scientific biases, their explanations of their scientific theories, and their personal interpretations of the universe or some ultimate principle behind the universe. Their humanity, sometimes painfully and awkwardly, sometimes clearly and maturely, comes through their words revealing, over and beyond their intelligence, the state of development of their consciousness. Too often the almost embarrassingly sophomoric level of their philosophizing presents itself in high contrast to their intelligence and scientific acumen.

Certainly we are seeing these people through John Horgan's subjective lens. But, taking his perceptions and, sometimes verbatim and sometimes summarized, quotations of their words with his own caution that he is ". . .fond, overly fond, of mocking scientists who take their own metaphysical fantasies too seriously" as at least journalistically accurate and fair, consider his concluding words:

> . . . we humans, even as we are compelled to seek truth, also shrink from it. Fear of truth, of The Answer, pervades our cultural scriptures, from the Bible through the latest mad-scientist movie. Scientists are generally thought to be immune to such uneasiness. Some are, or seem to be, Francis Crick, the Mephistopheles of materialism, comes to mind. So does the icy atheist Richard Dawkins, and Stephen Hawking, the cosmic joker
> . . .

But for every Crick or Dawkins there are many more scientists who harbor a profound ambivalence concerning the notion of absolute truth. Like Roger Penrose, who could not decide whether his belief in a final theory was optimistic or pessimistic. Or Steven Weinberg, who equated comprehensibility with pointlessness. Or David Bohm, who was compelled both to clarify reality and to obscure it. Or Edward Wilson, who lusted after a final theory of human nature and was chilled by the thought that it might be attained. Or Marvin Minsky, who was so aghast at the thought of single-mindedness. Or Freeman Dyson, who insisted that anxiety and doubt are essential to existence . . . And then we hear the voice of Horgan himself, expressing his own emotional response,

"But after one arrives at The Answer, what then? There is a kind of horror in thinking that our sense of wonder might be extinguished, once and for all time by knowledge. What, then, would be the purpose of existence? There would be none."

Horgan is not immune to the kind of fantasies he mocks.

The Scientific Method Cannot Encompass Our Consciousness

To approach this fundamental problem of consciousness with any kind of scientific methodology is proving itself quite difficult if not impossible. The usual tri-partite nature of the usual experimentation is composed of the phenomenon to be observed/measured, the observing/measuring device and the conductor / measurer / recorder / interpreter of the observation / measurement. Even this setup, regardless of the rigor and control of conditions, caused sufficient problems to force the recognition of the participatory alteration of the experiment by the experimenter. In the case of our consciousness, the phenomenon to be observed, the observing device and the observer are all one. Not a pretty picture and certainly almost impossible science according to the rules of the scientific method. We may take this as an insurmountable dilemma or understand it as a positive indicator of how we must expand and modify our approach to attain the goal.

The reason David Chalmers says no one knows "the underlying fundamental laws" of consciousness is, apparently, that he is still working from the assumption – or hope – that they must be physical laws and cannot be different than the laws of physics. Because the laws of consciousness are part of the same universe as the laws of physics, I hold no doubt that the final laws of consciousness can be unified with the laws of physics and that the two are, at very least, analogous and complimentary. And I am also quite certain that, once the laws of consciousness have been completely codified, the scientist will find some way to interpret them as physical laws regardless of how far that definition has to be stretched or even warped. If so, no problem, if it makes the scientist feel better. As Murray Gell-Mann has observed, almost laconically, in *The Quark and The Jaguar* "If something new is discovered (and reliably confirmed) that does not fit in with existing scientific laws, . . . we enlarge or otherwise modify the laws of science to accommodate the new phenomenon." (p. 282 W. H. Freeman, 1994, 1996 second printing) The term "physical" is a study in itself. Far more than most would like to admit, the reaction against the "spiritual" and "supernatural" concepts of Christianity caused science to swing to the other extreme and to posit the universe as only physical. Physical is defined variously as materiotemporal, temporal-spatial-material, matter and energy, phenomena and epiphenomena. If a new realm is discovered, perhaps consciousness is such, (and reliably confirmed to satisfy the scientists) then, inevitably, it will be subsumed – perhaps aggrandized is a better term – into the realm of matter/energy regardless of how bizarre a performance that may require.

Sharpening the Outline of the Problem of Consciousness

It certainly is consensual that there is a phenomenon we call, generically, consciousness, and that information and data can be recognized and gathered about our collective and individual experience of that phenomenon. **There seems to be a real problem only with two related concepts**: physicists find it difficult, under the conditions they have set themselves, to derive the "existence of consciousness" from physical laws and, therefore, to arrive at a grand unified field law, the "law of everything." It is also insisted that, so far, there are no laws of consciousness that we have discovered or can recognize.

The first problem may be clarified by making a distinction of kind between consciousness, as such, from the most primitive to the most complex just short of us, and our kind of peculiar consciousness. That there is generic consciousness of any kind seems to be a problem for the physicists in that, having limited the solution to physical explanation, they are frustrated in their attempts to account for consciousness in the first place, and to write its laws in physical terms in the second place.

Eliminating an Unnecessary Handicap

But it seems patent that our self-aware, self-referencing kind of consciousness is the real focus and the cause of the problem. We are unnecessarily handicapping our efforts by lumping all kinds of consciousness together. The consciousness of a reptile is not as much a problem in these considerations as our type of consciousness. The reptile can perceive and react but it does not, as far as we can tell, grasp its cosmological environment, determine its relationship to it, consciously attempt to alter itself or even the nature of that environment. We, having read out the entire genetic code of the reptile and determined its range of actions and responses, will eventually write the laws of serpent consciousness, perhaps even without having arrived at the LOE. But our consciousness is capable of considering itself considering itself and therefore we call it self-referential – and, pretty certainly, *that* the reptile ain't. Our type of consciousness, specifically our peculiar psychology, is the reason why the final theory, the ultimate "law of everything" will, by definition, have to explain us explaining the universe and ourselves, as part of that universe, to ourselves in the only way and in the only terms by which we are capable of comprehending and explaining that universe. Can't pin any, even precociously conscious, reptile with that. The fundamental problem is a function of the nature of our consciousness. It cannot be emphasized too much: even if we were to derive consciousness from the laws of physics as we understand them now, i.e. completely in terms of matter and energy, we would still have to deal with the way our specific type of consciousness operates as the agent reciprocally thinking about, developing the concept of and formulating those laws in the first place.

The problem of the existence of consciousness is exacerbated because we are the observed, the observing device and the observer; the measured, the measuring device and the measurer, not only when we attempt to analyze ourselves but when we attempt to determine the grand Final Theory. The problem's intractability is made even more acute by the uniqueness of the LOE. The LOE, by definition, determines our consciousness to be the way it is in the first place: the measured, the measuring device and the measurer are all one and the same and completely determined by the LOE. It determines the convoluted, solipsistic nature of our consciousness with which we must explain ourselves explaining ourselves explaining ourselves to ourselves in terms that are a function of that same consciousness and its modes of perception. Even if we allow that its most convoluted, highest activities were governed by complexity theory, there is a dimensionality to human consciousness that would still have to be dealt with. The LOE would still have to account for and predict a (our) consciousness which can conceive of itself determining its own evolution to be a type of consciousness that can modify the LOE(!). If human consciousness is to be completely predictable – and the LOE, by definition, will have to be able to predict everything and every action of every thing – then it will have to predict how our consciousness can determine its own determinism. The painfully cute degree of convolution involved in these word games only points up the painfully acute problem of an inescapable higher order of self-reference involved. This is a problem of observer-participancy exponentiated to the absurd ultimate. You're right, human thinking doesn't get much more convoluted and complicated than that but it may get a bit more complicated, however, on the way to the LOE, because, in defining our consciousness, our consciousness will evolve in a feedback loop because the concepts involved are a product of our consciousness, as are the sciences, laws and means of investigation they engender.

Reality and The Laws of Physics as a Product of Our Consciousness

We should recognize that we have, in actuality, discovered, determined and formulated the laws of physics as a function of our peculiar type of consciousness. Ultimately, even the ostensibly consensual concept, "consciousness," is a set of individual, subjective, private conceptualizations based on each individual's experience of what that might be. At least that is what it seems to be, subjectively, to me . . . Einstein was speaking as a perceptive psychologist when he enunciated the principles of relativity referenced to a human observer. The use of the concepts of observer-participancy, complementarity, and "brought to a close" by an "irreversible act of amplification" did not make Bohr, Stueckelberg, Heisenberg or Schrodinger any less scientists – although they often acted more like philosophers and psychologists. I am not saying that we should attempt to derive the physical realm from the laws of consciousness any more than we should attempt to derive consciousness from the laws of physics. I suggest that we go about systematically determining the laws of consciousness, whatever they may be, and then unify them with the laws of physics to arrive at a final LOE.

A Better Way to Look at Reality

Before we all get too deformed over this there is still an option on the event horizon that I submit is the key to both a resolution of the problem and to the LOE itself. If we can lighten up a little we will be able to applaud David Chalmer's clear perspective when he recommends finally acknowledging consciousness as a fundamental phenomenon which is not amenable to derivation from the known physical laws. To do so immediately allows us to rephrase the problem from "deriving the existence of consciousness from the physical laws" to "discovering the laws of consciousness." Discover the laws of consciousness? Well, sure. Not just the neural correlates, not just principles of organizational invariance, not just some fancy footwork semantically juxtaposing consciousness with awareness or information states. If we can reach an attitude state where we can allow consciousness even tentative status as a fundamental entity without having to preemptively assign it to a category such as physical, non-physical, spiritual, transcendental, informational or whatever, we will free ourselves to recognize the operational laws of our consciousness – most of which we already know – and to further clarify those we are so close to that they are hard to distinguish. Having done so, we will be able to identify how the very nature of the operation of our consciousness inexorably and inevitably takes us to the point of unavoidable inescapability at that metaphorical (pun intended) Swartzschild radius where our consciousness – at least the third-dimensional component of it – can no longer anticipate itself. If we cannot escape "outside" of our consciousness to be able to consider it without direct involvement, objectively, then we had better become very adept at dealing with our consciousness very consciously, to make a rather bad pun.

We have discussed, in the previous chapter, how this most fundamental of problems may be called, precisely, the problem of self-reference. Indeed, we define ourselves in terms of self-reference; we think of ourselves most essentially, in comparison to other organisms of which we can be aware, as unique in being self-aware, self-referential. We, generally, classify any species, dogs and cats for example, as inferior to us in intelligence and consciousness if they appear to lack self-reference. As Stan Tenen has pointed out, Koko the gorilla is self-aware (looks in a mirror and signs "Fine gorilla me") and, perhaps, elephants and dolphins may also be. We would identify any alien species which was recognizable at all in the first place, regardless of how bizarre they may appear, as like us by the criterion of self-reference.

Transition

If you are not yet sufficiently aware of these concepts or too boggled by them to be comfortable with them then please set this book aside for a week and then go back and reread this chapter. You will most likely be amazed at your comprehension of these startling concepts. Our minds do a great deal of what is called, in computer jargon, "background processing": while we are going about our daily routines, we are processing concepts and ideas, making syntheses, working out logical consequences and coming to conclusions. Those sudden flashes of understanding, many times, are not as sudden as they seem but are the result of background processing having gone forward for hours or weeks.

Don't throw up your hands because you can't get outside of yourself far enough to deal with yourself "objectively," it isn't going to happen. **But do take that fact that you can't as the critical positive clue** that, because you are that self-reflexive way, as an integral, embedded part of the universe, the universe is also that way and that must be the final clue to writing the Law Of Everything in terms of the meta-metaphor of self-reference. How will you know if what I am suggesting here is true? By referring to yourself, to yourself as the universe, (no longer to yourself as some, somehow inherently evil, error prone, out of place, perhaps even detached from the universe, defective god slave who must look anywhere but to self as critic or authority) which is both formed by and, in a real sense, influences the Law Of Everything. That is the sum and substance of this chapter.

I am not about to tell you that this topic has always been a piece of cake for me even though I seem to be genetically programmed to be preoccupied with its kind. What I can tell you is that the gymnastics involved in contemplating these enormous ideas will stretch your mind, open up dimensions of freedom and potential and, once you have assimilated them enough, can be a bunch of fun. Fun? Certainly.

So, if, Reader, you are comfortable in your understanding of the basic concepts of the GUT (what a gross acronym. . . Grand Unified Field Theory) and you have the guts (even grosser pun) to deal with the mind-boggling notion of a Law Of Everything and the way it determines our consciousness to be the way it is, and to conceive of a Law Of Everything in the way we do here, and you don't need to take Advil for the pain of putting your mind through the monkey bars of shifting dimensions, then we can begin to talk about a Law of Everything and consciousness in a different way that may make things a great deal more interesting and profitable.

Some Speculation on Theoretical
Characteristics of the Law of Everything

There are two difficulties with the way science is attempting to "discover" the LOE. It will have to be in terms which encompass science, philosophy, psychology, mathematics, and every other discipline, but we are trying to get to it in the same way we solve a physics or mathematics problem in the terms which we use in science today.

The second problem is a result of the first: we limit the questions we ask about what the LOE could or should be and limit the guesses, conjectures and theories about it because we a priori force ourselves into a far too narrow mode. We have to begin to think, literally, in the broadest terms we can conceive of to even begin to get close to a LOE. The speculative ideas advanced here range far beyond the notions of the basic forces, superstring theory, the dissimilar physics of a multiverse(s), even beyond a great deal of science fiction. Is there justification for such seemingly wild speculation? I do not think that it can get too wild, and we actually have no other alternative.

Some Hypothesizing about What the LOE Should be Like

When one begins to contemplate such a radical concept as a law of everything and its power and effects, things become rapidly quite strange, if not weird. Try imagining what a law of everything might be like. A unified field law is a strange animal; it can't just explain the laws by which the known types of force are unified, it must explain even how we must perceive it, even explain us to ourselves explaining ourselves to our selves. It may seem, after only a little play with the notion, that it is even quite disconcerting. Let us consider the concept of a law of everything from the broadest and most comprehensive view, with the fewest preconceptions and assumptions possible.

It would seem that the LOE must be absolutely true, absolutely obvious, absolutely simple, absolutely universal by its nature and by definition. Perhaps the LOE must be so simple that it is not possible to write a mathematical equation of it or a statement of it in the form of a scientific law.

The law of everything, at least by definition, must be absolutely simple: to give rise to all else in the universe it must be a principle of initiation, of propagation, expansion, contraction, explanation and of cessation all in one. Therefore, whenever the law of everything is experienced or comprehended by any kind of conscious being it is, at least potentially, experienceable and comprehensible in its totality.

Yet is seems that it must be a statable, stipulatable law in the sense that it is the ultimate determinant of at least all the universes that we can know. The LOE could be completely fixed in the sense of generating fixed, unchanging laws or it could be a meta-law that determines all laws to change, a law of evolving laws. The LOE must be a law even though we could extend it to an extreme by conceiving of it to be a law of transmuting laws: a law that determines that there are no fixed laws, that even itself is constantly transmuting. Because it is the ultimate determinant of the universe of universes it therefore is a law by definition, at least relative to our conceptualization and consideration of it, regardless of how un-law like it may seem. Even if we were to hypothesize that the universe is completely un-determined, then we would still have to acknowledge that there was some inherent law that determined it to be that way in the first place. (The concepts become very strained and preposterous here, obviously, due, even more obviously, to our mind's limitations of conceptualization, but we have to at least consider these extremes at some point.) Even if it is a law that says that all laws change, or that there are no laws, or even that itself is a law in constant mutation, it would still be a final law. Even if we postulate that the universe is a consciousness that evolves and constantly determines itself on an ad hoc basis, as some have done, we can hardly do so without conceiving of a "rule" that says that's the way it must be. It might even seem to some that we are determined to subjectively judge that ultimately reality is lawful regardless of how unlawful that fundamental law determines it to be. All of which I am sure are simply pitifully tiny constructs of our type of consciousness. Perhaps, in this search for the obvious, we are reaching a point where we may see what is beyond the elastic limits which we find our consciousness beginning to strain against.

It would seem that the LOE must be above proof, self-derivatory, self-referential, self-evident and self-explanatory by its very nature and definition. It must surpass Godel's limitations in that it must explain itself completely because it is intrinsically totally self-referential. By explaining itself completely it will prove itself. It would seem that a true Law Of Everything should be totally self-evident. In fact, by definition, it should be hyper-evident and inescapable for those consciousnesses which have reached the level of self-reference.

It would seem that the LOE is unavoidable. If the LOE, at least for the sake of this groping discussion, is understood, by definition, to determine everything including us completely, the very perception, thinking, hypothesizing that we are carrying on here is completely determined by its rule. It, in a literally real sense, causes me to think that I cannot avoid operating according to it.

It would seem that the LOE must not be just a general rule of how things happen, but also an initiatory principle in itself. It may be more difficult to answer the question of why things go forward than it is to answer the question how things go forward. It is at least possible to conceive of the LOE as both the impulse to, and the rule as to how, things go forward.

It should be derivable beginning from any thing or any phenomenon at any time and place in the universe since all parts and phenomena in the universe should refer to all other parts, and any time refer to any other time or place or thing. The key, the "clue" to the LOE should be embedded ubiqitously in the universe and discoverable as a self-revealing "cryptomarker." It should be as self-evident and discoverable to any self-referential intelligence as to any other since all intelligences are the product of that LOE. It should be as self-evident and discoverable at any scale of size as at any other scale. It should not be lost at a certain level of size like a hologram's information is lost. It should be complete and robust at all levels.

In a strong sense the LOE, besides explaining and precisely allowing us to explain and predict everything, all events and phenomena, should be self-revealing and explain itself.

With regard specifically to ourselves: we should anticipate that the discovery of the LOE, affording us the ability to predict ourselves completely, will cause us to go through a philosophical crisis because we will feel trapped in a completely deterministic universe with no choice or free will. But being totally determined by the LOE still does not negate the strong possibility that it may determine us to determine our own determination without contradiction.

The inevitable question is: How would we write the law of everything? The LOE will be stated in a form that is comprehensible to us and a function of the way we perceive and understand, which is a function of the LOE. By definition the LOE will determine how we are, how our consciousness is and functions, how the universe is and operates. We hypothesize here in terms which our consciousness can comprehend, and our considerations will be a function of our type of consciousness. But since we are determined to do so by the LOE, that may not be a problem at all, only a clue. We will have to express it and its operations in terms of our type of consciousness but, since that consciousness is a product of the LOE, the entire explanation should be contained in our consciousness.

We have been considering here some conjectures and hypotheses about what a Law Of Everything should be. We have been talking about its hypothetical nature and characteristics. But how would we state such a law of everything? How should we apply it or use it to predict events and phenomena?

Some Hypotheses about How the
LOE Should Manifest and Work

It is ironically strange that we should be talking hypothetically about the LOE. If it is only half of what we have described above, it is working all around us all the time and we are operating according to its mandate in this very discussion. So, in a sense, this discussion is a process of raising the intensity and scope and nature of our consciousness to be able to perceive and understand what is too obvious and too close to us. We need a precise concept to express what we have been guessing the LOE should be. I have suggested above that I believe that this concept, used as a meta-metaphor for the LOE, is self-reference. We should take the LOE to be a rule which determines the entire universe from energy and matter to consciousness, to initiate itself self-referentially, propagate on the basis of self-reference, as a self-referential universe, with self-referential sub-systems all the way from self-referential strings, to particles and atoms, to molecules, to organisms. We should understand the evolution of the universe as a gradual complexification of self-referential systems in all its parts and as a whole. Our approach, reiterating, is to hypothesize that the universe "works," fundamentally, in a self-referential mode and produces self-referential systems from the simplest to the most complex we know, ourselves, as a result. How do self-referential systems seem to work? They work according to an interdependency, through the inter-referencing, inter-identifying and reinforcing of all their parts by all other parts. An idealized self-referential system would be totally closed, self-contained, with all its parts interdependent and referencing all the other parts of the system, but nothing outside the system. We do not seem to find any ideal, totally closed, self-referencing system in the universe. Our solar system might be taken as an example of a self-referencing system. The relationships of distance, gravitational attraction, orbital periodicity, etc. between the sun and each planet and between the individual planets makes the entire system a self-referential system. Change any of the variables of the physical relationships and all the other relationships are effected and the character of the solar system itself is altered. The individual parts, the sun and the planets, all reference each other. If one discovers the laws that govern these intra-referencing relationships then a system of this type can be recognized also as self-correcting: if one were to visualize the solar system in miniature as a sort of child's puzzle and a particular planet were missing, the laws of interrelationship would tell one what size the missing planet should be and what its orbit should be like. It will tend to restore equilibrium if there is a disruption. But our solar system is not an ideal, totally closed and independent system; it is related to other systems like itself and other bodies in the universe also. It is self-referential with regard to it members taken as a system but it, as a system, is also a part of a larger self-referencing system beyond it. This self-referential inter-dependency between self-referential systems, I suggest, should be the key to the LOE.

The Universe as a Closed, Self-Referential System

The universe itself, in the classic conception of it as "all that is" would be an ideally self-contained, closed system. But even that concept is changing; there is speculation about the existence of other sub-universes, such as ours, all part of a meta-verse or, sometimes, multiverse. We simply do not have enough information at this point to know. But, in the event that that hypothesis proves to be true, then the LOE must encompass the meta-verse, and sub-universes would be self-referential systems also part of an all-encompassing, fully self-referential metaverse.

We see ourselves, as human beings, as organized systems integral to the universe. The self-referential self-consciousness we experience in ourselves we take as a defining characteristic of that human nature. The degree of that self-reflexive awareness – we are aware that we are aware that we are aware – we take as unique, particularly since it is of such a degree that it affords us the ability to control, alter and direct it and its future evolving trajectory. It is this added quality of control and self-direction that seems to us to put our self-reflexive consciousness in a class of organized, self-referential system by itself, a full magnitude beyond any other level of system from particle to atom to solar system – at least in the tiny portion of the universe known to us so far.

If all we were to establish is that the universe works according to self-referential law, it would hardly be worth the bother. As with current science, it is the predictive power of any law that constitutes its value. If we can predict an outcome we have the power to control or avoid. Obviously, the power is, unfortunately, misused and abused far too often. But it is the same power that allows us to develop cures for disease, non-polluting sources of energy, or to attain immortality and means of travel to other star systems. But the practical applications of the principle of self-reference as a "law of everything" is doubly powerful because it allows us to predict not only all events and phenomena in nature in the universe, but in ourselves.

The LOE as a Mathematical Formula?

Can we state or write the LOE as if it were a mathematical statement? An equation for the LOE? In mathematical terms the LOE would have to be able to predict/derive/generate the equations in which it is stated. I do not believe, by the nature of the LOE, that it will be completely expressible in mathematical terms. I believe that those areas and phenomenon of reality which are amenable to mathematical expression will be expressible mathematically as special cases of the application of the LOE. The mathematical modality is designed as a language to precisely express quantities and measurable phenomenon and their relationships; it is not meant to express a primary initiating and determining principle. Clearly the concept of the LOE as a mathematical equation or set of equations is the product of their a priori assumption that all that is may be cataloged as physical and therefore quantifiable. It would seem that the LOE must be a simple universal principle that says how the entire universe "works," and each discipline must be seen as a specific and special application of it.

Isn't This Modality the Same as
the Scientific Method Now in Use?

The objective of science is to explain the universe by discovering and codifying written laws by which it operates and applying those laws to extrapolate and predict phenomena and events for the sake of greater and greater pragmatic control of the future. The scientific method currently taught and used demands that we assume that there must be a fixed objective order of lawful reality to discover, independent of our perception of it, before we can even consider attempting to discover its laws. The assumption that this assumption of objective order is necessary is based on the perception that any kind of reality which is a product of anyone's subjective judgment would be unique to each individual's perception and evaluation, ultimately not effectively correlatable to anyone else's subjective reality. Any reality of which the laws themselves changed over time would also be effectively unpredictable – and, perhaps, undetectable because the perceiver, the investigator, would be changing as part of that metamorphosis. How would we go about using self-reference as the Law Of Everything? Simply take the LOE as the ultimate self-referential principle. If the universe or meta-universe is, by definition and scientific determination, a completely closed system then the LOE, the grand unified field law can only refer to the totality of the universe, to itself. This is just another way of saying that the LOE is a principle of self-reference from micro to macro. If we take the universe as a closed, self-referencing system governed by the ultimate law which is essentially self-referential, and understand ourselves as essentially a self-referential consciousness and integral part of that universe, the critical advantage is that the LOE will be in the only terms which we can understand, operating according to the only way we can know the universe in the first place, and giving us the immediate advantage of having our consciousness explained and its laws written as a function of the LOE.

The LOE is, by concept and definition, the most comprehensive self-referential law we can conceive of, literally. Every part of the universe, explained by the same law, by that very fact refers to and gives a clue to the nature of every other part and signals its relationship with those other parts and their relationship with it; in addition it also gives indication of the nature of the law itself, the way it initiates everything, and the way everything propagates in all dimensions in the trajectory it itself is.

The reason why the LOE is, by definition, a self-referential law in our projection, comprehension and gradual development of it is because it is a function of our consciousness, of the way we have evolved to survive/understand/deal with the universe on this planet reciprocally as a product of that most fundamental principle. Its formulation and eventual implementation and application is relative to our peculiar sensory impute devices, information handling and communication modalities, but is universal because our nature and the way we conceive and formulate it is a function of that very Law. The topic is so difficult to get outside of and to deal with, to articulate and communicate about, because we are so self-referentially close to the subject (pun intended). We may be preventing ourselves from any progress by assuming that the brain's neural processes give rise to consciousness. If we think strictly in three-dimensional terms we may have no other choice. But consciousness, essentially, may be a hyperdimensional phenomenon and conscious experience may be a three-dimensional manifestation of the entire phenomenon. We have some possible examples of analogous phenomenon in other areas. Topologists have theorized that, if physical bodies we normally perceive as three-dimensional were actually existing in hyperdimensional space (four or more dimensions) and therefore were subject to the physics of those other dimensions, we might actually perceive the effects in three dimensions under certain conditions. On a very large spinning sphere like a planet, one effect would be some sort of hexagonal pattern at a pole of the body showing up as a sort of "shadow" in three-dimensions of the physics operating in four or more. When the Voyager II took pictures of the north pole of Saturn a clear hexagonal pattern was revealed in the atmosphere, rotating with the planet – the winds inside the design were blowing backwards at 300 miles per hour and making the precise angles. It may be controversial, but it looks like pretty good evidence to me. We need to face the question Is it true that "physics provides a complete catalogue of the universe's fundamental features and laws?" How would we know that it doesn't unless a phenomenon was encountered which physics could not explain?

If we accept conscious experience as a fundamental feature, irreducible to anything more basic, then how shall we go about determining the associated fundamental laws as we did with electromagnetism by introducing electromagnetic charge? By introducing self-reference as a fundamental principle. If we do so, I submit, we will have the key to the explanation of consciousness and the means of unification of the laws of consciousness, conscious experience, with the laws of physics and, therefore, the LOE.

We should concentrate on writing the laws of consciousness rather than trying to derive consciousness from the laws of physics and see what they turn out to be without making the a priori assumption that everything can be explained by the laws of physics –including even the physicist who claims that.

Either we are going to recognize consciousness as a phenomenon in its own right with laws that apply, or we are going to have to a priori try to muscle it into a classification that it doesn't fit. To say that we can't derive it from the laws of physics and then go right ahead and assume that we can is a bit startling. Either we should state the situation as: we have not been able to determine the laws of physics that determine the subjective experience of consciousness and we are going to do so, or say that the laws of physics cannot explain consciousness which may be of a nature that is not physical. We would have to relinquish the attitude that a "nature that is not physical" is an oxymoron to allow ourselves to do that in the minds of many.

We should clearly differentiate between the statements that consciousness as a phenomenon cannot be explained by the laws of physics and the other, quite different, statement that the subjective experience of consciousness cannot be explained by the laws of physics. Not just because of the questionableness of the unquestioned assumption that the entire universe or meta-universe is material-energy-physical; not just because of the classic philosophical debates over whether there is an "objective" order of the universe independent of any mind or whether the universe, as we know it, is pure "subjective" experience; not just because the, residually theocultural, attitude toward the "mental" (which borders on that unthinkable horror, "spiritual") on the part of the scientist and the philosopher alike presents a major obstacle to considering conscious experience as an irreducible, fundamental lawful phenomenon. I am proposing here that we unite the laws of physics and the laws of psychology into a final psychophysical law which will be the LOE. To do so we must push to the extreme until we determine the point of unanticipatable inescapability. It is not that the laws of physics cannot account for consciousness or that the laws of consciousness cannot account for the physical world. It is that neither can account for the universe. The ultimate enigma is not that we cannot prove whether the universe is all consciousness or all just physics, but that we can't prove the existence of the universe by either.

We cannot derive the universe from the laws of physics or from the laws of consciousness. So what do we do to arrive at a LOE? Put it the other way around: as individuals, we apparently can't know any way but subjectively, which is a type of self-reference, i.e. we refer ultimately to our own experience to determine the truth or falsity of all we perceive and experience. We arrive back at that place where we cannot get behind, our subjective judgment, to judge the truth or falsity even of our subjective judgment – much less the truth or falsity of the laws of physics and/or the laws of consciousness. We cannot escape the circular loop at that point– or at least we cannot get outside of the situation at that point to see if there is another dimensionality of the universe which might explain or generate it. No one seems to have a problem with conceiving any attempt to physically exit the universe as impossible because it would simply mean extending the universe. So the only thing that we can do at that point is conclude that we cannot arrive at a LOE or, positively, take that as the ultimate clue as to what the LOE is. To paraphrase Leon Lederman's famous question, If you can't beat the universe, join it. As I repeatedly suggest herein, the way to arrive at the LOE is to approach it as a fundamental single principle in terms of the meta-metaphor of self-reference.

A "Partial Law of Everything?" The Ultimate Oxymoron

Since the LOE, as we shall come to understand, formulate and apply it, will be in terms of a function of our consciousness, which it has literally orchestrated in the first place, an interesting question is prompted: Would it be possible to formulate the LOE in terms that would be so comprehensive as to literally predict and explain everything, every entity, phenomenon and action that we could possibly perceive and conceive of, yet not cover everything?

Let us suppose that there are phenomena in the universe which we could not possibly perceive, sense, detect or measure with our technology. If these phenomena did not manifest indirectly as variables in some way, influencing the phenomena and objects we can perceive or detect with our technology, would we not believe that we had arrived at a LOE if it predicted and explained all that we could perceive and detect? I would say that this would not be the case: If the LOE is truly so it would also predict phenomena and entities, sentient and non-sentient, which we could not perceive, through their indirect effects and interaction with things we could perceive so that indirectly we would know of their existence. One of the first clues that we had not gotten the LOE right would be a loose end of some sort showing up as an anomaly which was not predicted or which was indirectly indicated as something missing from an explanation.

I suggest that this situation is already confronting us. One way of understanding the current situation where consciousness is the obstacle to arriving at the law of everything is to see it as a problem of incompleteness. The concept of the Unified Field Theory, in this perspective, may be seen to be that of an incomplete and partial law of everything – which, in this case it would not be. We may take the fact that consciousness cannot be explained by or derived from the basic forces as a sign that something is missing, there is a loose end manifest. Consciousness may be taken as a significant anomaly pointing to laws, phenomena or dimensions which we are either not aware of or not taking into consideration – or are ignoring or denying because of prejudice.

A practical consideration is immediately suggested: Should we settle, or could we even settle, for a "partial" LOE of our universe? It certainly seems adequate if it was able to predict nearly everything, or at least everything we consider important. The prediction and control we seek would be satisfied. But the logic of the situation would indicate that we would not be able to realize a complete explanation and prediction of even our sub-domain universe because there would be some unknowns, some discrepancies due to the interrelation with other sub-universes even if they manifest only as boundary interactions between them. We would be forced, perhaps, to seek a LOE for the meta-universe and understand our domain's peculiar law of physics as a sub-set. The assumption here is that, by definition, the LOE would have to apply to the meta-universe and to all possible sub-universe domains. Otherwise, if we indeed attained what seemed to be a true LOE for our sub-domain universe, it would eventually be proven to not be such since anomalies would begin to manifest when we examined areas, theoretical or physical, where our domain interacted with other sub-universe domains whose laws of physics were different. To extend the notion further, if we do intend to determine the LOE of the full multiverse then we should, at least, give some thought to the self-referential condition automatically imposed by the inherent logic of that situation. To be absolutely self-reflexive it must, to be a complete LOE, predict and explain itself with reference to itself and nothing "outside" of itself, i.e., something "outside" the universe, which would be contradictory. In this way the LOE would transcend the principle of Godel's theorem that no system can explain or prove itself totally.

It is interesting to consider how this general principle applies to the recent theory concerning a self-generating, fractally propagating universe consisting of many domains which are actually sub-universes in which the laws of physics are not the same. The conclusion seems unavoidable that the LOE would have to be a law of the meta-universe of sub-universes. Each sub-universe domain would have a set of physical laws and laws of consciousness peculiar to itself, but the LOE would have to determine all those various sets of physical laws and laws of consciousness in all sub-universes. It would have to determine how the various sub-domains of the multiverse interact in general and at their contiguous boundaries in particular.

Certainly, being able to alter our environment for better or worse is one thing, but actually altering the physical laws of the universe is another. It is awesome enough to unleash the power of the atom or to achieve anti-gravity capabilities, but to actually alter a fundamental law of the universe would be quite another extreme. To do so is, obviously, fantasy now but, by the very fact that we can conceive of it, it could be a possibility in the future, particularly if our universe is a sub-universe domain of a meta-universe. This consideration leads to another fascinating question.

A Self-Altering Universe?

While we are at it, we might also try to determine the LOE of a meta-universe which is self-determined to be self-altering or constantly transmuting, since we will not be satisfied until we have at least thoroughly explored that concept. That should keep us preoccupied for a bit. . . we certainly will not get bored with outmoded and inadequate metaphors.

Is it possible that the LOE, intrinsically, may be self-altering? We have no direct evidence for that of which I am aware. From a purely logical point of view it is possible to project that, even if it were, since we are embedded in it, we might not be able to tell since even our memory of the universe being one way in the past and different now, would be altered along with everything else in the universe. But that can only be the most speculative kind of conjecture. What we do know is that our consciousness has now reached a point where we can actually begin to determine our own evolutionary trajectory. Each individual can evaluate the history of the race, the history of our ancestors, and decide how SHe envisions HIrself in the future and work toward that goal. These individual visions of oneself as evolved in a certain direction in the future contributes to the collective racial evolutionary direction. Obviously, we, individually and collectively, are restricted in the range of those choices at any given time. Our current physical makeup and understanding of the potentials that the universe offers limit what we can envision for ourselves. But, as we make a step, more potential opens up in front of us. No evolutionary development of this type constitutes an alteration of the physical laws of the universe. But the interesting facet of this process is that we, as self-modifying entities, are acting according to, and have been generated by, the LOE so we can safely say that the LOE has the potential to be and do whatever we discover is and happens in the universe which it governs.

Could We Alter the LOE?

The most complex area of reality the LOE will have to predict and explain is the way it modifies itself – if it does – and, consequently, how our consciousness can modify the universe around us. We can even conceive of being able to change the laws of physics, literally modify at least our domain of the multiverse. We can conceive of an evolving multiverse. If we can self-modify through our self-referential awareness and we are part of the universe/multiverse, then the universe/multiverse can and partially does self-modify. We generalize that anything that we or other or even more advanced consciousnesses can do, the universe can do. Even more generally: whatever we can conceive we can achieve.

The Grand Unified Law of Everything and "Unreality"

The most interesting facet of the LOE is that it must predict and explain the phenomena we call the irrational, the imaginary, the impossible, the illogical, "evil," the contradictory, the insane, the unreal. From one perspective, all these phenomena are explained easily by saying that the universe is limited, that reality, therefore, is limited to certain parameters, certain laws which determine it to be in certain ways and not in others. The argument would hold that, although our minds require the ability to make mental associations and consider possibilities which cannot exist in this universe in order to have the degree of freedom to be self-determining we enjoy, it does not mean that these things could exist as such, at least in this domain of the multiverse.

Now it would appear that, if there is a LOE, then it would seem to determine just such an objective order of the universe. But, if so, the subjective nature of our consciousness would have to be determined as part of it. It would force us to say that the objective order of the universe determines reality, for consiousnesses like ours, to be subjective. Could this be so without contradiction? I think so. When one considers the classic philosophical argument between those holding for an objective order and those holding for a subjective reality, the debate seems to have a clear resolution when we pay strict attention to the way the conflicting statements are formulated. Too often the opposing views are stated thus: the objective view says that the universe is an objective, independent reality which exists outside of any mind, and the subjective view says that the universe only exists as we perceive it in our minds, giving one to believe that somehow we create the universe with our minds. The consideration of this theory is also a good example of the reciprocal relativity of our concepts of the LOE to our consciousness. If that which we call universe may quite possibly be only a kind of self-generating, fractally propagating, sub-universe domain among sub-universe domains of a meta-universe where the laws of "nature" are different in each domain, then our specific and peculiar kind of consciousness would not be probable and, perhaps, not possible in another domain since we assume at this point that it is a function of and adaptation to the laws of its domain. But it is conceivable that a generic form of self-referential consciousness might be possible in another radically different domain due to the fact that other domains would be a part of the self-referentially, self-generational and fractally propagating meta-universe, or multiverse, as some have already begun to call it. It is further tempting to speculate that self-reference as a generic phenomenon may be possible in any multiverse or domain of such, but that is not determinable at this point and seems, if anything, improbable. And it goes almost without saying that substantially all our speculation on this topic will probably seem like embarrassing kindergarten babbling in fifty years, but we must work it through.

We could, and probably should, therefore, immediately expand our preconception of the LOE and consider attempting to arrive at the LOE of a meta-universe which determines each domain and its respective type of consciousness to be different.

We should consider here whether we can agree that our consciousness could have a fourth-dimensional, space-time component. If so, whether our self-referential consciousness, obviously aware of itself, could be, currently, aware of itself only in its three-dimensionality due to its reflexive awareness being dependent on the current state of development of its three-dimensional neural correlative component, the brain. If so, whether the three-dimensional correlative component should be taken as the brain itself or specific functions of it; whether there is a hierarchy of causation in that the three-dimensional component of our consciousness is a sort of topological "shadow" generated by the physics of the fourth-dimensional component or whether the three-dimensional components generate the fourth-dimensional component or whether there is simply an integral unity of four-dimensional, self-referential consciousness, any imputed hierarchy being only

a result of the relative difficulty of that type of consciousness being as self-referentially aware of its space-time dimensionality as it is of its three-dimensionality;
if so, whether, we may further take the problem of determining the nature of our
consciousness as an indicator of an evolving process, that we are evolving toward
an expansion of our consciousness that will make us self-referentially aware of
even our fourth-dimensional component; whether the very recognition of that
possibility is both a key to the direction and reciprocal determinant of what we
shall, indeed, achieve; whether that evolution will demand and involve a further
development of our brain to accommodate; whether, on the contrary, we have
already begun to achieve that expanded self-reflexive awareness and are already
working our way through the expansion of our concepts and articulation and codi-
fication of it; whether the most fundamental nature of consciousness is that it is
determined by the LOE to determine itself in a process of constant expansion.

One can see, from that perspective, that criteria will always be a problem
since we tend to use the highest form of expanded consciousness/awareness we
can consensually recognize as the criterion of all other types of consciousness.
Syllogistic reasoning, whether philosophic or scientific, is the consensual criterion of truth currently. But, clearly the leading edge of our conscious evolution is
already far beyond the consensual binary security of true-false, yes-no. Yet reason judges ecstasy suspect, intuition naive, transcendent experience incomprehensible, the metasyllogistic meaningless or, worse, sometimes, insane. But, since
the consensually comfortable and familiar criterion is always behind the leading
edge there are conflict and "problems." But the LOE will have to account for our
most expanded consciousness as well as the more pedestrian consensual forms in
which it will be formulated. The recognition that that is where the most fundamental problem lies with regard to the topic at hand brings us around again to
square one: how do we arrive at the laws of our (entire) consciousness so that we
may unify them with the laws of physics to arrive at the LOE? I suggest that we
proceed as follows. Although we use the terminology of relative sophistication of
Einsteinian physics and even indulge in quantum-speak when the party gets rolling, we still tend to do our difficult scientific theorizing and philosophizing talking down from the security of the Cartesian monkey bars according to the Gospel
of Newton. But dimensionality seems to be at least a factor of high importance
with regard to bridging the putative gap between the "physical" and the "psycho-
physical." To arrive there we need to differentiate clearly between the "existence
of consciousness," consciousness as an observed phenomenon, conscious experience as a set of data, and the laws which govern conscious experience. We need to
acknowledge the difference between types of consciousness and that it is our
specific type which is the focus and cause of the problem itself. To bring it right
down to bare basics, if we take the laws of reasoning, logic, as one of the operational laws of our consciousness, then physicists will not allow that the physicist
can be derived by the philosopher, but the philosopher can be derived by the
scientist – if he could just find the right formula. This is not to say that the scientist is intrinsically illogical: it is meant to point out the humorous things we have
been calling each other over the criteria fence. The scientist should realize that the
philosopher does not think that he is either unreasonable or derivable by the laws
of reason.

The greatest contrast between knowledge and experience manifests when
the question is addressed regarding what is possible and what is not. Reasoning
from even the broadest and deepest factual knowledge base can only tell one what

is possible, and that in a circumscribed way when compared to actual experience of a phenomenon. The reasoned anticipation of some doctors, when automobiles were just invented, was that traveling over thirty-five miles per hour would be physically destructive to the human being. Very soon the actual experience of those who habitually traveled at velocities well over that limit without harmful effect prevailed.

The universe is at least partially consciousness simply by the fact that we are conscious and are part of that universe. From another perspective of logic, the universe must be potentially conscious if it determines us to be so. The laws of that part of the universe which is conscious, by consensual definition, are conscious(ness) laws. Various aspects and types of awareness are explainable by basic laws of physics and some kinds of consciousness are explained only by their own laws: we can predict the reaction of an amoebae to a specific stimulus, but it strains complexity theory to even begin to model the way a human will react under complex choice at a high level of abstraction.

A More Fully Human Way of
Coming at It: Consciousness Revisited

The new human must anticipate living in a known universe, known in the sense that we may reasonably anticipate that we will know how all things "work," including ourselves.

The consensus of scientific expectation is that, even if we do arrive at a final law of everything, we will not be able to totally predict everything, including our own actions and fate, indefinitely into the future; chaos and complexity rules will allow prediction a very short span into the future. This may well be shortsighted: the LOE should explain chaos and complexity theory. But the ability to predict oneself, even on preliminary consideration, is a startling and, for some, a very disconcerting possibility. When we try to write the LOE that determines us to write the law of how we work, we resist the notion that we are absolutely determined by that law and tend to write the law to encompass the maximal freedom for ourselves of which we can conceive, trying to preserve our comfortable sense of freedom, self-determination and free will. But once we have arrived at the LOE (I almost said "ultimate LOE" which would be an automatic oxymoron – the worst kind, actually. . . the puns also get funnier and funnier the closer we get) we will also have arrived at, and in, a completely known universe. Now that, for some, is just very damn scary. You can predict yourself predicting yourself predicting yourself? Very cramping!? Unless we "discover," on the way to the LOE (sounds almost ignominiously prosaic) that the LOE is so trickily accommodating as to allow that it is continually self-modifying by nature and we, as part of that busy process, are determined by it to determine our own determination. Let's hope so or, better yet, "determine" right now that that's the way things are going to be for certain.

Toward a Possible Resolution

Taking David Chalmers' terminology and statement, quoted at the beginning of this chapter, as a challenging launch point, I suggest that the resolution of the problem of consciousness will not be a bridge theory but a unification theory. A bridge theory somehow cobbles up a connection between consciousness and the laws of physics. A unification theory, as I am using the term here, determines the laws of consciousness and unifies them with the laws of physics as laws, rather than trying to reduce consciousness to a point where it can be derived from the laws of physics.

A Critical Redefinition of Terms

To be explicit, whereas David Chalmers would define his term "psycho-physical" laws as laws which will tell us something about how some of the physical laws are associated with conscious experience, I would respectfully suggest that we reserve the term "psycho-physical" law as the unification of the laws of physics and the laws of consciousness. If we discover and codify the laws of our consciousness, the most complex we know, we will be writing the laws of all lesser consciousnesses on this planet which are subsumed into ours. If we can write those laws we can then attempt to unify the exciting electricity of the laws of physics (which, by that time, hopefully, may be all unified, gravity included) with the sophisticated magnetism of the laws of consciousness, thereby arriving at a true psycho-physical law – which should be the LOE. If we make the least pre-judgments about what the laws of consciousness are or can or cannot be, we will free ourselves to be able to recognize them, whatever they are. If we look only for physical determinants of consciousness and they are not physical in any way we currently define physical, we simply will never find them.

To resolve this problem, to unify across the explanatory gap, I submit that we need to do the following (notice that these points effectively summarize what I have been saying in so many ways to this point in the chapter):

1> recognize and come to terms with the historical prejudices, religious and philosophical, that influence our thinking in regard to this topic.

2> recognize that it is our particular type of consciousness and, specifically, a very particular characteristic of our consciousness, subjectivity, self-referential awareness, that is its own self-caused problem.

3> recognize that the physical laws are conceived and written in the form that they are as a function of our psychological laws of consciousness.

4> that self-reference is the clue and the key to the solution of the problem.

5> that the laws of our consciousness are at least partially known to us and we should recognize them and codify them.

6> recognize that the final formulation of the LOE will be available to us when we determine the laws of our consciousness, of our self-referential (subjective) experience and unify them with the laws of physics, formulating the LOE in terms of self-reference as a meta-metaphor.

7> initiate the process of writing the LOE in terms of self-reference.

To this purpose I present some admittedly highly speculative conceptualizations of self-referential systems, taking the largest, the universe, first. I simply lay down my opinions as flat statements for your consideration, criticism or agreement since they are, inherently, not open to empirical experiment.

A true Grand Unified Field Law of the Universe or a Law Of Everything (if the GUT doesn't turn out to really fill the bill), by definition and by nature, must be completely obvious and self-evident.

If it is truly the unified field law or the Law Of Everything then it must explain itself, since it must literally explain everything, which is what Universe is defined as in the first place.

All but gravity have been unified to date to the satisfaction of many physicists. The basic presumption of consensual scientific thought is that the entire universe can be explained in terms of those elementary forces. But the effort is so great to achieve just these fundamental unifications that there is little attention paid to phenomena such as consciousness. If the universe as we conceive of it – and what other universe can there be for us than one which we can at least conceive – is a totally self-contained, closed system then it is, by that very fact, intrinsically what we call self-referential. (It may also be that there are other universes. Our science fiction and even a few latter day scientists have spoken of parallel universes dimensionally shifted from ours. It could be speculated that there are neighboring universes like ours, perhaps, which could conceivably interact with our universe. But these putative universes would all seem to be closed systems, by definition at least, and therefore self-referential by nature.) A totally closed, self-referential system, if it is dynamic at all, would seem to have to generate its own movement, be its own engine. Since all parts of a self-referential system are self-referential also, all parts, from smallest to largest, contain the key to the self-referential law by which the whole operates.

A self-referential system is self-correcting. Since all parts are a product of each other it is redundantly rich in information and all parts refer to each other, thereby giving a signal if a part has been used incorrectly. A complete unified field law must also be able to predict and explain the generation of all deviations, innocuous or harmful, the "incorrect," and the not-yet-realized, conceptualized, manifest, i.e. all possibles within the context of the system.

Our Consciousness as an Unique Phenomenon

We should not preclude, however, the possibility that our consciousness may constitute a realm essentially not the same as the physical in any way we define the physical realm currently. Just as we had to posit electromagnetic charge as a fundamental entity and then determine the laws by which it operated, so should we posit consciousness as a fundamental entity and determine the laws by which it operates – without making any limiting prejudgment that consciousness must be completely determined by the physical laws of the universe. We have already reached an impasse, the focus of the discussions concerning consciousness at this point in both the scientific and philosophical communities, in attempting to account for consciousness on the basis of the known physical laws. I suggest that we should proceed without delay by hypothesizing that consciousness may be a fundamental entity, a phenomenon of the universe with laws that are unique to itself. We should then go about the process of determining the laws of consciousness, letting the laws which we discover tell us what consciousness is.

We should, besides keeping an open mind as to what consciousness is and what laws it operates by, also keep open the possibility that consciousness is a law of the universe itself. By that I mean that consciousness may be – I believe it is – a unique kind of fundamental entity in that it is the law by which it operates, that it can determine its own evolutionary development and its moment to moment operation. This is due to the fact that consciousnesses like ours that have reached at least the primary level of being self-aware, and those that may have developed beyond even our level of self-reflexive awareness, can obviously orchestrate their own evolutionary trajectory.

This line of reasoning forces a critical question: Even if it is obvious that our kind of consciousness is capable of determining its own evolutionary direction and its day to day action and development, does that not mean that there is a law that determines us to be just that way? I think the answer has to be yes. We should, at least, seriously investigate the possibility that not just self-reflexive consciousness but self-reference itself is the most primitive orientation, the fundamental character and law of the universe. It is, I suggest, the most fundamental metaphor by which we understand ourselves and, therefore, is going to be, inevitably, the most profound metaphor by which we are going to comprehend the universe and in terms of which we will formulate the LOE. It is easily conceivable that this fundamental entity can eventually be derived from some even more fundamental entity(ies) as we learn the laws of consciousness and the nature of consciousness. So be it. But, by the uniqueness of its relatively unanticipatable position, I don't think that will soon be the case.

If the determinants of consciousness and even consciousness itself are not physical, then what could they be? Consciousness could be a pure non-material, non-energy, effect. Can an effect exist completely "on its own" or does an epiphenomenon always require some carrier, some physical entity, whether matter or energy configuration, in which it can only exist? Even though it required a physical carrier, consciousness, essentially, could be a transcendental effect of the interaction of physical elements from simple perception to synthesis to concept and reflexive awareness of itself. If it is an epiphenomenon generated by a complex of physical elements it could be constantly and instantaneously changing as those elements changed. But it could also be related to all those elements in a feedback system so that every change in consciousness would also alter the complex of elements generating it. This is a good working description of the psychosomatic processes and the concept of "mind over matter." But could a pure effect be aware of itself and self-modify? I can see no intrinsic reason why it could not if, indeed, it was generated as such. It does not matter whether we say consciousness, our self-reflexive consciousness in this case, evolved that way as a matter of adaptation and survival on this planet, or that consciousness will develop in this generic form throughout the universe, or that it is a chance development among many in the universe. Is it possible that physical elements could generate a purely non-physical effect such as self-reflexive consciousness? I can see no reason why not. Could physical elements generate a non-physical effect of such complexity that it could, by its self-referential nature, become semi or completely independent of the physical elements by which it was initially generated? I can see no reason why not if, indeed, it was generated, according to whatever rules, to do so. At this point we are at a primitive point of the reasoning process where we have to take the reality of what we experience as possible because we can get no further or farther behind it.

Giving the Physicists' View Its Due

Consciousness can certainly be inspected from the opposite perspective. We tend to take a bottom-up approach and some physicists anticipate that consciousness will be finally determined to be derivable from superstring theory. I suggest that we take a different perspective. Even if consciousness can ultimately be seen to be derivable from superstring or an even more fundamental theory, we will still find that it will have more than just three-dimensional components. The problem as the physicists see it now may well resolve into determining the full dimensionality of our current consciousness. I anticipate that in that process things will continually become more and more complicated by the very nature of our consciousness: we know that, on full recognition of its current dimensionality, our consciousness will inevitably begin to expand to encompass greater dimensionality. Lack of recognition of the fourth-dimensional component of our consciousness – or reluctance or lack of potential to do so – I suggest is the primary obstacle to both recognizing the laws of our consciousness and its full nature. I believe that we are already recognizing the fourth-dimensional component of our current mode of consciousness and I am oriented to exploring beyond that. We hypothesize that consciousness comes about in three-dimensional physical organisms such as we, through the action of fourth-dimensional physics manifesting as a specific effect in three dimensions under certain very specific conditions, in the same way that higher dimensional physics inform three-dimensional phenomena in general.

Whether Consciousness is a Purely
Physical Phenomenon or Not as a Moot Question

The final determination as to whether consciousness is beyond dimensionality and matter and/or energy or whether it is an epiphenomenon of matter and energy although following laws of the universe peculiar to itself, for all practical purposes, may be moot. As Murray Gell-Mann has pointed out quite fairly, physicists are pretty much of one mind that the entire universe is just matter and energy and that there is nothing else. If a phenomenon or epiphenomenon is claimed to be beyond either, or operates according to laws which have no known counterparts in the laws of physics, it will be denied – until someone has the nerve to prove it rigorously. At that point the physicists will then graciously extend the laws of physics and the definitions of matter and energy to subsume or include it.

This event horizon of proximity to our consciousness and its epistemic limitations is what is causing the myopically fuzzy distortions in our perspective. We can't back away far enough to get a decent view of the situation to see an easy resolution. If we clearly define those high-level laws we will see how they are the very determinants that have brought us to this event horizon of circularity that may be stated thus: our consciousness has evolved as a function of the conditions on this planet and may be totally unique in the universe. The psychological laws and laws of logic under which it operates are a function of that type of consciousness. When this type of consciousness interacts with the universe self-referentially, as it has been determined to do by the universe, it inevitably reaches an epistemic barrier where it is confronted by the fundamental problem of subjectivity: we cannot get "outside" of our subjective, self-referential perception of what is real, including this perception that we cannot. Any final law of everything, therefore, must predict and explain not only our peculiar consciousness as such, but the apparent self-referential modality by which it operates.

Do We Really Know Nothing
of the Laws of Our Consciousness?

When we speak of the "laws" of our consciousness we must carefully distinguish between the law or laws by which our consciousness operates or manifests in operation, and the conceptualization of the term as held by those who believe that the "laws" of physics will eventually explain with mathematical precision how consciousness is generated by the physical universe. By way of analogy, we have discovered and put into mathematical formula the "laws" of how gravity manifests and operates even though we did not know what gravity is in its fundamental nature – or how it was generated by the universe, if it was. I submit that we know a great deal of the law by which our consciousness operates, as we will discuss below, although we are only now beginning this discussion about what its fundamental nature is and how it is generated by the universe, if it is. I also submit that we will probably have to recognize that consciousness is a "fundamental" of this particular universe just as we have had to recognize that gravity is.

There is a widespread belief that we know nothing of any laws by which our consciousness operates. I find that amazing. I set aside any theological bias which would see any notion of a law or laws of consciousness as deterministic and therefore disputing free will as ulterior prejudice. I hold in abeyance any scientific bias which would claim we do not know the laws by which our consciousness operates because they cannot, at least so far, be expressed in terms of known physical laws as prejudiced by that assumption. When we examine the general situation with regard to our consciousness, I submit that the belief is held in error. All of the schools of psychology that claim to be scientific are based on operational principles which are considered to be objectively fundamental and universally applicable to human nature. That is a good definition of a "law". . . Even the Phenomenologists, who claim that all we can know of reality is our subjective stream of consciousness, are implying that that is the way reality is, that is the "law" that governs reality. Freudian and Jungian psychology, even though not advancing strict deterministic rules, are built on the assumption that there are certain mechanisms of the subconscious and the conscious, some negative and some positive, which are identifiable operating in the human psyche. Even though these operational principles are usually utilized to analyze and predict mental, emotional and social phenomena for the treatment of psychopathological conditions, they can also be understood, regardless of whether they are correct or not, as, at least, proposed rudimentary systems of laws of our consciousness.

When one revisits the philosophical schools of thought from the beginning of known history to our time, whether they fall in the time honored categories of Rationalism, Subjectivism, Existentialism, Phenomenology or whatever, a fundamental assumption is that consciousness can be understood, that it has structure, boundaries, identifiable modes of operation, potentials and limitations. Although they may radically disagree as to what it is, all these approaches and perspectives work against some definite conception of what consciousness is essentially, and can do and cannot do. Although often very informal, these assumptions of rules of one kind or another may be taken as rudimentary descriptions of the laws of consciousness.

More narrowly, the structure of syllogistic logic has been developed on the assumption that accurate conclusions can be drawn if the proper rules of thinking are followed. These "rules" of logic are considered to be objectively embedded in nature and our consciousness as part of nature. Again, from another perspective, these are purported minor laws of consciousness. There are the clinical operational principles at basis for therapies which make a fundamental assumption about what is normal and right and, therefore, lawful. There are the stiff rules of formal logic and the precise rules of mathematics. There are the generalized conceptions of consciousness and its operations embedded in the schools of philosophy in the west. We find, turning East, that there are venerable and robust systems for the positive development of human consciousness which exhibit rather elaborate rules of "right" consciousness. We need only to mention those three systems rediscovered with a good deal of excitement and delight by the Sixties generation: *The I Ching: The Book of Changes*, *The Book Of The Dead*, and *The Book of the Tao*.

Consider the following analysis of the I Ching as an example of a well developed, holistic system of the operational law of human consciousness.

Exhibit One: The I Ching, a Sixty-Four Gestalt Hologrammatical Analog of the Dynamical Field of Human Consciousness

Is there a simpler way of saying that? Yes, The I Ching is a well developed model of human consciousness and how it constantly changes.

For some six thousand years of continuous history we have possessed, from the ancient Chinese tradition, a well developed system of human psychology called the I Ching, the Book of Changes. It purports to be an advanced method for mapping and getting in touch with both the ordinary, more profound and higher dimensional elements of our consciousness. It is explicitly based on "chance" as its operational modality. This, by inspection, may be understood in "modern" terms as being based on what we now call chaos and complexity theory. Terence McKenna has been able to create a computer program (Time Wave Zero) which models the I Ching accurately and can be used as a predictive algorithm for the future. My conviction as to its power and accuracy is based on my own experience working with the I Ching for the last thirty six years. It may be easier for the Western mind, now beginning to free itself from the godspell mentality, to appreciate and accept its validity as the injunctions against anything the opinion of the Church held to be "occult" or paranormal or "magic" are ignored or forgotten. I suggest that it is a highly refined and deep well of insight. It also is a very strong, coherently integrated self-reflexive system – thereby mirroring the mind of which it is a dynamic model.

The Philosophical Basis of the I Ching

The I Ching, based on the philosophical premise that the only permanent thing about the universe is change, has been used as an oracular system, teaching one how to live most consciously and well on the level of opposites (good/bad, beautiful/ugly, true/false, etc.). It purports to predict, if one pays attention to the positive elements and negative elements in any given current situation, how that situation will transmute into a particular situation in the future. This is its most important feature: it can accurately determine, predict what will be the future resolution of any given situation from the elements which are internally undergoing transmutation within the present one. This presupposes that there is a law which determines and governs our consciousness, that it is identifiable and that it can be comprehended and practically applied. It specifies that there are 64 hexagram/gestalts and not some other number of identifiable gestalts which are fundamentally adequate and sufficient to differentiate and specify the states of the human mind. (If we evolve to habitual four-dimensional consciousness then, I speculate, we would have to expand the I Ching system, perhaps adding additional components to the hexagrams to mirror the additional dimensionality of our consciousness – although the "secret" of the I Ching is that it already integrates the effect of the fourth-dimensional component implicitly.)

The I Ching as Oracle

The oracular nature of the I Ching system is employed by using the ancient method of division of yarrow stalks or by tossing three coins six times to arrive at the sums (6, 7, 8, or 9) that determine the nature of the six lines of each hexagram, built up from the bottom, line by line. There are only two kinds of lines: a yang, solid line is either an unchanging 7 or a changing 9; a yin, broken line, is either an unchanging 8 or a changing 6. Any line that is changing simply changes into its opposite: a solid line changes to a broken line, a broken to a solid. The hexagram is determined by the chance of the yarrow stalk or coin toss methods and is intended to indicate the current status of the situation that is the subject of the user's question to the oracle. The changing lines of the hexagram indicate the elements within the situation that are active and influencing the situation, causing it to mutate into a second hexagram, gestalt, that will be the future, resultant situation or outcome.

The inevitable question prompted by the nature and operation of the I Ching system is, obviously, How could the originators of the I Ching even conceive of the chance division of a bunch of yarrow sticks or chance throws of three coins being able to describe the precise situation in answer to a question posed by the user, much less predict the future outcome of that situation? I do not claim to be able to answer this question definitively. I can, however, say what I have learned that has proven out over time.

There are several interrelated principles fundamental to the I Ching which can be expressed in modern terminology:

1> The system is built up from the concept of primitive distinction. By primitive is meant the most fundamental operation of the human psyche at basis of all perception and comprehension. By our nature, unless we discriminate one thing from another we do not comprehend anything. Even discrimination itself must be discriminated from other things and actions for us to recognize it and other things. We comprehend by distinguishing, by discrimination. The I Ching, working on that basic assumption, takes it a step further: it says that, as soon as one discriminates any thing as such, its opposite is immediately discriminated. If one discriminates cold, hot, as its opposite, is automatically discriminated, distinguished. If one distinguishes truth, so false is automatically distinguished. Since a thing can have only one opposite, its logic goes, things may be seen to come in twos. This is a foundation of the I Ching, based on empirical and experiential observations over thousands of years.

2> The I Ching is a closed, self-referential system; each hexagram contains within itself every other hexagram and each hexagram expresses the nature and dynamics of the entire system. This structure mirrors the image of the human psyche held by the developers of the I Ching. They had a full appreciation of the universe and the human being as part of that universe as lawful, operating according to the laws of chaos and complexity. But they also realized that the most fundamental characteristic of human nature was that we identified ourselves as being self-referentially aware and, ultimately, had to rely on our subjective perception and comprehension as criterion of truth. A system that modeled our mind's operation therefore would be adequate if it modeled its essential self-referential characteristic.

3> To reiterate, the I Ching is explicitly based on chaos and complexity theory and, since it is a dynamic model of the human psyche in operation, this means that it assumes the human psyche operates on the basis of the rules of chaos and complexity also.

4> The I Ching assumes action at a distance is a fact. The most adequate metaphor we currently possess for what the I Ching calls "Influence" is that of the morphogenetic field as theorized by Sheldrake (*A New Science of Life*) and the mental influence of the human mind on electronic machines as documented by Jahn and Dunne (*Margins of Reality*). Both studies present evidence, controverted and resisted by our current scientific establishment, that information is transmitted to other minds and objects at a distance, effecting information modification and increase and physical change. Although we are now only beginning to tentatively even experiment in this area and tend to deny even the possibility of such phenomenon, it has been a robust concept for thousands of years within the Chinese ethos.

5> The reason that action at a distance could be taken for granted is that the Chinese scientific view also assumes the existence of higher dimensionality. In fact, it is this higher dimensionality, conceived of as both higher intelligence and higher geometry, that the I Ching is designed to enable the user to contact through the self-referential modality using chaos theory technique.

6> There is a certain hologrammatical structure inherent in the I Ching. Due to the fact that it is essentially a self-referential system, it is fractal and hologrammatically expansive so that it contains temporal cycles, the length of which (three days, ten days, three years, ten years, etc.) are natural periodicities arising directly from its structure.

As a result of having recognized these cycles within the written commentaries on the I Ching and programmed them into his Time Line Zero computer program, Terence McKenna has been able to overlay them on human history and has found startling and precise correlation with the major events and turning points in human history. If we put together all these elements then an answer to the fundamental question as to how the I Ching could conceivably work at all is at least possible. The user, in throwing the coins at this moment, under this set of circumstances, with this specific mind-set and intention and question, altering the physical conditions immediately present in just this specific way, is, in terms of chaos theory, effecting the initiating conditions of the situation with the coin tosses in such a way that it is at least conceivable that a kind of attunement with one's own and general higher consciousness and with a higher, probably fourth, dimensional geometry is achieved with a degree of relative accuracy so that the hexagram produced resonates from and with all those factors. To put it in a much simpler form: the I Ching is a psycho-physical mechanism intentionally structured to put the user in resonance with the space-time dimensionality of his or her current intention/question. From the space-time perspective, although we think in linear, sequential cause and effect, all events and phenomena are "simultaneous" and mirror all other events fully – just as the I Ching, in slow motion in consideration of our "slow motion," sequential thought processes, demonstrates how all 64 hexagram/gestalts flow one into the other. As a result, the hexagram produced as an answer to the user's question, even if one only grudgingly allows subconscious anticipation on the part of the user, can trigger the proper understanding and act as an entry into the system itself. The hologrammatical richness of the self-referential nature of the system, in which each hexagram contains the potential to become all the other hexagrams, can take one to the transcendent level of the perspective of four-dimensional consciousness regardless of through which door one enters. Is the I Ching an accurate mirror of our consciousness? Based on my own experience with it, I must answer Yes, on the level of opposites. And its nature and structure makes it seamlessly linked to a psychic tourist guide like the *Book Of The Dead*, and for the transition to more expanded states of consciousness as dealt with in the *Book of The Tao*.

Is it comprehensible immediately to almost everyone? I would have to answer No – although the inherent attunement with the human psyche and the almost poignant names of each hexagram dictated by the system itself make it maximally resonant and intuitive. But, because the system is as rich and deep as it clearly is, it must be learned as a whole, the principles not only understood but patiently observed operating in one's own psyche. It has been important for me to also experience what the limitations of the I Ching are. There is a certain level of consciousness at which it drops out. It is structured and fine tuned to express the laws of the psyche operating at the level of opposites and to mirror the subtle, constant changes as one moves from gestalt to gestalt, situation to situation, at that level of consciousness. But when one moves to expanded, relativistic awareness, the I Ching, as expressed in line four of Hexagram One, The Creative, enters into realms that are uncharted – at least by its context. At that point one moves into the flux of the *Book Of The Dead*, the guidebook for the transition past opposites to the highest levels of consciousness we can attain at this point in our evolution – although, by the very fact that we can at least conceive of there being levels of consciousness beyond the most expanded we can attain now, we are on our way there. The important point here is to recognize the limitations of the I Ching: it is as correct and powerful in the three-dimensional context of our consciousness as it is because it recognizes the involvement of a fourth-dimensional influence, but it does not address our consciousness operating directly in a spacetime mode as such. *The Book of the Tao* does that.

To summarize: I submit that the I Ching is a well developed, comprehensive schemata of the human psyche's range of possible actions and reactions at the level of three-dimensional consciousness, the habitual, usual level at which we comprehend and communicate by discriminating and naming discreet phenomenon in terms of opposites. Even more succinctly: the I Ching is a powerful, dynamic system expressing the operative laws of human consciousness. Since it explicitly deals with the three-dimensional realm of discriminated opposites in lineal time as its primary focus, it may be seen to be analogous to the Newtonian laws of physics. Are there laws of human consciousness that apply in relativistic or quantum mechanical terms? Yes. And we have indication of those already. We will speak of them later, but the point here is that we have laws of consciousness already known to us. I suggest that the I Ching actually encompasses the basic operational laws of our consciousness quite well. Not only can and have they been computerized in a robust system, but they are amenable to empirical verification through rigorous psychological studies: assuming that we have an objectively oriented tester who knows the I Ching in depth, is familiar with its system of metaphors, knows its cycles, and understands its basic philosophy and operation, numerous subjects would be trained to use the I Ching as an oracle and the hexagrams received as answers to their questions, and the advice, admonitions and praise received through the changing lines in the hexagram would be compared to the person's situation as seen and agreed on by the user and the tester.

The resultant hexagram would then be compared to the outcome of the user's situation in question to see if it really mirrored the outcome as the user and the tester agreed it to be. I would predict that the correspondence of the hexagram received as an answer and its internal changing elements would seem to grow as the tests progressed, especially if the user was new to the I Ching. This phenomenon would manifest since the I Ching, as a feedback type of system, would also be instructing the user as the test progressed, and the learning would allow for greater subtlety of comprehension and relation to what would probably be a somewhat foreign metaphor at first. Obviously this test scenario only suggests a format and would probably require a much more rigorous approach. But the possibility of rigorous testing clearly is available. The I Ching is but a single example of a rich and developed system that mirrors human consciousness in its full operational range.

We Are Not Completely Ignorant
of the Laws of Our Consciousness

We are so "close" to the laws of our consciousness that we cannot generally see them as such. We may recognize that there are principles of nature by which our consciousness operates, but we may not want to acknowledge them as such because we know that we might have to face questions concerning whether we had free will or not, or that we might have to face some conflict with our "religious" beliefs. Parenthetically, we might note the position of someone who claims the Bible is an adequate expression of the laws of human consciousness. I maintain that it is almost diametrically the opposite. It embodies the rules which were laid down by a Nefilim master on his human slaves. They are not the fundamental principles by which human consciousness operates, but the oppressive, coercive rules enforced for the convenience and service of Yahweh.

It is most amazing that any intelligent human being, who is the very product of the rules and parameters of human consciousness, can say that we know nothing of the laws of human consciousness since we are prompted by, operate by, are limited by, understand and judge by those rules unceasingly. We cannot escape those laws of human consciousness. Just because we have not definitively codified them in their last fine detail does not mean either that they do not exist or that we are not aware of them. That the laws of our consciousness, quite obviously, involve chaotic behavior and elements of complexity difficult to identify is trivial. But, just as we have identified and worked through chaos and complexity theory in other areas, we will work through those elements operational in the nature and operation of our consciousness. If we know the laws of human consciousness or, more precisely, that the laws of consciousness are already substantially or partially patent to us, then we should vigorously commence to determine their full extent.

I have said above that the laws of our consciousness cause us to hit an epistemic barrier beyond which we cannot pass, preventing us from deriving a LOE unless we also explain ourselves explaining ourselves and predicting ourselves to ourselves as part of it. What is that situation at the epistemic barrier?

We see that the way we conceive the universe and its laws is peculiar to our type of consciousness; we realize that we cannot determine the truth or reality of anything except subjectively including the LOE and that the LOE, as we conceive of it, is a function of our type of consciousness; yet the LOE must, by definition explain us and why we conceive of the LOE in the way we do; that, at the extreme of circularity, we can no longer use linear logic to reach a conclusion, we have to give in and accept that it all comes down to our individual, subjective perception and comprehension and judgment of even logic's validity itself. It is so extreme that even our understanding of our consciousness and awareness – even our understanding of our understanding is subjectively determined. And that's where we are blocked. Our consciousness by its very nature is meant, apparently, to be where the buck stops for us: one would probably not be able to operate, at least under the conditions on this planet, if consciousness was an unrestricted infinite regression or progression of judgment calls on each item discriminated from perceptual impute.

If we use reason or any modality which, by definition of the LOE, is a product of the LOE, to arrive at the nature of the LOE, what does that indicate? Is it possible? Is it possible, indeed, to use any other method than what is determined by the LOE? Apparently not. It might be safe to assume that whatever method we use by the nature of things, gives us a clue to the nature of the LOE. The ultimate method or recourse, perhaps we should call it, that we are forced by the nature of things to fall back on is our own subjective, self-referential, awareness and judgment. Our self-referential judgment pronounces after (hopefully) it has considered the consensus of the senses, the input of intuition, the vectors of the emotions, the arguments of reason. We should, therefore, consider that self-reference is the most significant key to the nature of the LOE since we cannot get beyond it.

Self-Reference as the Key to the Law of Everything

Is it correct to assume that we have learned just enough about the universe to impose an exponentiated version of our self-reflexive type of consciousness on all-that-is as an adequate model of its nature and the "laws" by which it operates, or are we just rolling over psychemorphism on top of psychemorphism? When would we be able to say that we had used self-reference so perfectly that we were able to be completely certain that self-reference was the most fundamental character of the universe? So far, the only way I can see is that only self-reference will tell. . . The reader should not conclude that I am deliberately attempting to toy with her or his mind: as I stated at the beginning of this chapter, I repeat the same inescapable conundrums over again in many forms here to thoroughly acquaint the reader with the nature of that strange point where we cannot get behind or outside of our consciousness, we cannot anticipate ourselves, we cannot even be aware without being aware.

The Clues in the Web

All of this thinking is transparently a product and function of our type of consciousness, our type of perception, our type of awareness. We might conclude that our consciousness may be so limited, so primitive, so inadequate that we can literally not even come close to forming any kind of concept of how the universe works. But we can form the concept of a law of everything and a consequence is that we, sooner or later, will register the thought that our consciousness is a direct function and product of that law. If so, then our experience of that law is, in a real sense, an experience self-referentially of ourselves. That is the most direct experience and conscious awareness of anything that we know. Rather than conclude that we might not even be able to conceive of anything close to the law of everything, we can see that we cannot escape a direct experience/awareness of it, ourselves. Although this might easily be judged to be a partial awareness of a law of everything because our consciousness is clearly limited, again we can only go with what we know.

A simple point should be clarified here. I know of the infra-red portion of the light spectrum but I have no sense organ to "see" it directly like a snake does, yet it seems clear that the full perception of the "reality" of any physical body should include this particular radiation – as well as all others in the electromagnetic spectrum – as an integral component along with the portion of the light spectrum available to us. This tells me that, at least, my perception of any physical object is limited and is a function of the type of sense apparatus I possess, literally am. So we can understand, apparently, that we have limited and relative perception. But I mention this here only to point out that the problems do not lie on that relatively superficial level. They lie at the very root of our consciousness, in that non-space where we cannot get behind or outside of our own immediate awareness, where we push and probe the thought about our thought, get caught in the self-referential, infinite regressions, attempting to find a more fundamental criterion for the criterion of our criterion. We cannot, apparently, (I don't seem to be able to tell, except subjectively) get outside of our own perception and conceptualization any more than we can get outside the universe. Having dealt with the above mentioned epistemological problems over my life in a very intense way, I do not see a way through or around them in any way that we have tried to this point in the last two thousand years. Although it is trivial to postulate an order of reality as objective, existing in a precise way independent of any mind, the problem is how to determine what it is and to know it except in terms of one's general subjective experience and specific neurological impressions and evaluation of it. Especially when one's general subjective experience tells one that, indeed, there seems to be an objective order outside one's mind.

The Law of Everything and Human Freedom

It is with this last point that we come to the most fascinating potential of self-referential systems. It would seem that a completely closed system for which one can determine a unified field law, a principle which explains everything about the system including itself and its operation, would imply an absolute determinism and no freedom, at least as we think of it in terms of free will and creativity. But it would seem, after some reflection, that, although the complete system itself, the universe in this case, would be quite absolutely determined, there is still potential for its parts, though determined, to be determined in a paradoxical way. Our self-referential consciousness gives us the potential, individually and collectively, to evolutionarily determine our own determinism. Having said that, it is necessary to recognize that there seems to be an inherent contradiction in that statement. But, as in the case of the circular logic above, we may take this as a clue to the nature and degree of freedom of the system rather than be stopped by some traditionally inadmissible syllogistics.

At Play in the Unified Field

What do we do when we have deciphered the universe, written the full unified field equations for the physical realm, determined the laws of consciousness and integrated them with the fundamental physical principles and realized self-reference as the meta-metaphor of the LOE? It's inevitable: we will begin to play with and within the universe. At first, it is also probably inevitable, we will sometimes play harmfully and destructively. But the power suddenly in our hands will be far greater than any atom bomb in destructive potential and we will realize that even one mistake could be our last. Our psychology will change profoundly, we will mature as a race rapidly and those in the universe who have already attained understanding of the Law Of Everything will have no choice but to warn us – as perhaps they already are beginning to respectfully do – as we warn children in advance about the potential danger of a powerful tool used incorrectly. But mature and learn we will, becoming responsible citizens of stellar society. Learning and living with the knowledge and potential of the Law Of Everything is the second most profound god game we will play for some time to come.

CHAPTER 15

THE EDUCATION OF REALITY CREATORS

"The most successful tyranny is not the one that uses force to assure uniformity but the one that removes the awareness of other possibilities, that makes it seem inconceivable that other ways are viable, that removes the sense that there is an outside."

Allan Bloom
The Closing of the American Mind

Some teachers have extraordinarily flat personalities because they have spent their entire teaching careers pressed between the pages of the syllabus.

Ema Klinger
The Schoolroom Soaps Yesterday and Today

We still matriculate our young, these amazing star-children, semi-illiterate and naive for fear of them questioning our shambling senilities.

John Brokencrock
Diary of a Crazed Principal

With the demise of godspell religions and the dawning of a generic definition of human nature, the educational process will undergo a radical revolution.

In the profound shift we are undergoing to the new paradigm, it is quite easy to see that we are in the process of redefining ourselves as humans. Our children will be redefined in that process also. They are critically important enough to require a discussion devoted entirely to them. They are ourselves in the future; they are, literally, the future of the race. It will follow that, as we redefine the children, we will have to redefine the entire educational process meant to prepare them for life in the context of the new paradigm. In order that we redefine the educational process correctly we need to revisit and understand the process as we know it today.

The Educational Status Quo

The educational systems extant in our time are the products of thinking that is already far out of date. The religious and sectarian schools educate and program their charges into the old godspell syndrome. True, we have education for all rather than for just the elite or those who are wealthy enough to buy it, but the public school systems supported by taxes is rapidly degenerating because of two major flaws.

The most basic flaw is sourced at the top of our society. The slow agony of the American educational system is Constitutionality.

Jefferson and his friends worked out a set of rules under which those holding completely opposite understandings of what a human being is and what purpose there might be to human life could live in proximity with the least ideological and physical mayhem. Jefferson et al, it appears, never conceived or anticipated a time when a common consensual definition would be arrived at, thereby superseding the Babel factor with a new paradigm of generic humanity. And that was no mean feat; the Constitution stands as an evolutionary landmark for the planet and we salute it.

Consider, however, how adherence to the principles upon which this country was founded effects the performance of the teacher every moment in the classroom. The teacher who is conscientious realizes, at least subconsciously, that whatever approach is taken to almost every major topic dealt with in the classroom may bring criticism from some parent or group as contrary to their personal belief system. The Constitution has succeeded as obviously the most advanced attempt to create the most comfortable way disagreeing groups can coexist in a pluralistic society. But that does not mean that it has succeeded in the precinct of the classroom where the parent members are not usually present, their charges of interest are not expected or allowed to have a say, and the single individual hired to preside over this amazing phenomenon is by contract expected to exercise routinely consistent neutral genius to intuit and never transgress on any and all parent member's private principles – while educating in depth – for a modicum of pay. The President of the United States is expected to act in an unprejudiced way regardless of his or her religion in matters of Church and State. That's a piece of cake compared to the day to day teacher's delicate dance.

The focal problem for the teacher, particularly the good teacher, today, is the fear that whatever significant SHe teaches will be perceived as infringing on or threatening the religious views of some parents of the children in the class, leading to complaints, charges, harassment, and unpleasant confrontations. One parent's religion is another parent's curse; one parent's philosophy is another parent's psychosis; one parent's science is another parent's abomination. There can not even be any fundamental definition of what a human being is within that context and you are supposed to nurture the flowering in the child of the fullness of that which may not be defined. That's stressful pressure on particularly the good teacher, pure and simple, and the most creative ones too often succumb and fine creative ways to leave the educational system. That is the child's loss. In the earlier times in this country when there was not quite the diversity or the rejection of institutionalized religion or, lately, the intimations of a resolution with the advent of the new paradigm, the pressures on the teacher were, perhaps, not so acute. But the godspell ethos was and is at root of the conflicts based in authoritarian theological absolutes that require a powerful and ingenious Constitution to regulate the situation.

The public educational system that has evolved over the history of our country is based on a concept of democracy that has taken the form of primacy of equal opportunity under the powerful influence of John Dewey, one of the best known educational philosophers in American history. The result has been a system that focuses on the basics, tends to be geared to educate the average student while the below average and the advanced student tend to be dragged up or down to the middle or neglected. Students that have not done the work or qualified are often passed through the system rather than being made to repeat a year in school because it is believed that it will traumatize the child. The curriculum is often quite pragmatic, teaching the child the skills that will be needed to read traffic signs and fill out applications, read instructions, do the calculations necessary to handling money, getting paid, paying bills, making change, using a tradesman's rule. Social studies teaches one to vote and participate in society as a docile citizen. If, currently, industry needs scientists, then science is encouraged in the schools and scholarships are awarded accordingly. Students are shaped into new members of the military-industrial complex, hopefully with the basic skills to participate competently. Philosophy or, more accurately, the history of philosophy is taught only at a higher level: the school does not teach the student to think independently for HIrself and avoids anything that could be called "valuing" for fear of being called into question by an objecting parent or institution. Neutrality of this degree is a straightjacket.

In addition to neutrality, the second basic flaw is that the educational system is still working according to the rate of information transfer we knew in the eighteen hundreds – linear, packaged, slow to change and include the new. Pedantic and authoritarian, divided into isolated categories, the system presents information to the student far too often in an unappetizing and meaningless fashion. Ultimately, the unspoken but implicit purpose is to educate the child at least to the minimum requirements for functioning in a pluralistic society.

Conflict over what a human being really is has given rise to the proprietary conflicts of sectarian interests and the constant effort of those parties to control or restrict whole segments of information not deemed true, sometimes considered evil, or, at least, not in their interest. The conflict between the Creationists and the Evolutionists in our time is a classic example, as these two factions battle for the minds of our children in the school textbooks.

Genetic enlightenment will cancel that general situation. Certainly there will be disagreements over policy in any area but they will be incidental, not fundamental. The managers of a corporation building computers may argue and disagree over how to perform a certain operation but they never disagree on the fundamental nature of the company and its product and what its function is. If we know what and who we are, then the proprietary control, suppression, destruction, or withholding of information on the basis of a religious or sectarian belief system will be eliminated. In contemporary society, even in America, information will be treated as an essential need of the individual and valued as such. There is a faint glimmer of this coming world situation in some of the contemporary philosophizing of a segment of the computer elite. In the minds of some of the more enlightened students of advanced media, information should be free and available to all, not treated as proprietary or as a commodity. But that will not come about until we also have created the true leisure society on the basis of an unlimited supply of wealth.

The school and the teacher will be unshackled. Teachers will again be free to deal with fact, truth and conjecture, with science, philosophy and profound questions as well as sex education without fear of a parent protesting that their religious belief and value system has been attacked and their child harmed. The school will be able to deal harmoniously with parents not only on subject material but on sickness, neurological problems, learning disabilities and social problems of children without stigma attached or parent protest because of their private value system or religious philosophy being brought into question. The school, unshackled, can be a place where all share the same definition and understanding of what a human (child) is. An unshackled system can work toward common goals, will be able to identify social problems, neurological problems, insanity and criminal behavior because a common definition of these aberrations and attitude toward them exists. Once identified they can be separated from the school as such and treated or dealt with. Students who do not learn or perform for whatever reason will be identified and worked with specifically because the idea that a child is traumatized when singled out, held back, corrected or challenged will have died out.

The quality of teaching will exponentiate because gifted teachers will again be found in the school operating in a comfortable freedom of creativity. In general, teachers will be older than they are now and will continue working because the good ones will be recognized for their wisdom, as well as their teaching ability, and will be sought out. Will any of this come about soon? Probably not, realistically, although it may sooner than we would imagine at this point.

The new human society will not be plagued by the theological absolutism that causes these current problems, and the good teachers will again be found in the classrooms from which they are gradually retreating in our time because they become discouraged because they cannot teach effectively.

Education of the New Human in the New Society

The entire education of the new human child will be toward freedom, independence and self-determination, preparing the child fully to be HIr own evolutionary artist, HIr own reality creator. Children will not look on school as a dreary wasteland that they must tolerate but the vital and exciting arena where they can expand their minds because the informational content will not be filtered, sanitized, manipulated, or restricted to an impossible and artificial package of desiccated platitudes specifically designed to not infringe on any belief system. Education will not be confined to producing humans who can at least function in supermarkets and on highways that require a minimal literacy or who can qualify at a certain job skill level. Those desperate gambits on the part of our educational system as it reflects our societal demands will be unnecessary. Education will be positively oriented to consciousness expansion in all dimensions. An indication of that concept is found in some of what has come to be known as New Age thinking, although that philosophy is cast in terms of what humans and human limitations are today and is only a glimmer of the robust things to come. This kind of education will require very efficient, intelligent and very high speed means of information transfer.

A significant educational development will be virtual reality simulations of an extremely realistic nature through which the individual or groups will be able to explore any area of known reality and even invent realities to investigate and experience. An advantage of this technology will be to afford the individual or group the opportunity to explore and experience areas and topics which are dangerous, generally forbidden, harmful or lethal, harmlessly, while sophisticated programming presents real world psychological, physical and ethical consequences of these scenarios to be demonstrated and experienced. The individual will be able to contend, argue, even attack in this virtual reality and the "computer" will play the part of the race and the social context or whatever designated role the player assigns it. The news from the science community is that this technology is already very close to being immediately available.

We already have before us the eventual refinement of electronic technology that will lead to bio-electronic capabilities to make micro information data bank implants linked to the brain. This kind of "instant" education may well make the individual receiving the tiny device the equivalent of a Ph.D. in a given subject, an expert in the field in five minutes. The educational process may never be remotely the same from that time forward.

These are only two examples of technological advances we can anticipate in the very near future. These scenarios I have sketched are inadequate and will rapidly be outmoded. What seems to be only the most advanced science fiction today clearly will become reality, and then antiquated, more and more rapidly. We should view these future possibles against the background of the known rate of development of the common consciousness and technology we have witnessed to date, and attempt to estimate the way that rate will accelerate rapidly as we go into the next century. We will rapidly see the development of technologies in the first part of the next century far beyond even what is frequently shown on the television documentaries as still all but science fiction. But technological advance will only be a tool, the focus on consciousness expansion and being in charge of one's unique personal evolution will be primary for the new human child.

The curriculum courses in the new society's schools may carry titles such as Anthropohistory, Genepool Navigation, Experiential Evolution, Virtual Reality Creation, Exopsychology. The concept of conscious genepool participation and navigation is particularly significant relative to the education of developing evolutionary artists.

The Gene Pool as the Primary Social Unity of the Post Godspell Social Order

The primary collective social unit of the post-godspell social order beyond the family will be the gene-pool. "Gene pool" is defined in Webster's as "the collection of genes in an interbreeding population that includes each gene at a certain frequency in relation to its alleles." "Allele" is also defined as alternating Medelian characteristics (such as smooth or wrinkled pea seeds)." Which does not really tell the layman who has not formally studied genetics very much, actually. It simply means that a gene pool is a collection of actual genes in a person's DNA that determine that person to exhibit certain characteristics similar to other persons who have the same collection of genes in their DNA. In this discussion, however, I am going to use "gene pool" in the same sense but with the focus on the person and their characteristics, rather than the genes themselves that determine those traits. In popular speech this has come to be an accepted terminology. So, when I use "gene pool," herein I will be referring to the individuals who naturally congregate, associate, and celebrate together – "birds of a feather flock together" – because they are programmed similarly.

Am I saying that genetic programming totally determines us, that we have no real free will, that we couldn't do anything else but that which our genes determine us to do? No. Certainly not: and even if we thought they did, part of that programming clearly is a mandate to determine and alter our own programming. The notion of being determined by one's genetic makeup should not be taken to signify a seemingly mindless "hive" mentality: just the opposite – although some humans seem to act that way in spite of themselves.

The new human society will be characterized by conscious, aesthetic, easy, natural associations of those who recognize each other on the basis of genetic programming, common life goals and interests, and common functional orientation. I believe we may see a gradual geographic relocation of groups that will supersede the artificial boundaries of nations. Obvious examples of major gene pools are those whose genetic proclivities prompt them to go off planet into space as contrasted to those whose genetic disposition prompts them to remain on earth to turn it into a park, a great place to visit and a great place to live. Another gene-pool is now forming of those who will choose to die in their time, and another of those who are oriented to immortality.

The Genepool Golden Rule

These are not just divisions of groups with different worldviews on primarily a philosophical basis. The natural developmental processes on a planetary scale tend to create various functions and focuses analogous to the beehive and, therefore, distinct types. There are and will be philosophical differences between gene pools but, in the new human society, those differences will be recognized by all as a function of genetic orientation and honored as different perspectives and as superficial relative to fundamental genetic orientation. These different perspectives will be understood as analogous to schools of artistic approach, each of which emphasizes a certain facet of human nature's and nature's potential. There will be a mutual respect and positive recognition between gene pools beyond just tolerance, and the only rule necessary will be a consensually agreed prohibition of any gene pool doing anything that will harm or obstruct the function of another gene pool. The gene pool golden rule.

The major difference between the way gene pools are viewed now compared to the way they will be understood in the new human society will be that individuals will be very conscious of their being of, or having an affinity to, a particular gene pool rather than other ones at a given time in their life, that they can and may and will change evolutionarily genetically, and that moving from gene pool to gene pool will be a natural function of their unique personal evolutionary process. They will understand that those of other gene pools are not strange or wrong, only of a different orientation.

The godspell mentality could only lead to artificial groupings of humans on the basis of ideology in the form of dogma, theological belief systems or intellectual philosophical convictions. The boundaries and characteristics of nations developed over long periods from tribal tradition and regional control. But conscious gene pools will gradually dissolve all these ancient divisions as the individual becomes a citizen of the planet and seeks out those genetically similar to HIrself. When the godspell is broken by an individual the artificial ideological ties are broken and the individual is faced with a complete reevaluation of not only HIr associations but the interpretation of life in general. This is the reason why institutional religion is decaying by attrition so rapidly in our time. Divorce in our time is prevalent because of the gene-pool disorientation so widespread on the planet. Once the individual breaks the godspell, all too often SHe realizes that the ideological religious and cultural reasons for choosing a partner are false and SHe is prompted to separate from a partner that may now obviously belong to another gene pool. The partner is often seen not as wrong, inferior or bad but of another orientation to life with whom one has little in common. One's "trip" is realized to be different, at that point in time, from the other's "trip," to put it in common slang.

There is a misunderstanding to be avoided in this regard. It is already possible, on the basis of ontogeny following philogeny (an individual's personal physical and mental development recapitulates the evolutionary phases of the race), to trace the development of the human being recapitulating the stages of primitive organisms that have been thought to have marked our evolutionary ancestral development on this planet. The human fetus does go through stages where it has gills, is covered with heavy hair, has a tail, and exhibits physical characteristics which resemble typical primitive stages of development. And the child after birth does exhibit some developmental phases which are hardly different than some of the higher primates.

On that basis, some would compare some gene pools of adults to earlier developmental stages. But we must carefully differentiate between truly primitive traits and adaptive behavior on the part of earlier humans. More and more of the evidence both from the new paradigm and traditional anthropology and paleontology points out that we, humans as we know ourselves, have been the same from our very beginning creation by the Nefilim with regard to our general mental and physical capabilities. When we were a new species without a substantial historical experience of the planet, if we hunted to live it does not mean that we were de facto primitive predators at that stage any more than the sophisticated Native American hunters of the North American plains in the seventeen hundreds – or the theoretical physicist who hunts elk in season seriously for food.

On the other hand, as example, if an individual is *psychologically* fixated at the mammalian muscular stage where all social interactions are in terms of the stronger, the physically dominant, and the solution to all social conflict is through violence or conflict then we may begin to talk of psychopathology rather than retrogression. It is when an individual becomes fixated at a developmental phase long after experience of that phase should be finished that we begin to differentiate. But, with the recognition of the difference between gene pools, also comes the realization that one has the potential to move between gene pools as one evolves over one's life span. An antelope will never move from its gene pool but we, being self-aware, can conceive of ourselves as members of various gene pools and follow our genetic tendencies until we find the one we relate to at any given time. As we develop in tune with our own unique complex genetic makeup's we may find ourselves moving from gene pool to gene pool as we consciously create our own genetically influenced realities. It is this freedom and complexity of personal development that makes us more than any gene pool, and ultimately capable of creating an entirely new one. It is our DNA which gives us that potential and prompts us to do so.

The problems that arise between parents and children may be understood better in terms of these genetic tendencies. It seems evident that children, even from the time that they are born, may not be of the same gene pool as their parents. This can be sometimes a real shock, a very difficult thing for parents to understand – even if the same problem existed between them and their parents. A child can be a very different creature than HIr parents and misunderstood by them, and them by the child. This problem is more acute under the old godspell mentality. With arranged marriages and marriage between individuals in older times and cultural and superficial bias in more modern times and marriage on the artificial basis of common religious affiliation rather than a true knowledge of the other's genetic direction and preferences, or even the other's true sexual orientation and attractions, a great deal of tension and cross purposes and misunderstanding can arise. But we will eventually reach a stage of human social understanding where the genetic basis and nature of the gene pool will be thoroughly understood and respected. We will recognize it as a survival mechanism to maintain diversity and balance of functions in the population and that we should be very cautious about manipulating it into a too narrow, uniform or monochromatic phenomenon for our own sake. The genes know what is needed to be robust through diversity. But those genes also are the prime movers directing us toward consciously determining our own evolutionary path.

The Gene Pool as an Educational Medium

A further refinement of the gene pool concept will be its incorporation into the educational process: the stages of the developmental process of the human child will be mapped onto the various gene pools, but without assigning hierarchical ranking to them. A child will have the opportunity to experience the best features of the various phases of human evolutionary experimentation. The quintessence of the hunter, the gatherer, the agriculturist cultures presented by those steeped and expert in the context will put the student with affinity to those gene pools in tune with deep genetic roots. The crafts, industrial, mechanical, engineering, construction, architectural gene pools; the arts, literary, musical pools; the military, police, and civil servant gene pools all have some essence of their orientation to reality to offer as insight into a facet of human nature. The domestic, child-bearing and rearing, cooperative, social, political, educational, pools will certainly all contribute. The futant gene pool will also welcome its own.

A person's education may be said to be fully matured when there are no facets of existence on this planet that could be a complete surprise or, even more importantly, disconcerting. Bizarrely strange creatures from the depths of our seas, perhaps groups of Gigantropithecus or other early humanoid types still living in remote areas of our planet, perhaps dinosaur species supposedly extinct millions of years ago still extant in some deep jungle region, bacteria living in boiling springs, proof of our supposedly solid universe being multi-dimensional and self-generating, clear evidence of action and the influence of thought at a distance, all perceived as amazing and fascinating but not surprising or disconcerting. If some new thing encountered or learned disconcerts, it is a clear sign that one has no information base with which to correlate it or one's information base is too limited or perhaps even incorrect. One's convictions are such that they are threatened by the new information. If information is accurate and true then it should integrate with all other true information. If it does not, either the new information or the previous information is incorrect or one's evaluation of either the old or the new is incorrect even though they are both correct. We define "correct" here as in harmony with all other "correct" information.

Philosophy is a symbol system, a language, a metaphor through which we communicate, manipulate, and explore the nature of what we consider reality. Inherent in the discipline of philosophy is the fundamental idea that logic is its primary mode of operation. Philosophy takes as its subject the entire universe. Since the universe includes phenomena such as human nature, science and even philosophy, then these become fair game for philosophical analysis. But, unlike mathematics, we do not teach philosophy as such to small children and to teenagers in the public school systems because the differences between the variety of views of human nature conflict to such a degree that there can be no common public policy, position or platform from which to proceed. On the other hand, we teach small children and teenagers about the human situation and about reality dogmatically. We foist on them parentally dictated, religious, philosophic and scientific information, often as if it were some absolute, unquestionable truth, in the home or sectarian schools. But we do not teach them, in general, to question or to wonder. As if that were not enough of a ridiculous web of contradictions, the children are now introduced, at a very early age through an electronic media in the form of video games, cartoons and fantasy programming, to a variety of virtual realities. This type of consciousness is prelude to the total relativity of realities as a fundamental premise. There soon will be arcades where they will be able to play extremely realistic games in virtual realities and eventually create ad hoc realities of their own in which to explore and operate. The new human will move past this web of contradictions to an educational modality based on a generic concept of human nature that will afford the child full unfoldment of HIr potential.

Philosophy as Child's Play

The new human child will be taught the mechanics of generic logic early, educated in philosophical thinking commensurate with each stage of intellectual growth and capabilities. But even the discipline of logical thinking and the philosophical processes will be taught in the larger context of the entire spectrum of the potential of human consciousness. Although, formerly, we kept Philosophy from the children until they had reached legal adult age, in the new society it will be taught from the time the child is able to reason because we will have a planetary generic consensual view of human nature. It will be taught as a form of approach, among others, to the nature of humanity and reality, integrated with science, psychology, mathematics, history, etc. Its essential nature will be understood as an expression of a specific type of awareness, self-reflexive consciousness. It will not be restored, however, to the royal position is was considered to possess in the olden times when it was understood as the all-encompassing discipline that subsumed all other disciplines. It will be considered as child's play.

We will eventually teach the child to be HIr own philosopher from the earliest age according to HIr capacity at any age and stage. We will no longer indoctrinate the child with artificial and false theologies, moral, ethical and value structures based on the godspell notion of sin or eternal punishment or reward. They will be taught the nature of reality in the most expanded context we know, introduced to the concepts of action which enhances others while it enhances us, encouraged to make their own decisions about reality and to explore and follow their own unique evolutionary development and to respectfully make their unique contribution to the ongoing conversation about the ideal trajectory of our racial evolution.

I suggest that what is perceived as a problem in the educational system manifesting as "hyperactive" children, often classified as "attention deficient," is a sign of the readiness of the newest models coming off the assembly line for a much more intense and information-dense kind of experiential learning. Many of these children exhibit high IQ's and no neurological damage and, as Tom Wolfe pointed out in a recent college lecture, can zero-in on a Nintendo game for four hours at a time. This is a significant clue. It appears that they can often absorb the slow, plodding, linear and too often mediocre transfer of information in the classroom with a few neurons doing background processing, and the rest of their minds can go ranging on at a much faster speed. Teaching them to use these high-powered capabilities should be a major focus of the educational process.

The Education of Potential Immortals

We should keep in mind that all this will eventually take place in a context of potential or real immortality. The judgments we make from day to day will not be implicitly or explicitly influenced by the inevitability of death. The universal and unremitting pressure that influences our every decision, our every choice, and forces our lives into a few decades of melancholy activity will be gone. Immortality will be a profound turning point; even the potential for it at this early stage is already that for some of us. So the education of the new human child will not be a remote preparation for a short life span in a family and social context conditioned by the random possibility and long term depressive inevitability of death. It will be focused on the well-rounded education of apprentice, potentially immortal, evolutionary artists.

CHAPTER 16

WHAT DO YOU DO FOREVER?

In the satisfactory afternoon
Of bicameral integration
We become our own
Genetic credentials,
Mythic dimensions,
Theopolitique
Merging our planetary genius
Into positive unity.
The godspell is broken;
Let our god-games begin.

Neil Freer
Neuroglyphs

To describe the situation when we are free of the godspell restrictions and limitations I have used the phase "going one-on-one with the universe." "One-on-one" is a sports term, actually in the dictionary, used when a player plays directly against a single opposing player, and it is also applied to a direct encounter between one person and another. I have simply extended it to describe the direct interaction between an individual and the universe and it is very precise in this sense. A person who has broken the godspell is no longer compelled to understand the universe through the interpretive screen of religious absolutes, no longer must take the universe as defined by theological dogma, no longer must take the universe as owned by some peevish and domineering deity. SHe is free to not only interpret the universe of which SHe is an integral part for HIrself, but to interact with the universe directly and immediately, free to seek, inquire, demand of the universe as a part of that universe that which seems reasonable and just for HIrself. This includes material things as well as information, and extends to such things as perfect and perpetual health, the ability to remain at whatever age SHe wishes and, ultimately, immortality.

What Then?

If relatively immortal, in control of our own evolution, living beyond war, want, ignorance and disease, in an enlightened leisure society in a known universe (possessed of a science based on a law of everything) the question becomes not just What do you do as a new human? but **What do you do forever?** That is a profound question and – at this point in our development – a startling threat to some and an exalted challenge and opportunity to others.

We have already examined the potential of immortality, what it will be like

293

to take charge of our own evolutionary direction, what living in a full leisure society might be like. Although we have caught sight of the vision of the new human in the new human society, we have not fully considered what it might be like to live like that forever. We might well ask ourselves, anticipating that that option may be ours in a much shorter time than we would now even guess, Could we envision ourselves living under the most optimal conditions conducive to excellent human living, as we usually conceive of them currently, forever? I believe that it is quite obvious that we would not be able to do it. Regardless of how ideal conditions might be, things would simply get boring and impossible to tolerate. A static situation is not something that humans can take for very long. So far, we have been accustomed to struggling through a short life span without the critical information we need. We are born with only the most basic hard-wired responses for survival, without information about the human condition and how to understand it, dependent on parents who seem to have little information and are almost as mystified about the whole mess as we are and who cling, as we quickly learn to do, to some theology or half-baked philosophy for the direction of our lives. The vast majority of us muck through, rationalizing awkwardly that death must be nature's way, without committing suicide, by focusing on trying to suck some pleasure and meaning out of the immediate, and betting on some borrowed notion of heaven, reincarnation, longevity pills and a quick and merciful death. We can rationalize, avoid, and deny that situation for a short life span, but an immortality of that nature would be a horror. But the change coming will be radical and almost completely opposite to those conditions. That realization prompts us then to investigate what we will really do when immortal, what will be our truly fulfilling and most profound activities. What, then, shall we do forever?

Reinventing Ourselves: Reinventing the Universe

Continual expansion of experience and knowledge is the only thing that does not satiate. We shall not only continually expand our consciousness and information base but, based on what we experience and learn, constantly reinvent ourselves. That may be taken as an accurate, basic description of how we will go about the process of our conscious evolutionary exploration in its fullest sense. Focused transcendental experience, as we have redefined it, will be the primary activity of the new human. Some will proceed slowly, cautiously experimenting in small incremental steps with potential evolutionary directions and possibles over a protracted time frame. Some will not be inclined to much change, content as they are or accepting only changes that have been thoroughly investigated and declared safe by others. Some will seem more adventuresome, exploring more experimental and less certain areas, and the scout futant types among all these will tend to feed back information and evaluations for the consideration of others. In general, all humans will pool their personal experimental knowledge consciously for the good of the race. Some humans are involved fully in this process already. It simply is more difficult at this time because of the lack of consensus about what a human is in the first place, and what the generic human process is in the second place. But that it is already possible is attested by those who go quietly and consciously about reinventing themselves continuously. What forms does such "reinvention" take? Generally two forms: the reconceptualizing of what a human could be and could do with appropriate exploration and experimentation with those potentials, and the reconceptualizing of the universe and experimentation with reinventing it – beginning with the universe we know and extending to other possible universes.

This latter notion of reinventing the universe(s) at first seems like an outrageously arrogant and, perhaps, preposterous idea to even suggest. A bit closer consideration, however, shows that it is simply a full-blown description of what we have already been about in a timid and disconnected way for quite some time: it is called science, currently. If we are already going about trying to improve the human being in general, can talk theoretically and quite casually about genetically inventing a modified human with gills who can live comfortably underwater, are experimenting with physical immortality, have engineered new bacteria, have already conceived of enclosing our sun in a shell of appropriate material to utilize all its energy, among many other amazing things, we should frankly acknowledge that we have already begun to modify/reinvent at least the immediate local sector of the universe. Taking our science fiction as a projection of ourselves into the future, we will continue to inevitably expand this activity and its scope into the universe.

Practical Methodologies for Making the Transition

Real expansion takes place only after we have eliminated the major negative aspects and elements in our psyches. Any constructive change must address the body and the mind. I only mention some tentative ways to get oneself through the changes because, the closer to real freedom one approaches, the more unique and personal the needs and methods that apply become. The entire thrust is to become one's own person, achieve one's unique genetic potentials, one's own identity. So the less anyone can say about how to do so outside of oneself. Only general ideas are advanced here because of that fact.

Various therapies have their place in this part of the process although most, like psychoanalysis, can help one see the problems, but have no technique to get down far enough into the level of neurological imprint to allow for effective change. The fundamental base of all expansion is a focused intent to constantly grow and experience and understand. To this end we have developed a number of techniques over time. Classically, shamanic practices, meditation, yoga, and, more recently and controversially, psychedelic experience, are all tools we have and do use to get beyond our current mind-set and limitations to allow ourselves to expand our awareness and our conceptualization of potentials. All are based, in one form or another, on the understanding that the human being's consciousness can be effected in various ways by chemical induction of various states because it is bio-chemical in makeup. Even the early Christian monks who changed their consciousness, their bio-mental stasis, with fasting and sensory deprivation were operating according to that principle. Some techniques employ the ingestion of chemical compounds that alter the chemistry of the mind and, therefore, its perception. Fasting is an ancient technique; the use of a flotation tank a modern one.

The techniques of Buddhist yoga train the practitioner to control the autonomic nervous system (the part of our brain-neurological system that operates the involuntary responses and systems) and to focus the mind to achieve very expanded states of consciousness with the goal of reaching a steady state of nirvana, a stable rest in the highest state of awareness.

The techniques of Taoist Tai chi and chi Kung meditation train the practitioner to attain a focus in high consciousness beyond the opposites through which we usually conceive reality and to stay focused in a universe in which there is no steady state, where change is the only permanent thing.

The use of higher dosages of pure LSD (lysergic acid diethylamide) in proper set and setting can be much more rapidly and profoundly effective in achieving expanded awareness. LSD is currently illegal in the United States because of the socio-political backlash to the profound threat perceived by the academic, medical, political, police, military and religious power elites. Its effectiveness, nevertheless, is proven and it is inevitable that, when the public and the power elites have matured sufficiently to reconsider it, it will again come into its own. Phases of the LSD experience clearly take one into areas where consciousness can only be said to be directly four-dimensional and the best metaphor and language by which to express and communicate the experience is that of advanced physics. Now, all these techniques and practices are certainly beneficial and effective and take one in the right direction, but they are only a beginning, a gateway to habitual four-dimensional consciousness. Once habitual, the consciousness of humankind will be radically altered and the benefits great. The perspective from space-time extends one's vision so that the evolution of the individual and the race can be viewed as it extends into the future.

Reinventing the Universe?

There is a reciprocal relationship between reinventing ourselves mentally and physically and reinventing the universe around us, the primary reason being that we are integral parts of the universe. It seems very curious that it is necessary to emphasize the fact that a human being is literally, both physically and mentally, an integral part of the universe. One would think that it would be unnecessary to even mention such a fact because everyone should take it for granted. But the real situation is that, although everyone would take it as a given, many do not, in fact, act as if it were true. This is an effect of the doctrines, philosophies, teachings and theologies of godspell religions, the notion of the mind/spirit/soul as an intangible substance that is not physical, gradually sublimated over the millennia to a sort of implicit sense that spirit/soul is not only not of the physical realm but not of the universe at all.

Part of this thinking is the doctrine that the "soul" is "in" the body, leading to the various interpretations that the body, is, after all, inferior to the soul, which should not be tainted by any material thing. Part of this thinking is due to the interpretation of "heaven" as being a state of being, a being "in" and "with" God the Supreme Being, with Spirit conceived of, almost always implicitly and inferentially, as independent and separate from the universe. The reasoning, usually subliminal, seems to be that, if the spirit/soul can be and is completely separate from the universe when "with God" in "heaven" – and that is its most fulfilled state – then it can be considered somehow independent of the universe and its laws even here and now.

What is the Immutable Essence of an Individual?

There is an interesting question raised by this kind of discussion. Some of the latest theory developed by advanced physics is that our universe is one among many sub-universes of a metaverse. If we assume that we will eventually arrive at a profound enough understanding of our particular sub-universe to be able to discover where and how we could somehow travel into another universe, even though it was governed by different physical laws than ours, would we still be a part of this universe intruding into another universe? Would we have to somehow become a part of that other universe and no longer a part of this one? If we did have to become a part of another universe would we still be ourselves, the same identity principal, in order to go there? If we did have to become "someone" or something entirely different would we be dead in this universe even though we still existed in another universe? By the logic and physics of this universe we would not only be dead but non-existent if, indeed, we were to move into another universe, at least in the terms and considerations of our current thinking. What we would be, if we could be, in another universe is purely speculative – even "other" sub-universes are purely speculative at this time – but, again, according to the logic and physics of this universe, we would either be or not be at least the same identity principle, even if in another form, according to a different set of physical laws. This brings us back, perhaps with a bit more perspective, to the question of what that identity principle is in this universe, if it "is" at all.

If we consider the question in the form: What are we, as an individual human being, essentially? we find ourselves in the very center of a number of different controversies, none of which advances an overwhelmingly convincing and definitive answer. We know what we experience ourselves to essentially be: there is that elusive principle that is "I," regardless of whether my body is gradually and totally replaced by new cells over a seven year period, whether I am fully aware or in coma, whether I had gone through a sex change, or had all four limbs amputated, or had suffered complete and prolonged amnesia and recovered, had died and my body put in cryogenic suspension and I was restored to life, or all of those thing together – "identity crises" notwithstanding. We have seen that science is stymied in arriving at a law of everything because consciousness will not yield to a definition in terms of the physical forces and fields that physics will acknowledge. Although it is not usually put in those same terms, the "I" falls into the same category whether we, as do some schools of thought, equate consciousness with the identity principle or not, as do some other school of thought.

We return here to the primary focus of this discussion with the question: If we can and will reinvent ourselves, what, precisely will get reinvented?

We can answer that question easier if we cast our speculation in more practical and concrete terms. Let us theorize that, at some stage in our future development it becomes possible to be what is popularly known as a "shapeshifter" in the science fiction literature. A shapeshifter is an entity who can take on, change into the physical form ("morph" in the language of computer graphics) of almost any creature he or SHe wishes for either pragmatic purposes or simply for the sake of directly acquiring the experience of a particular type of creature. At least in our current thinking, modifying one's body to the degree that it could shape-shift is a fairly extreme capability. As much as this seems like an extremely remote possibility only realizable in the far distant future, consider that a biologist recently claimed in a television documentary that, within five to ten years, we will be able to have any (kind of) body we wish once a month through genetic engineering and nanotechnological advance. One month one could be female, the next month male, the next month American Indian, the next month Scandinavian, etc. The anticipated way of bringing this about is through the development of a benign virus, perhaps even synthetic viruses which would carry the pertinent genetic instructions into the body's cells to effect the transformation. It is amusing to think about what kind of industry would produce and market such a product. ...perhaps a pharmaceutical company, or perhaps even the Avon lady would bring it to our door. . .? Such technology is surely only the very crude beginning of capabilities that we will inevitably develop. Eventually, instead of relying on a constructive infection by a benign virus and having to bring the changes about by such methods, we may perhaps re-engineer our genetic code to give us the ability to shape-shift just by thinking about what we wish to be.

The very process of considering all of these possibilities, physical and mental, weighing the attractiveness, feasibility and relative value and significance and determining to implement one and not another is the way we shall go about reinventing ourselves. The conception of possibles, the sorting of them, the choosing, experimentation and implementation require the acquisition of facts, data, evidence. The experimentation itself is part of the process of expansion and enrichment. As we learn and sort we learn to recognize what is most beneficial, what is riskier, what is in keeping with our nature and what potential we have to become something beyond. We may go on from there to afford ourselves the ability to bi-locate, use telepathy as our primary mode of communication, the ability to view at a distance, and perhaps the ability to heal wounds to our bodies instantaneously by an act of will – among other things. All indicators point to the time when we will have achieved mastery of the physical level by mental control, a stage on the evolutionary path on which we are learning to be able to reinvent the physical universe.

I anticipate that the general process will progress from the gradual modification and reinvention of the physical part of human nature through the modification and reinvention of the psychical/mental part of human nature. Experimentation with the modification of the human body, from immortality to shapeshifting, will teach us gradually what is the true essence of the person, what identifies the individual when just about everything has been so radically modified or altered that even a partner might not be able to recognize the person at all and, indeed, the nature of that person might not even be human, but something else. Can one be so changed in nature as to not have a human body as we know it, but perhaps some kind of body that is still beyond our imagination at this time and yet be a "human" person? I suggest that we already acknowledge that a "person" may have any kind of body – or no body as we know it – and still be that unique identity, the same person. I would even extend the notion further: A person should have the right to become literally another person by choice, another completely different identity principle, just as routinely as now one may change their name legally. Obviously, these considerations bring up all sorts of legal questions about identity, legal status, responsibility, individual rights, and definitions of just what constitutes the essential person, just who a "person" is under the law. And, just as inevitably and inexorably as the law has adjusted to such radical notions and changes in other areas in the past and present, so will it gradually adjust and be forged to these new conditions and potentials. It may be easier for the law to adjust once the new paradigm has taken hold because we will be free of the obstruction of the contradictory pluralistic definitions of human nature that have necessitated the kind of law under the Constitution which we have had to work with to this time.

The Real Meaning of Contelligence

It is in the reinvention of ourselves and our ongoing reinvention of the universe that the concept of contelligence, a term invented by Timothy Leary, finds its full meaning and significance. By "contelligence" Leary meant both the difference between and the fusion in operation of intelligence and consciousness. Consciousness taken as a manner of awareness, and intelligence taken as an understanding of information operate in a reciprocal interaction with each other. One should compliment and reinforce the other. An expanded consciousness facilitates the assimilation and integration of data and a broad and deep informational content facilitates the expansion of consciousness. This expansion of consciousness into greater and greater dimensionality and its concomitant acquisition and assimilation of more and deeper information is the basic mechanism by which we reinvent ourselves. As we realize more and more possible ways we can understand the universe, so can we expand our consciousness to realize more and more ways in which to interact with the universe and to acquire more and more information. Current New Age jargon often calls the beginnings of this process, as we witness it in ourselves, "reality creation." The concept needs to be thoroughly discussed for clarity.

"Reality Creation"

For some time we have been speaking in terms like "reality creation," "our minds create their own realities," "reality tunnels" and "reality creators," sometimes literally, sometimes metaphorically. It has been a popular notion within and without the New Age culture with a great latitude of meaning. The fundamental question that arises with regard to such a concept is: Can and do we create previously non-existent real things or conditions literally with our minds?

On first consideration it would seem to be a completely straightforward question amenable to a very simple positive or negative answer, even determinable by rigorous scientific analysis: set up precise experimental conditions and test for results. Even one completely verified positive case in which a subject created even a simple physical object by the power of mind would prove the first point. If only one subject demonstrated the ability beyond scientific question to directly influence a current or future event, the second point would be proven. Not all subjects would have to be able to do it. One would prove the point.

It is not as if we do not have any evidence at all of such kinds of things, at least on the very basic level of action at a distance. The work of Jahn and Dunne (*Margins Of Reality*) has shown statistically significant mental influence on the output of an electronic random generator. There is quite a bit of documentation of certain persons who apparently have the ability to move physical objects without touching them, purely through the exertion of some force they are capable of summoning by mental concentration. But these are examples of influencing physical phenomena at a distance, not of the creation of physical objects out of "nothing." But note that it is an influence at a distance. It is amusing that, paradoxically, we tend to think of the creation of a physical object out of "nothing" as more radical than being able to move a physical object across a table top without touching it. It is not inconceivable that same force would be exerted by a person manipulating the elements – quantum particles, atoms, molecules, force fields – of space-time into a new physical object as exerted to move a physical object in space-time. Be that as it may, the heart of the matter here is the basic question, What is the nature of the force or forces exerted by a human being in doing such seemingly extraordinary things? It has been variously described as a "mental," meaning non-physical, force; an "intention" in the sense of non-physical; "influence," a mental/physical concept, as it is called in the Chinese tradition in the I Ching which does not entertain the radical Western separation between mental and physical, mind and matter; a physical effect of and in space-time but involving higher dimensions, i.e. quantum mechanical. All these various aspects and connotations are and have been investigated, tested and experimented with, denied, ridiculed, argued about, approved, reinforced and rejected by various authorities, scientists, mystics and meditators over time. Action at a distance was suggested as a logical possibility by the Einstein-Rosen-Podalski "thought experiment" some sixty years ago. Various individuals, the world over, have demonstrated under controlled conditions the ability to move physical objects without touching them. Experiments in ESP and distance viewing under rigorously controlled conditions have yielded highly significant statistical evidence of their reality.

We can, it seems, be reality creators at three different levels. We, being self-aware, can, through intention and proper information handling, determine our individual trajectories of development and make our own choices, thereby modifying circumstances, altering events and influencing the events to come in the future. This is certainly an interactive mode, but it is limited to a radical responsibility, potential and freedom to determine what kind of reality one intends for oneself and others subjectively. Everyone has some sense of this first and most obvious level, which can hardly even be avoided, if only that things are not turning out as they intend.

A more fascinating reality creation, or reality intervention, is the capability we have for influencing events at a distance. We have always been handicapped by the vague theological orientation that would hold that only God could do such a thing, the "common sense" view of even a mind such as Newton who said that action at a distance was, obviously, nonsense, and the related idea that non-physical things like mind could not exert any influence, physically or otherwise, directly on physical objects, persons, or even other groups of minds. It is not difficult to see the mind-set of the modern scientist already at work here: Newton thought of physics in terms of physical objects and forces, and the mind as non-physical. His logic, according to those premises, is that a non-physical entity, mind, could therefore have no physical effects. It is not his logic that we reexamine here but the basic premise about the universe being nothing but objects and forces in Newton's sense and the sense in which the modern physicists explain it. For most contemporary scientists the discussion – if, indeed, they will even allow themselves to consider it a valid one – of this kind of topic is reduced to the question What medium, what carrier energy or field, what matter or energy of what type carries the (assumed physical) effect from the alleged initiator of action at a distance to the recipient of the effect of the action? This is a very reasonable question from the point of view of contemporary science, which knows all space to contain something, some energy, some matter, however rarefied in some places, and which works from the basic assumption that cause and effect operate through physical objects and force fields and that there is no such thing as a perfect vacuum. Because of these basic convictions any effect, whether it be ESP, viewing at a distance, or any kind of action at a distance, is usually rejected out of hand as impossible. This is the other side of the coin of the discussion about the nature of consciousness, which is thwarting the physicists from arriving at a Law Of Everything since they cannot find a way to explain consciousness in terms of the known physical forces. Consciousness has been abandoned or, at least, neglected by physicists for as long as the scientific method has been employed. Consciousness is subjective, the scientific method's goal is objectivity. But the cycle has come around full circle in that consciousness, as a phenomenon of the universe, has to be dealt with sooner or later and particularly in the development of the Law Of Everything.

It appears that this two sided problem of action at a distance and the derivation of consciousness will probably be resolved in the not too distant future. Resolution will come by either a conclusive demonstration that consciousness is, indeed, a measurable energy field of a type that must be considered as primary itself, of the class but not derivable from the other primary forces (strong, weak, electromagnetic and gravity), or that it will be determined to be of a type and class that is essentially different from the known primary forces of current physics, probably in terms of it having an inherent fourth-dimensional, at least, component and being able to operate at least partially independent of the known laws of physics.

It should be noted that other cultural orientations do take action at a distance by mind or other energies for granted and develop such proficiency on a disciplined basis and practice such skills routinely. With the freedom from the cramping dogmatic postulates of theology and Newtonian type thinking, we will accelerate our exploration of this area of human activity. It is amusing that, while some humans even in Western culture have been doing such things for as long as we can remember, those who are supposed to be "experts" in the area of psychology, but trapped in the billiard ball universe of the Cartesian three-dimensional coordinate system, are only now finding it barely respectable to begin to investigate the topic. But one should not wait for the "experts" to verify what is already quite evident to the common consciousness.

How Do You Act at a Distance?

If the possibility of action at a distance is a reality, how do we learn to use to that potential? The most straightforward answer is to just do it, try it and, in the trying, one will learn. This recommendation may sound rather simpleminded but we are working at the leading edge here, where the rules and laws are being discovered by the experimentation. The unique character of this kind of action is that we are, in a real sense, creating the potential, the possibility and the fact as we attempt it. This is a general principle that one discovers at the leading edge of evolution: when we develop and activate a new ability or faculty that we never possessed or never knew we possessed before we are, in the process of realizing it, literally making it real, acting as reality creators in the process of discovery. The principle seems to hold universally true: what we can conceive, we can achieve. It is in the reciprocating interactive relationship between conceiving and achieving, between the assimilation of information about new potentials and their realization, that we find our consciousness and the dimensionality of our experience expanding and evolving. Isn't this, after all, just another way of describing the ordinary learning process? Although there is a basic correspondence of learn and implement, we are focused here, primarily, on the actual discovery process and activation of new potentials in the human being, literally, in some phases, in the human neurological system. These potentials may well require or, indeed, are causing augmentative developments in the human brain and nervous system and require us to invent new conceptual systems of psycho-philosophy, psycho-physics and psycho-mathematics to express and communicate them. With new perceptual potentials resulting from expansion of our consciousness into more dimensions will come the pressure to develop the bio-psychical capabilities in our brain and neurological system to accommodate such direct perception and expanded conceptualization. The multi-dimensional information we receive through this expanded perception will force us to invent communicatory systems robust enough to process and convey such rich data and to extract expanded meaning from it.

In the trial and error process of teaching ourselves the bio-psychic proficiency in action at a distance of any kind (distant viewing, telepathy, physical actions, etc.), unfamiliarity is the obstacle and in the experimentation one learns both how one's mind works and how it works successfully in this mode. One learns what kind of intent and attitude works and doesn't work, what kind of physical conditions are helpful and what not, what kinds of phenomena and events are more influenceable and what ones less, what persons are open to communication and influence and what ones are not. One quickly learns that the most successful communication and influence is one that is based not on power but on respect, empathy, compassion and with the intent of accomplishing the maximum good of all – simply the bio-psychic application of the golden rule. Learning to move objects at a distance seems to be more difficult – I have never been able to accomplish it under any circumstances – because it seems to be dependent on the level of a natural force more concentrated in some persons than others, similar to the variations in physical strength among individuals. But that it can be learned, as well as the ability to set objects on fire through a concentration of heat in the hand by mental focus, the effecting of photo emulsion to create images on film, etc. seems to be borne out by the example of adepts who have claimed to have developed it through long practice or have been born with the talent.

Although these abilities seem mysterious and exceptional at this time, I believe that, in time, they will be understood as the ability to manipulate forms of energy or fields that we will be able to identify and quantify, and learn to manipulate easily. I think that, once we expand our science to identify and thoroughly investigate the chi force taken as fact by oriental physiology, we will have made a positive beginning. As I have speculated above, the chi force seems to be completely describable and quantifiable only in four dimensions, although its effects can be manifest in three-dimensions. I think the evidence already shows that this force is as real as gravity or electromagnetic energy or the weak and strong force, can interact with those forces and is both a "mental" and "physical" force when inspected from the perspective of three dimensions. To put it the other way around, mental energy appears to involve both bio-physical energy and a fourth-dimensional energy component. If this were so then we would only have to expand our physics to include the fourth-dimensional component to resolve the problem of consciousness in physics. The solution of the "problem" of consciousness in physics and the integration of the fourth-dimensional energy component into physics, I suggest, are two sides of the same coin. This is to say that consciousness is an organized energy field, fulfilling the requirement that it be explainable in terms of the forces of physics but, in order to do so we must add another force or field, that of the fourth-dimensional component. In these integrations the intrinsic self-reflexive relativity of our thought and the physical universe, understood through that same subjective perception, give us the clue to the final realization of the Law Of Everything expressed in terms of self-reference.

The Status Quo: Barely a Beginning

There is always the danger of misuse of these and any extraordinary abilities and powers we have and will develop. One only has to reflect on the criminal misuse of powerful psychoactive substances by the military and intelligence agencies, the less than enlightened way remote viewing capabilities have been exploited, the perverted and sometimes hideous experimentation with genetics that are on record, to know that the potential is always present. But, just as with everything from gunpowder to atomic energy, from lipstick to birth control, which have been predicted by someone at some time to destroy civilization, so we will see the doomsayers and righteous godspell mongers proclaim that immortality, extrasensory powers, action at a distance, space exploration and alien contact will be the downfall of all that is human. Although we are only at the beginning of the conceptualization of such activity, a sort of early adolescent playing with the birthday chemistry set kind of naive trial and error process, we shall not destroy ourselves or our planet – although we will scare ourselves badly more than once with meltdowns, toxic spills, dangerous radiation, genetic mistakes, misuse and abuse of any and all of these new powers – and religious backlash. But, more powerful and overriding by far will be individual and racial self-respect and, therefore, survival and right action. The godspell broken, we shall treat ourselves as if we deserved to continue in existence, deserved to live on a planet which was ecologically in harmony and supported us and our children, deserved to be treated and to treat others for mutual maximum benefit, and act accordingly.

Matriculating into Solar and Stellar Society

There are two events, contact with alien cultures from outside our solar system and re-contact with the Nefilim, that demand serious consideration in relation to our near term evolutionary progress. From one perspective they are both political events and, initially, do not seem to be related to the way we will reinvent ourselves. The profound nature of these anticipated events will, however, force us to rethink human society, ourselves and our place in the universe and, therefore, our evolutionary trajectory.

I take as a given that entrance into whatever stellar society is out there – or intelligent interaction with one here already – is contingent on our reaching racial maturity. I define racial maturity as reaching an integration of the various facets of our racial consciousness into a subtending consensual reality. Most fundamentally and simply this means arriving at a clear and obvious common factual understanding of our beginnings and our nature. We need to do so not primarily to meet some projected criterion of admission but for our own sake, for the maturity to be able to handle the jolts of radical relativistic paradigm shifts from alien reality views that do not overlap with ours. Is it possible? Certainly, by the fact that you or I can conceive of it. Whatever we can conceive we can achieve.

In a time when we are challenged by possible contact and communication with aliens with potentially radically different perceptions of the universe (universes?), the new paradigm empowers us to expand our awareness exponentially. By explaining our metamorphic psychology it highlights just how much that reality is a function of and relative to our unique metamorphic development at any given time. Knowing our own reality from the "unassailable integrity" of our true history, we can become the confident jugglers of alternate realities with the maximum flexibility to consider an acceptable alien reality and the maximum certainty and firmness to refuse an unacceptable one; to make as certain as possible that our rights of passage be respected in the rites of passage into stellar society.

Is there a more evolved modality to which we can move to adequately facilitate our own species development and enable ourselves to confidently interact with an alien species? Whatever we can comprehend we can transcend. How do we go about it? We need to reconsider and redefine our modes of understanding as relativistic – at the very least. Being able to operate relativistically from that base buys a great deal even from the practical point of view.

We can compare the developmental pattern of another species with what we call evolution on this planet, including our own special case of it, to see if indeed the politics of evolution as we know it even exists on other planets. Being able to see clearly that our concepts of god, gods, God, and Avatars, are a direct product of our original relationship to our humanoid creators and sublimations thereof, we can move beyond the crucified West and the karma-dharma-diddle of the East. We can compare our most profound concepts of whatever ultimate principle we can conceive and project as the ground of universe – or universes – with the concepts of other species without loosing our center. We can compare our view that our "unassailable integrity" dictates that the direction of our evolution is toward greater and greater benevolence, tolerance and respect for the importance and rights of the individual, toward unlimited knowledge and freedom with the relative "unassailable integrity" of an alien species. Not on the basis of some religious concept or the idea that there is an objective order of reality in which those rights are embedded, but solely and precisely on the basis of the expediency of our survival and evolutionary development. It will someday be pertinent and interesting to compare our situation with both the development of a linearly evolved species and that of another engineered species.

We can see already that intelligence alone is not wisdom. Even wisdom is relative to species reality. At least – and I am frankly guessing here – it will be relative until we reach some understanding of a true common ground more profound and comprehensive than the shallow level of binary code or the format of the message scratched on the side of a space probe. There may be certain common elements such as survival, but even that should remain an open question. Benevolence may not be a potential in some species? Violence may not be a potential in some species? Is the universe recursive enough to allow the "contradictory" process of cyclical improvement through time travel to the extent of going back in time to invent something of which you are the present end product? We no longer have to handle these possibilities at arms length in our science fiction like children telling stories to frighten each other. From a position of "unassailable integrity" we enable ourselves to move beyond the gossiping, sightings cataloging mode, the disastrous docile passivity of waiting for daddy. We enable ourselves to avoid the regressive tendency to make a cargo cult center out of Arecibo, to far surpass Kirk's blustering macho insecurity, Spock's lumbering dehydrated logic and McCoy's alcoholic fearful tendency to just avoid the whole thing. Maslow may not have realized how correct he was when he said that we fear most the godlike in ourselves – like the adolescent rejecting the parent in her or his own personality. So far we have just been waiting for some nebulous Martian Express as earth warriors dreaming luminous dreams under the godspell, the old ways of the old gods, listening to the sky, yearning for some word of return, of direction, of knowledge, of command. But, if we have come far enough to recognize that we shouldn't be looking any more for some biblical god, savior, or divine being with dominion over us, we should rethink our attitude and expectations and prepare ourselves as a mature race for contact with mature races. Instead of dragging our dried larval shells across the sands of an imagined metaphysical desert under the ancient godspell, we need to step out of species adolescence, separate from the ancient parentage and confidently assume responsibility for ourselves from the base of a known history with a mind as open as the universe.

This realization and unassailable integrity will also prevent us from missing a grand opportunity. The tendency will be, when any official recognition of alien presence is finally made, to lump all alien visitation and actual presence in the past and present together. We will tend to let blur the difference between the Nefilim from within our solar system and the greys from the backside of Zeta and other truly foreign types. We will marvel at the several types of greys and miss the significance of our own alien genetic half. We will tend to assume that the intentions, objectives and actions of the greys and the Nefilim were and are somehow the same since they are "alien." There will be great confusion and discussions ad nauseam about whether the greys have actually manipulated, even engineered our history and that of the planet in general from even millions of years ago like something out of a cosmic Burpee's hybrid DNA seed catalog while we neglect the systematic investigation of our genome to sort out Nefilim components, Homoerectus components, residual conflicts between the two resulting in genetic defects, etc. We will tend to glaze over the necessary psychological integration of the details of our own peculiar inception, history and the attainment of a profound racial identity and independence in the giddy media glut of the immediacy of the novelty of the greys' public presence. If we let these things happen to us we, already predisposed by the subservient master/slave programming of the godspell mentality, will tend to look to the aliens for answers about our history, about the universe, about just about everything, and will be worse off than before. On the contrary, if we take advantage of any alien revelation as an opportunity to carefully distinguish our own half-alien history and nature from the history and nature of aliens that are very different from both the Nefilim and us, we will gain a tremendous advantage. We will be able to deal confidently and gracefully with true aliens from a base of self-knowledge and racial unity.

A Lesser Alien Problem

Even though Sitchin's work has shown clearly the fact that the Nefilim came here from the last planet in our solar system, it does not exclude the possibility that the Earth has been visited by races other than the Nefilim from other places who may have influenced events on Earth. It is quite possible that an advanced alien race or races could have visited Earth even millions of years ago. Assuming that they had advanced scientific capabilities relative to our current abilities, they could have influenced the Earth's biological and physical development in many ways, including seeding or manipulating species genetically, crossbreeding with indigenous species or the opposite – in the way of wiping out species accidentally or intentionally. There is a very definite possibility that someone had visited here millions of years ago, evidenced in high-tech objects like ooparts, found in ancient coal strata, as a single example. This is very different from ooparts that can clearly be dated and associated with the Nefilim. Cremo and Thompson, in *Forbidden Archaeology*, have presented a collection of impressive finds of this sort which are inexplicable in any other way, unless we allow that high indigenous civilizations had arisen even millions of years ago on Earth and then vanished, or such material had arrived here as space junk from some far distant unknown source, perhaps cast into space by some catastrophic explosion. Again, arriving at a true racial identity will facilitate our evaluation of any such information and put it into a complete historical, biological and genetic perspective.

The discovery, recognition, presence of or contact with an alien species, the return or reestablishing of contact with the Nefilim, or the recognition and evaluation of traces of ancient alien visitation or presence, or even the discovery of very ancient vanished human civilizations is having and will have very profound effects on us and the way we understand and view ourselves evolutionarily. But even these events, if and whenever they come, will still be just others in the process of our evolution into higher dimensionality of perception, consciousness, experience and expansion into the universe. As mind-boggling and novel as many of these possibilities for the present and even the far future seem to be, our consciousness has the potential for absorbing and assimilating and transcending them to our benefit. With the proper balance of respect, compassion, attunement, curiosity and enthusiasm we will find that the astounding freedom and potential of the universe welcomes us to contribute to its dynamic participatory formation.

Experience and our limited history shows us that even the near-science fiction potentials we have discussed here for becoming directors of our own evolution, for exploring our own potentials both physically and mentally, for playing in relativistic fields and even in other sub-universes, can only be child-like and even primitive glimpses of what we will be capable of and will eventually accomplish. We may well hardly recognize ourselves as human the way we do now in a few short centuries. We are on our way to the stars, to cross mating with other advanced species, to stellar adventures and dimensions of consciousness and physical forms which will make our current science fiction projections look infantile. My objective here has been to prompt us all to shed our slave rags of our collective image as a subject race by exposing the myriad ways we think and act subserviently to our individual and racial disadvantage. Unburdened of that dark subliminal guilt and the ancient compulsion to look to a now nebulous "god" for forgiveness and reconciliation, we may begin to act as if the universe was truly our home, this planet is our planet, we have the integrity and capabilities to treat it and ourselves respectfully and with mature compassion and love, and we have the freedom to transform it and ourselves in any way that does not harm it or others. I make no apologies for my limited vision; I hope that within a short time our collective efforts and individual gifts will allow us to far surpass ourselves in this regard. I encourage all readers to interpret everything I have mentioned or said here for themselves and to prosper and flourish as their own evolutionary artist. Any "conclusion" to this book can only be a contradiction by the nature of the subject. The message is that liberated human nature and existence is open-ended, forward creative, poetic, rhapsodic, and expanding. Rather than a conclusion, only a marker can be placed to show where the pages stopped just now, perhaps a flag to guide our amazing children if they come curiously back as psychic anthropologists, searching along the evolutionary path to see where we were now in the continuum. To them let us record Greetings, our certainty that they would reach such capability, and confidence that their evolved compassion would move them to do so.

Or we may come back as immortals from the relative future, possessed of four-dimensional consciousness, to revisit this place of our minds where the stars are just beginning to become truly real to us, where we are still only represented on another planet by a toy-like robot, when we are only now beginning to summons up our courage to acknowledge our true history, when expanding our minds into new dimensions so frightens so many that we condemn a champion of such an evolutionary gambit as a political prisoner. Let us record for ourselves the reminder, however, that we knew that it is inevitable that we would attain immortality, that we would come to play our own four-dimensional god games, become our own casting directors of our own personal cosmic movies, that we would be back to revisit this place of our minds. Let us mark this place in space-time with the traces of our humor, pay our respects to our Nefilim relatives and let our own god games begin. Whatever we can comprehend we can transcend. Whatever we can conceive we can achieve.

Printed in the United States
994600001B